Managerial Decision Analysis Series

Data Analysis, Regression, and Forecasting

Managerial Decision Analysis Series

Data Analysis, Regression, and Forecasting

Arthur Schleifer, Jr. • David E. Bell

Harvard Business School, Boston, MA

Course Technology, Inc. *One Main Street, Cambridge, MA 02142*

An International Thomson Publishing Company

Albany . Bonn . Boston . Cincinnati . London . Madrid . Melbourne . Mexico City . New York . Paris
San Francisco . Singapore . Tokyo . Toronto . Washington

Data Analysis, Regression, and Forecasting is published by Course Technology, Inc.

Managing Editor	Mac Mendelsohn
Production Editor	Christine Spillett
Text Designer	Susannah K. Lean
Cover Designer	John Gamache

For more information contact:

Course Technology, Inc.
One Main Street
Cambridge, MA 02142

International Thomson Publishing Europe
Berkshire House 168-173
High Holborn
London WCIV 7AA
England

Thomas Nelson Australia
102 Dodds Street
South Melbourne, 3205
Victoria, Australia

Nelson Canada
1120 Birchmount Road
Scarborough, Ontario
Canada M1K 5G4

International Thomson Editores
Campos Eliseos 385, Piso 7
Col. Polanco
11560 Mexico D.F. Mexico

International Thomson Publishing GmbH
Königswinterer Strasse 418
53227 Bonn
Germany

International Thomson Publishing Asia
211 Henderson Road
#05-10 Henderson Building
Singapore

International Thomson Publishing Japan
Hirakawacho Kyowa Building, 3F
2-2-1 Hirakawacho
Chiyoda-ku, Tokyo 102
Japan

Case material of the Harvard Graduate School of Business Administration is made possible by the cooperation of business firms and other organizations which may wish to remain anonymous by having names, quantities, and other identifying details disguised while maintaining basic relationships. Cases are prepared as the basis for class discussion rather than to illustrate either effective or ineffective handling of an administrative situation.

Library of Congress Catalog Card no. : 94-68360

1-56527-273-0

Printed in the United States of America

10 9 8 7 6 5 4 3 2 1

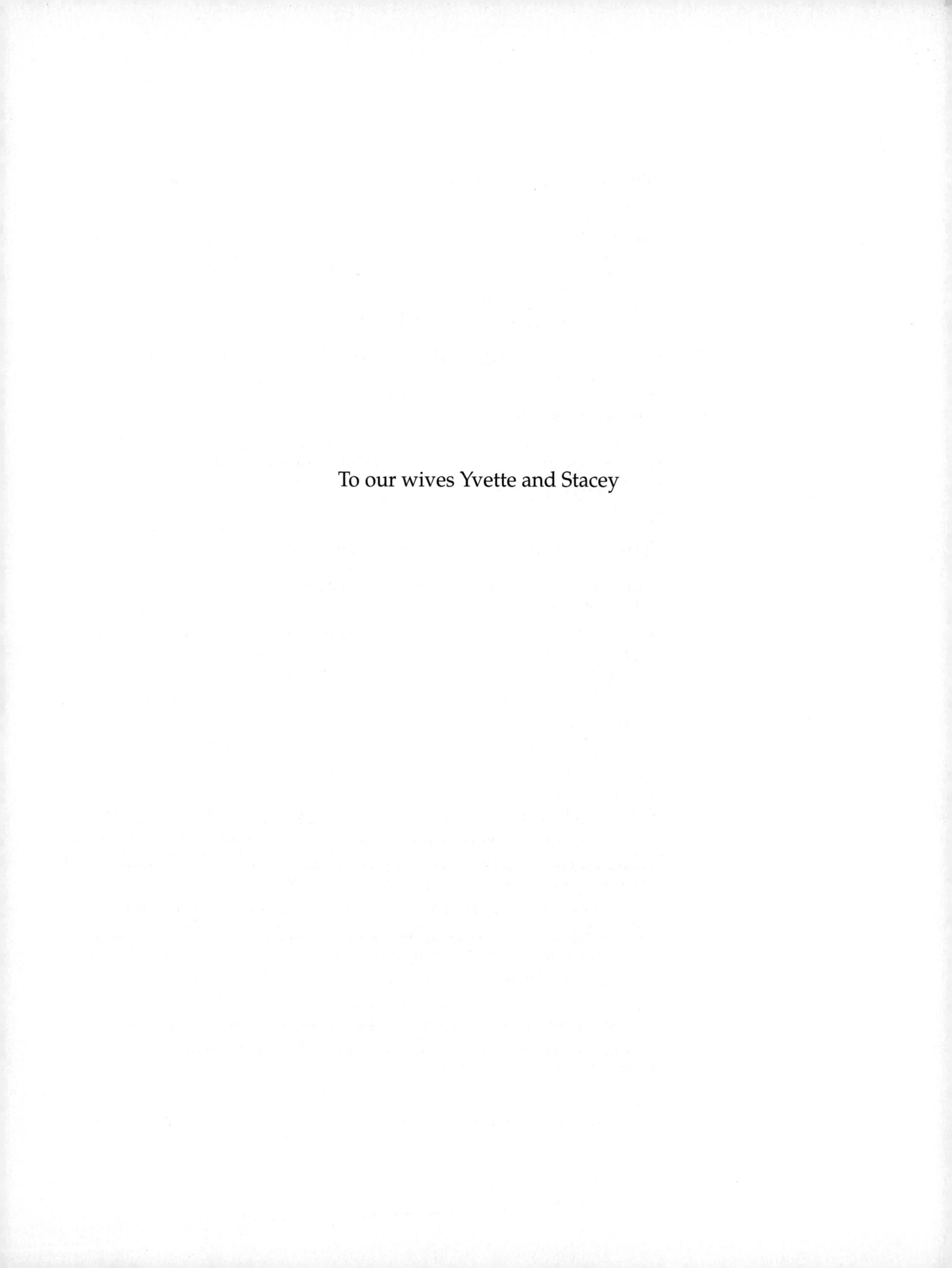

To our wives Yvette and Stacey

From the Publisher

At Course Technology, Inc., we believe that technology will transform the way that people teach and learn. We are very excited about bringing you, professors and students, the most practical and affordable technology-related products available.

The Course Technology Development Process

Our development process is unparalleled in the higher education publishing industry. Every product we create goes through an exacting process of design, development, review, and testing.

Reviewers give us direction and insight that shape our manuscripts and bring them up to the latest standards. Every manuscript is quality tested.

The Course Technology Team

This book will suit your needs because it was delivered quickly, efficiently, and affordably. In every aspect of our business, we rely on a commitment to quality and the use of technology. Every employee contributes to this process. The names of all of our employees are listed below:

Tim Ashe, David Backer, Stephen M. Bayle, Josh Bernoff, Michelle Brown, Ann Marie Buconjic, Jody Buttafoco, Kerry Cannell, Jim Chrysikos, Barbara Clemens, Susan Collins, John M. Connolly, Kim Crowley, Myrna D'Addario, Lisa D'Alessandro, Jodi Davis, Howard S. Diamond, Kathryn Dinovo, Joseph B. Dougherty, MaryJane Dwyer, Chris Elkhill, Don Fabricant, Kate Gallagher, Jeff Goding, Laurie Gomes, Eileen Gorham, Andrea Greitzer, Catherine Griffin, Tim Hale, Jamie Harper, Roslyn Hooley, John Hope, Nicole Jones-Pinard, Matt Kenslea, Susannah Lean, Laurie Lindgren, Kim Mai, Margaret Makowski, Elizabeth Martinez, Debbie Masi, Don Maynard, Dan Mayo, Kathleen McCann, Jay McNamara, Mac Mendelsohn, Laurie Michelangelo, Kim Munsell, Amy Oliver, Michael Ormsby, Kristine Otto, Debbie Parlee, Kristin Patrick, Charlie Patsios, Darren Perl, Kevin Phaneuf, George J. Pilla, Cathy Prindle, Nancy Ray, Marjorie Schlaikjer, Christine Spillett, Michelle Tucker, David Upton, Mark Valentine, Karen Wadsworth, Anne Marie Walker, Renee Walkup, Tracy Wells, Donna Whiting, Janet Wilson, Lisa Yameen.

Brief Contents

CONTENTS

▼ CHAPTER 1
Data Analysis and Statistical Description

▼ CHAPTER 2
Sampling and Statistical Inference

▼ CHAPTER 3
Time Series

▼ CHAPTER 4

Forecasting with Regression Analysis

▼ Chapter 5
Causal Inference

▼ Chapter 6
Multiplicative Regression Models

Overview

This book is one of a series entitled *Managerial Decision Analysis*. This series of four books represents the output of three long-term course-development projects with which we have been associated at Harvard Business School. Both of us spent several years, at various times, heading the semester-long Managerial Economics course, required of all 800 first-year MBA students. We have separated this material into two parts, according to whether it concerns a decision made under conditions of complete knowledge (Decision Making Under Certainty) or a decision made under some degree of uncertainty (Decision Making Under Uncertainty). The first of these two books covers topics such as relevant costs, net present value, and linear programming. The second covers material on decision trees, simulation, inventory control and cases involving negotiation and auction bidding.

Both of us have also taught elective courses to second-year MBA students. Schleifer developed a course on Business Forecasting that is the foundation of our third book, *Data Analysis, Regression, and Forecasting*. Bell developed a course that integrates approaches to decision making under risk, whether business, personal, or societal. This material is presented in our fourth book, *Risk Management*.

Together, these four books provide unprecedented case coverage of issues, concepts, and techniques for analyzing managerial problems. Each book is self-contained and can serve as a stand-alone text in a one-semester course (which is how the material was used in our classrooms) or as a supplemental volume for those seeking a set of demanding real-world applications for students in a more traditional text and lecture course.

Learning concepts and techniques through the case method may be a new experience for some. It takes time to adjust to the notion that problems do not always have neat, clear solutions, and, more profoundly, that learning is often greater when they don't. We believe that this set of material covers not only what a future manager should know as an intelligent user of quantitative methods but also as an intelligent consumer of analyses others have done.

As the reader will see, some of the cases and notes were prepared by our colleagues; we are grateful for the opportunity to use them. What will not be apparent is the debt we owe to those of our colleagues that have gone before us in heading the Managerial Economics course: Robert Schlaifer, John Pratt, John Bishop, Paul Vatter, Stephen Bradley, and Richard Meyer. We also thank the Division of Research at the Harvard Business School for their financial support of all the course development reflected in these volumes. Finally, we wish to express our appreciation to Rowena Foss and Laurie Fitzgerald of the Harvard Business School, and to Mac Mendelsohn of Course Technology, for their substantial efforts both keeping this project on track and contributing to its quality. Rowena Foss has been our secretary, at times individually, but for the most part jointly, since 1978. To her we owe a special debt of gratitude.

PREFACE

Managerial Decision Analysis Data Analysis, Regression, and Forecasting

This book is about the practice of data analysis. It teaches future managers what they need to know about statistics in order to make informed decisions. It illustrates the applicability of statistical methodology to real-world case situations, and provides students with opportunities to engage in analysis and to critique the analyses of others. It doesn't teach statistics for its own sake. Its primary goal is to provide students with the methodology needed to solve a variety of commonly occurring real-world problems, not to invent problems to illustrate methodological techniques.

Managers typically deal with *observational* data, not data derived from carefully designed experiments. The analysis of such data can illuminate their understanding, facilitate their planning, and improve their decisions, by providing them with answers to the following questions:

1. How can one variable, or the relationship among several variables, be succinctly described? For example, what are good ways of describing the relationship between monthly retail sales and monthly expenditures on retail advertising in the United States?

2. What can you infer about a population or process from a sample of data from that population or process? For instance, what can you conclude from a sample in which 320 out of 500 respondents to a questionnaire said they liked your product? And how do you modify your conclusions if (as is likely) the sample is not random and the responses do not necessarily indicate that those who say they like the product will buy it?

3. How can you predict future values of a variable from its past values and from past values of other variables? A typical example would be forecasting next month's sales of a breakfast cereal based on past sales of the cereal and various promotion policies used in the past. (As a by-product of the analysis, one might also try to judge the cost-effectiveness of the promotion policies.)

4. What will happen to a variable if you deliberately change the value of another variable with which it is causally related? For example, suppose you are the owner of a major-league baseball team, and are concerned about how many games should be televised. Given data on previous games, including attendance, whether or not the game was televised, and other factors that might affect attendance, what can you infer about the likely effect on attendance (and ultimately profits) of changing the number of games televised?

All of these problems have action implications and economic consequences. As a result, the focus is not on testing some abstract hypothesis, but rather on learning enough from analysis to be able to weigh the economic tradeoffs and decide on a course of action.

Through text, cases, data sets, exercises, and worked examples, this book seeks to give students guidance on the use of data-analytic tools, and to help students avoid their misuse. It focuses on real business problems with real data (but supplements business cases with cases from other fields, where appropriate, and provides simplified exercises for developing skills with new techniques). Its emphasis is more on helping readers understand how data analysis and statistics can aid managers, less on refining analytic skills by introducing the most powerful techniques available.

Because it relies heavily on cases, whose analyses can range from quite straightforward to highly sophisticated, this book is suitable for both introductory and more advanced classes, and can be used in classes where there is a wide range of student abilities. It can be used in conjunction with a more traditional statistics text, or it can stand alone.

What's Different?

Because of its focus on practice, this book differs from many introductory statistics texts in terms of emphasis. You may have had a previous course in statistics, and may be wondering whether this book covers ground previously covered. Here are some of the differences, and our rationale for them. (If you never have had a course in statistics, this section may seem intimidating. We suggest that you skip it for now, but read it when the course is over.)

The use of cases distinguishes this book from many others. The cases typically introduce a business context within which a forecast or inference or statistical summary must be made. As is true of similar situations that the student will encounter in the real world, there are seldom neat "solutions" that can be mapped from the text to the case. For example, forecasts are not always simple applications of regression or time-series analysis. Forecasters don't always have the right data at the right time, nor do they sufficiently understand the process that generates demand. Similarly, samples are seldom strictly random, and the sample mean does not necessarily measure directly what has economic significance for the manager. Cases expose students to these realities. The frustration they may experience parallels the frustration that managers often feel when the real world fails to conform to the models in the textbook. But the cases themselves serve as learning vehicles, and most students discover that the text, exercises, and worked examples provide them with sufficient background to engage in creative discussion of how to approach these less structured situations.

We concentrate on the inputs that are needed for analysis, and the interpretation of the outputs that statistical models provide. While complicated formulas and computational techniques may be important to more theoretically minded students, we rely on computers and spreadsheets to do the number crunching. Because spreadsheets have become so commonplace and powerful, we have not included statistical tables. For example, students who want more refined normal-distribution percentage points or cumulative probabilities than the 68%, 95%, and 99.7% benchmarks we supply can use the built-in Excel functions for these characteristics of the normal distribution (or a variety of other distributions as well).

While the debate between Bayesians and "classicists" may be absorbing for statisticians, it is not one that engages students who are interested in practice, rather than theory. As a result, we try to blur the distinction while remaining intellectually rigorous. We talk about confidence distributions (distributions of the parameter given the sample), rather than sampling distributions, and derive confidence intervals and tests of significance from them. We show that under most conditions occurring in practice, confidence distributions can be treated as probability distributions (posterior to a "uniform" prior) of uncertain population means or regression coefficients, an interpretation that is compatible with a decision-analytic framework.

We believe that using a rough approximation to a probability distribution is much better than ignoring uncertainty altogether, and that refinements to such rough cuts come at a high price in terms of student comprehension, while still failing to take into account all sources of uncertainty. For example, we use normal approximations where the Central Limit Theorem applies, but avoid problems of unknown variance or distributions based on small samples from nonnormal populations. And in regression forecasting, we use confidence limits that formally incorporate the residual standard deviation, but not the uncertainty in the regression coefficients. Our rationale is that even if these refinements are included, they do not reflect possible uncertainties in the values of the independent variables that are inputs to a forecast, or uncertainties due to nonlinear relationships or other possible misspecifications. We would rather have students sensitized to the many sources of uncertainty in a regression forecast than have them accurately take into account those that can be assessed mathematically but ignore those that depend on subjective judgment.

In statistical inference we emphasize confidence intervals, on the grounds that most managers are more interested, for example, in the range of values that some parameter can take on, less interested in whether the data in hand are or are not consistent with a hypothesis that the value of the parameter is exactly zero. Although we introduce the concepts of hypothesis testing and significance levels to make students aware of language and techniques they will encounter in practice, we avoid getting entangled in the details of null vs. alternative hypotheses, point vs. interval hypotheses, one-sided vs. two-sided tests, etc.

While marketing texts often distinguish between cohort and life-cycle effects, statistics texts seldom say what you can and cannot infer about the effect of age on a population from data in which age is one of the independent variables. We discuss the difficulty of sorting out cohort and life-cycle effects from a single cross-sectional sample, and show how these effects can be distinguished if we have two such samples separated in time.

We try to use techniques that are applicable to a wide variety of problems, not specialized techniques designed for one problem class only. In making inferences based on sample data about the difference between two population means, for example, we start by constructing confidence intervals for the mean of each population. Although not powerful, this approach sometimes reveals that the intervals don't overlap, indicating that the result is "significant". To use the data more efficiently, we don't introduce the concept of a standard error of the difference between two means; rather, we treat this problem as a special case in regression. A dummy independent variable is used to identify from which population each observation arose; as an important by-product, this approach permits us to perform the more relevant analysis involved in determining significance or confidence intervals about the difference between two population means while holding "constant" the values of other variables.

Managers who observe in past data that changes in price are often accompanied by changes in demand want to know by how much a proposed price change will affect demand. Such causal inference is fraught with pitfalls for the unwary. Nevertheless, to limit our interpretation of the relationship between price and demand in the above example to "statistical association" would render statistical analysis useless in practice. We provide text and cases that illustrate what can and cannot be learned about causal relationships inferred from observational data.

Although economics texts often introduce multiplicative regression models, such models seldom appear in texts on business statistics. Nevertheless, many of the variables involved in business exhibit multiplicative, rather than additive, behavior. It is more natural to think in terms of percentage changes in populations, prices, quantities, national-income accounts than in terms of absolute changes. When variables like these are incorporated in regression models, their natural values seldom behave linearly and additively, but their logarithms often do. We have included a chapter on multiplicative regression models because regressions involving such variables are pervasive, and may be badly misspecified without logarithmic transformations.

Outline of the Book

Chapter 1: Data Analysis and Statistical Description

The text is concerned with how to convey what is interesting about data using summary measures and graphical analysis. We look at both cross sections and time series. While the chapter is a self-contained introduction to what is commonly called statistical description, it also introduces you early on to many modeling issues that arise in regression analysis: scales of measurement, trends and seasonal effects, linear and additive relationships, multiplicative effects, transformation of variables to achieve simple structure, univariate and bivariate summary measures, life-cycle and cohort effects, and issues of causation. This "front-end loading" familiarizes you early on with a number of topics you will have to learn later, and simplifies the exposition when we come to regression. Graphical analysis and presentation are used extensively.

Exercises on Interpreting Data present three problems designed to "hook" you into realizing that even simple data sets require sophisticated analysis. The first problem is an example of Simpson's paradox (batting averages of two switch hitters are compared; one has a better average than the other both as a left-handed hitter and also as a right-handed hitter, but has a worse average overall). The second problem shows the importance of computing the right average: a frequency distribution of passenger-tire lives is used to compute demand for replacement tires. The third is an example of the regression effect: families with high incomes ten years ago had a much smaller increase in income than families with low incomes.

Worked Examples in Data Analysis Using Spreadsheets leads you through the steps required in Excel to produce the graphs shown in Data Analysis and Statistical Description.

Boston Edison vs. City of Boston is a case in which, because of a legal precedent, the property tax that Boston Edison has to pay the city hinges critically on the ratio of assessed value to market value, averaged across all single-family residences in the city. This ratio is based on a sample of properties—those residences that recently changed hands. The sample surely isn't random. How might it be biased, and how might the bias be reduced? And how should the average ratio be computed?

Hygiene Industries and The Stride Rite Corporation (A) are cases concerned with the problem of forecasting demand at the SKU level for items in a product line. Hygiene manufactures and sells shower curtains. They want to forecast demand by SKU to improve production and inventory planning and on-time delivery to customers. The case includes sales data by SKU from two retail chains. Stride Rite manufactures and sells children's shoes. The case concerns issues of structuring and revising forecasts for a variety of related products.

Chapter 2: Sampling and Statistical Inference

The text discusses what can be inferred about a population or a process from a sample. Point estimates, confidence intervals, levels of significance, and issues of determining an appropriate sample size are addressed. (These inferential procedures will be used later in the interpretation of estimated regression coefficients.) The biases that can arise through nonrandom selection of respondents or inaccurate measurement, and the effect of these biases on the inferences that can be made from a sample, are discussed.

Sampling Exercises provide some realistic problems in which it is natural to make inferences about a population from a sample. One exercise involves drawing samples (by a computer randomization procedure) from a population, calculating confidence intervals from the sample, and discovering that the theoretically correct fraction of such intervals comes close to covering the (known) population mean.

Chapter 3: Time Series

Although regression analysis is often used to explain the behavior of a particular time series in terms of related independent variables, the values of these independent variables are not always known in time to be of use in forecasting future values of such a series. In this chapter we introduce methods for analyzing univariate time series, showing what we can infer about future values of a time series from its past values alone. We introduce two simple time-series models: constant-average and random-walk, and discuss characteristics of the processes that generate them, how series generated by such processes can be identified, and how to make probabilistic forecasts of future values of such series.

Time Series Exercises introduces two time series—average annual temperature in Boston over a forty-seven year period, and monthly gold prices over a ten-year period, and invites you to analyze each series and make probabilistic forecasts of future values.

The Boston Gas Company: Winter 1980-81 chronicles the crisis that faced a local natural-gas distribution company during an unusually cold and unusually early winter. Could Boston Gas, by better forecasting and planning, have avoided the near miss in which they nearly ran out of gas, or was this just a once-in-a-lifetime run of bad luck?

Chapter 4: Forecasting with Regression Analysis

We then turn to regression as the methodology of choice when values of predictor variables are known in advance. We introduce the basic idea of using predictors, introduce the concept of regression, and describe some of the common measures supplied as output from a regression analysis. We discuss how to use the output of a regression analysis to generate point and probabilistic forecasts. Finally, we describe proxy effects and introduce some elementary transformations that permit us to use regression when the relationship between the dependent variable and natural values of the independent variables is not necessarily linear and additive.

The built-in regression function in Excel and other spreadsheets requires the user to move columns of data so that the independent variables are in contiguous columns. This impedes exploratory analysis of data. In addition, many such functions provide excessive output in formats that are difficult to read. For these reasons, we have developed a regression utility that runs as an Excel add-in. *Using the Regression Utility* documents its capabilities and gives instructions for invoking them. *Doing Regression Analysis* walks the student through a number of examples, showing how an analyst goes through the process of developing and interpreting a regression model and generating forecasts.

Exercises in Forecasting with Regression provide an opportunity to delve more deeply into proxy effects, and to learn the importance of graphical analysis as a prelude to regression. One of the data sets includes grade averages, admissions-test scores, college averages, age, etc. for a class of Harvard Business School students. How well can grades be predicted from these other variables whose values are known when a student applies?

Chemplan Corporation.: Paint-Rite Division provides data on a company's annual sales (which have been trending upward) and on two drivers of demand: a construction index and home loans. The two indices help explain past sales quite well. Can they be used to predict future demand?

Harmon Foods wants to use regression to forecast monthly sales of a brand of breakfast cereal whose sales are somewhat seasonal, have been trending up, and are affected by promotional efforts. An interesting problem in modeling, interpretation of output, and economic analysis.

Highland Park Wood Co. concerns a lumber wholesaler who has been asked to supply southern pine to a home builder for delivery in six months, but at current prices. What should Highland Park do? One option is to hedge with futures on a similar type of wood for which a futures market exists. Data on historical spot and futures prices are provided.

CENEX owns an oil refinery whose decision to shift oil purchases to crude oil with high asphalt content is under review. From data on asphalt prices and consumption and other indicators, students are invited to construct a price forecasting system as part of a long-term planning analysis.

CFS Site Selection at Shell Canada Ltd. gives the results of a model constructed by a consulting company that was designed to identify the best gasoline-station sites on which to construct convenience food stores. You will find a challenge in unraveling the mystery of what motivated the particular way in which the model was specified.

Firestone Tire and Rubber Company issued an industry forecast for passenger replacement tires that was substantially below Goodyear's forecast. Firestone's forecast decomposes replacement-tire demand into two components: passenger miles driven and tire life. Forecasts of passenger miles driven are largely based on DRI forecasts; tire-life forecasts at the time of the case were extremely uncertain because of the unknown wear characteristics of steel-belted radials. Could the discrepancy between Goodyear's and Firestone's forecasts be within the range of overall forecasting uncertainty, or was there something systematically different about their methodologies?

Data Resources, Inc.: Note on Econometric Models explains the methodology used by DRI, one of the large players in the macroeconomic forecasting business. The development of systems to measure national income and product, of statistical methodology to analyze the data, and of computers to crunch the numbers made this kind of service possible. The forecasts in the Firestone case, above, are grounded in DRI forecasts. How solid is the ground upon which DRI's forecasts are based?

Chapter 5: Causal Inference

Many books on statistics assert that one can establish association but cannot "prove" causation through the analysis of observational data. Nevertheless, managers and other decision makers base interventions on the analysis of data. Is this justifiable?

The text discusses what we mean by causation, and shows how, by holding certain confounding factors constant, experiments can sometimes reveal how one variable affects another. Just what is held constant, however, is critically important in sorting out what affects what, and the same is true when regression is used with observational data. Influence diagrams are introduced as a way of helping to decide which variables should be included in, and which excluded from, a regression model designed to reveal causation.

Nopane Advertising Strategy describes an experiment in which different advertising strategies were test marketed. The outcome of the careful experimental design is affected by the action of Nopane's competitors, who have varied their own advertising strategy across the test territories. The results of the analysis depend on whether the competitors will react to Nopane's national advertising the same way as they did in the test, or whether they will simply revert to the advertising policy that they had pursued prior to the test.

How much does attendance at a baseball game decrease when the game is televised locally? In *The Gotham Giants*, we have data on attendance and televising for each of last year's home games. Surprisingly, average attendance was higher for games that were televised. What is going on?

The directors of *Lincoln Community Hospital* are contemplating changing its status from nonprofit to for-profit. They have assembled a large data base of both kinds of hospitals, including a number of variables that affect cost. Can the data provide any guidance to the directors on the likely consequences of the proposed change?

An *Exercise on Causal Inference* presents data from a Federal Trade Commission study of whether test preparatory programs (such as Stanley Kaplan) really increase test performance. The data come from a "natural experiment" in which some students took the Scholastic Aptitude Test (SAT), then attended the Kaplan program, and then took the SAT again, while a "control" group of students simply took the SAT twice, without a test preparatory program in between. The gain in score of the Kaplan group exceeded that of the control group. Was the gain real or illusory?

Chapter 6: Multiplicative Regression Models

In this chapter we discuss models in which:

- Changes in an independent variable by a fixed factor are accompanied by changes in the dependent variable by a fixed amount;
- Changes in an independent variable by a fixed amount are accompanied by changes in the dependent variable by a fixed factor;
- Changes in an independent variable by a fixed factor are accompanied by changes in the dependent variable by a fixed factor.

We show how to specify these models and interpret their outputs. To do so, we analyze three data sets. The first concerns the relationship between average life span and per-capita income in 101 countries. The second deals with death rates of smokers and nonsmokers, classified by age, gender, and, for the smokers, degree of smoke inhalation. The third data set consists of selling prices of California avocados over a 25-year period, along with number of bearing acres planted, yield per acre, and various national-income and population statistics.

An *Exercise on Multiplicative Regression Models* provides data on titanium dioxide production and costs, from which you can investigate the applicability of the classical learning-curve model and other models for predicting cost behavior.

Barbara J. Key vs. the Gillette Company summarizes the positions of expert witnesses on both sides of a sex discrimination suit. Just looking at salaries, women were being paid only 70% of men's salaries, on average. But were women being paid less than men *with equal qualifications*? The two experts came up with startlingly different conclusions, based on what variables they included in their regression models. Who was right?

Other Materials

A **data diskette** is included with the text. It contains data in Excel 5.0 workbooks relevant to almost every case, some of the text, and most of the exercises. Hands-on data analysis is important; some of the cases and virtually all of the exercises demand it. The data diskette provides opportunities to work through structured assignments, or for students to explore the data and supply creative analyses of their own where appropriate.

In addition to the data sets, the data diskette contains two **utilities**. The first produces a graph of a "cumugram," or cumulative distribution function, from a column containing values of a variable. Cumugrams are useful for finding cumulative frequencies and fractiles, and are difficult to chart properly without this utility. The second utility performs regression analysis. It permits analyses in which the independent variables are not in contiguous columns; it allows you to exclude specified observations from the regression; it can compute estimates, residuals, and forecasts; and it provides easy-to-use graphics capabilties for examining relations between variables and performing residual analysis.

A **Teacher's Manual** is available to instructors. It provides analyses and teaching strategies for each of the cases, solutions to the exercises, a summary of the important points in the text, and some discussion of the pedagogy employed. A sample examination is included.

Read This Before You Begin

To the Student

You will need a copy of the *Data Diskette* to work through some of the material in this book. Your instructor may provide you a copy of the disks, or may make the files available to you over a network in the school's computer laboratory. If your instructor does not provide you with the disk, it may be obtained from Course Technology, Inc., by calling 1-800-648-7450 or by sending a FAX to (617) 225-7976.

To the Instructor

Data Diskette: The instructor's copy of this book is bundled with the *Data Diskette* which contains the Regression and Cumugram Utilities and Excel data files that will help your students to complete some of the exercises in this book.

Instructions for installing the files contained on the *Data Diskette* are found in the README.TXT file on the disk. The README.TXT file may be opened by using the Windows Notepad.

Data Analysis and Statistical Description

Sources and Arrangements of Data

Managers acquire data in a variety of ways. They draw upon internal sources of data, such as accounting or management-information systems, and external sources, such as libraries, trade publications, the United States census, public-opinion polls, market research, and consulting organizations. Managers sometimes commission surveys of customers, or conduct experiments to provide data on the effects of some contemplated change in a production process or marketing technique.

Raw data come in various forms: words, pictures, sounds, computerized bits and bytes, and ordinary numbers. In order to perform the sorts of analyses described in this text, raw data must be converted into numbers,[1] and arranged in a table or set of tables. These arrangements are called **data structures**. By far the most common data structures consist of **observations** on one or more **variables**. If the observations consist of people or objects examined at a particular moment of time, the data are called **cross-sectional**. Examples of cross-sectional data include observations of people in which the variables are their annual expenditure on a product, their age, their education, and their gender; or observations of personal computers in which the variables are retail price, amount of memory, number and type of disk drives, and hard-disk capacity. By contrast, if the observations represent periods of time, the data are referred to as a **time series**. Examples of time series include yearly observations, in which the variables are measures of economic activity, such as gross national product, unemployment, and inflation; or monthly or quarterly observations, in which the variables are sales of a product, advertising expenditures, and price.

Spreadsheets provide very useful ways to record observations on a set of variables. By convention, each row of a spreadsheet usually represents an observation, and each column a variable. Each cell contains the numeric **value** of a variable on an observation.

Harvard Business School note 2-191-114. This note was prepared by Professor Arthur Schleifer, Jr.

[1] Ways in which nonnumeric data can be recorded will be discussed later in this chapter.

Purposes of Data Analysis

Why do we analyze data? One reason is to condense a mass of data into **summary statistics** that succinctly characterize the observations and variables. This condensation is called **statistical description**. Measures of distribution (histograms and cumugrams), of centrality (means, medians, and modes), of location (fractiles), and of spread (standard deviations and ranges) constitute the primary outputs from statistical description.

Another reason for analyzing data is to examine the relationship between two or more variables. Usually, we want to know not only if a relationship exists, but how to quantify it. For example, we might want to know if product sales are related to advertising expenditures and, if so, how much sales levels tend to change with each additional advertising dollar. More ambitiously, we may want to determine what causes what. Does advertising affect product sales, do sales levels influence advertising expenditures, or does some other factor influence both sales and advertising levels? Finally, we may need to base decisions on the results of the analysis: to boost sales, should we lower price, increase advertising, or both? By how much?

Once we determine the relationships among variables, we can **forecast** the unknown value of one variable when the values of other variables are specified. For example, we could use a time series to forecast the future value of a product's sales based on a specified level of advertising. However, this forecast represents only an **estimate** or best guess, since the relationships obtained from a sample of data may not be the same as those obtained from much more data. For this reason, it is useful to calculate the probability of an estimate's accuracy using **statistical inference**, and then report this probability along with the estimate.

Subsequent chapters address forecasting and statistical inference, while this chapter serves as an introduction to data analysis. It focuses on statistical description and the kind of exploratory data analysis that should always precede the use of these more powerful and sophisticated methods.

Description of Data Sets Available for Analysis

You should browse through the following data sets on your Data Diskette to become familiar with their contents, as they will be referred to frequently in this and subsequent chapters.

Smoking and Death Rates [SMOKING.XLS]

A 1966 study of the effects of smoking and inhalation on death rates[2] tracked a number of people annually from late 1959 through September 30, 1963. For each year that a person remained in the study, his or her age, gender, smoking behavior (self-reported) and status (alive or dead) were recorded. Treating each person-year as an observation, there were 2,904,813 observations and 29,562 deaths.

What can you say about the effects of smoking, age, and gender on death rates?

[2] E. Cuyler Hammond, "Smoking in Relation to the Death Rate of One Million Men and Women," *National Cancer Institute Monograph No. 19.*

Forecasts of Consumer Price Index [CPI.XLS]

A number of econometric forecasting services provide forecasts of key macroeconomic indicators up to two years ahead. These forecasts are typically issued each quarter.

The file contains quarterly forecasts of the consumer price index[3] (CPI) made by Data Resources, Inc., (DRI) a prominent forecasting firm. Each forecast predicted what the CPI would be one year later. To facilitate comparison with actual values of the CPI, the time associated with a forecast is the time at which that value was predicted to occur, not the time at which the forecast was made. Thus the forecast value of 123.3 listed for 7201 (the first quarter of 1972) was actually made in the first quarter of 1971.

What can you say about the accuracy of DRI's forecasts? Do they predict inflation rates well?

First-Year Harvard MBA Data [HBSMBA.XLS]

The file contains data on 667 first-year MBA students at Harvard Business School in the early 1970s, including grade information, section, age, college average, and test-score results. The variables are more completely described in the data file itself.

What can you say about the factors affecting performance in the first year?

DESCRIPTION OF ONE VARIABLE

Measures of Distribution

A variable takes on different values on different observations. These values can be grouped into brackets of equal width. The number of times each bracket occurs constitutes the variable's **frequency distribution**. The frequency distribution is depicted visually by a histogram or a cumugram.

Histograms

In a **histogram**, the **relative frequency** of each bracket is plotted as a bar in a bar graph. Figure 1.1 shows these relative frequencies for the weights of 768 first-year students at Harvard Business School.[4] The frequencies signify the fraction of students that fall into each weight category.

Figure 1.1

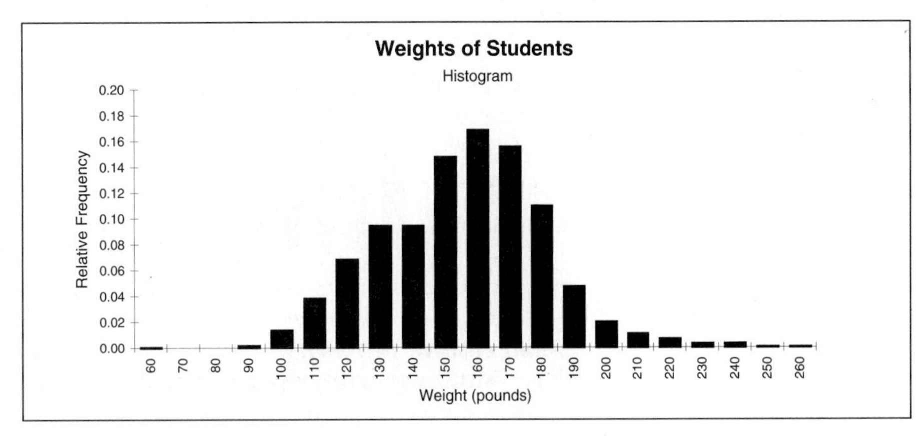

---

[3] The index measures the price of a standard selection of consumer goods relative to a base price of 100 in 1967.

[4] In a computer exercise conducted several years ago, students were asked, among other things, to supply their height (in inches), weight (in pounds), and gender. These variables are in file HTWT.XLS.

A first step in constructing such a histogram entails grouping the values into brackets. This step requires careful choices about the bracket width. If the brackets are too wide, disparate observations are grouped together. On the other hand, if the brackets are too narrow, each bar comprises very few observations; and the histogram is not likely to convey anything sensible about the shape of the distribution. In this example of HBS students, the weights are grouped into 10-pound brackets, and labeled such that the bracket denoted by 100 contains weights of at least 95 pounds but less than 105 pounds. For a discussion of how to produce histograms in Microsoft Excel 5.0, see the section called Worked Examples in Data Analysis Using Spreadsheets near the end of this chapter.

There are several important characteristics of a histogram. These include:

Mode. The **mode** of a distribution corresponds to the highest bar in a histogram. It signifies the most frequently observed bracket. In Figure 1.1, the mode occurs at 160 pounds, i.e., for the bracket that extends from 155 to 164 pounds. In this example, the heights of the bars also decline steadily as you move away from the mode in either direction; the distribution is therefore called **unimodal**. By contrast, if the heights first decline as you move away from the highest bracket and then get higher again, the distribution is **multimodal**; when there are two peaks, it is **bimodal**.

Symmetry. A distribution is **symmetric** when the bars on the right of the histogram mirror the bars on the left. An easy test for symmetry is to flip over the page containing the histogram and hold it up to the light; it is symmetric if its shape is pretty much like the shape of the original histogram. For example, the histogram in Figure 1.1 is fairly symmetric.

Skewness. Distributions that are not symmetric are referred to as **skewed**. Figure 1.2 presents a histogram with a skewed distribution. In this case, the distribution is skewed to the right, since it has a long tail to the right. In the same vein, a distribution is skewed to the left if it has a long tail to the left.

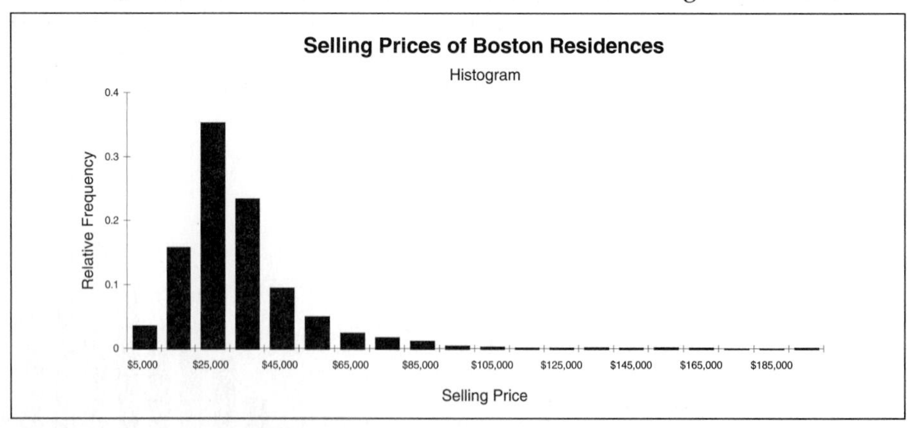

Figure 1.2

Multiplicative Symmetry. Sometimes a variable with a skewed distribution has a **multiplicatively symmetric distribution**. In the data from which Figure 1.2 was derived, 50% of the observations have values below \$28,000, 7.5% have values below half of \$28,000, or \$14,000, and approximately 7.5% have values above twice \$28,000, or \$56,000. Taking the logarithm of these selling prices generates a fairly symmetric distribution, as shown in Figure 1.3.[5]

[5] See the final section of this chapter for a short discussion of logarithms and logarithmic graphs.

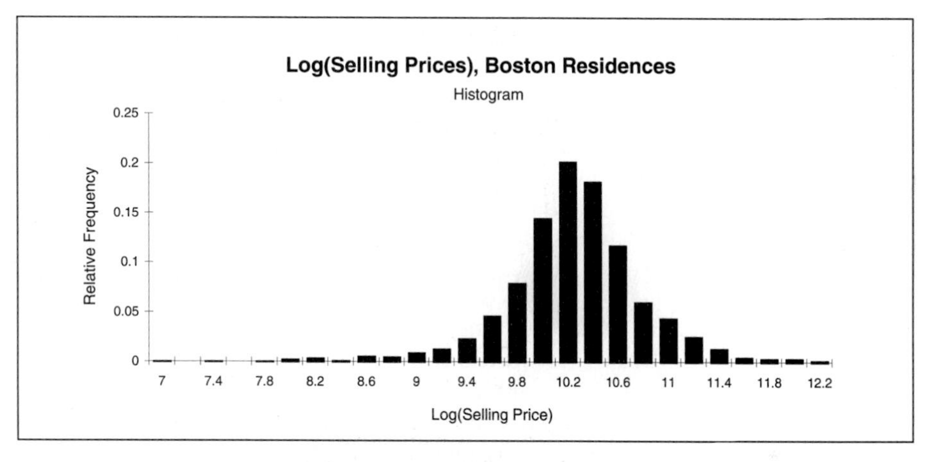

Figure 1.3

Outliers. In Figure 1.1, there is a barely visible bar corresponding to a bracket of 55-64 pounds, and no other bars until the 90-pound bracket. Both because it lies so far away from the other data and because this does not appear to be a reasonable weight for an HBS student, we might conjecture that this **outlier** was misreported and eliminate the observation.[6]

Cumugrams

In contrast to histograms, plots of cumulative distributions, or **cumugrams**, use the individual values of a variable, without grouping into brackets. They show the fraction of values that are less than or equal to some particular value. Cumugrams assume that the values of a variable have a natural ordering, or that the relationships "less than" and "greater than" are meaningful. Figure 1.4 shows a cumugram of the student weight data. It shows that roughly 10% of the students weighed 120 pounds or less, and about half weighed 157 or less. The jaggedness of the cumugram in Figure 1.4 stems from the fact that many students rounded their weight to the nearest five pounds.

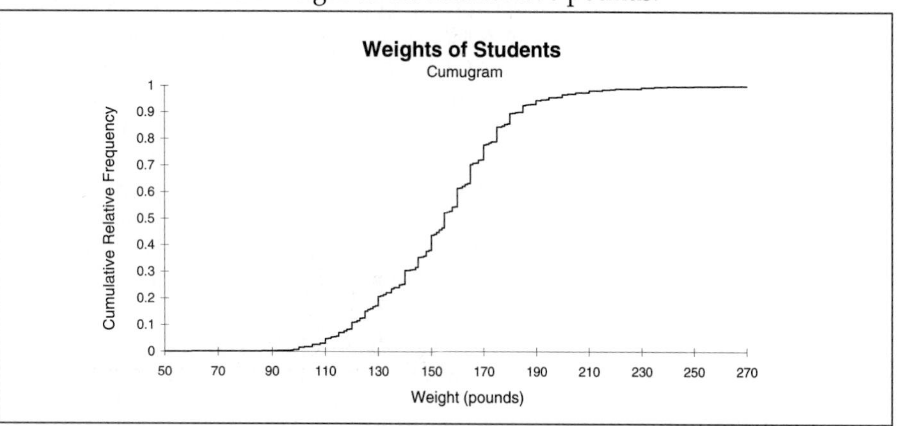

Figure 1.4

It is harder to detect modes and symmetry in cumugrams than in histograms. A mode occurs at a point of inflection in the cumugram where the graph stops increasing at an increasing rate and starts to increase at a decreasing rate. A cumugram is symmetric if its shape is about the same when viewed upside down as it is right-side up. In this way, you can verify that the cumugram in Figure 1.4 is roughly symmetric.

[6] If we check HTWT.XLS we see that the 60-pounder reported himself as 72" tall.

▼ Measures of Centrality

In describing a variable, we often want to specify a single value that is typical of the variable. Such a value is called a measure of **centrality** and indicates roughly where the middle of the distribution lies. The most common measures of centrality are the mode, the median, and the mean. We have already discussed the mode: it is the highest bar in the histogram. If the values are sorted, the **median** is the middle value: roughly half the values of the variable are less than the median and half are greater. In a cumugram, the median is the value on the horizontal axis that corresponds to a value of 0.5 on the vertical axis.[7] The **mean** of a variable is simply the average of its values, or the sum of its values divided by the number of values.[8]

In a symmetric distribution, the mode, median, and mean coincide. However, in a distribution that is skewed to the right, such as the one in Figure 1.2, the mode is less than the median, which is in turn less than the mean. By contrast, in a distribution that is skewed to the left, the mode is greater than the median, which is greater than the mean. The three measures of centrality for the student weight data and the Boston selling-price data are as shown in Table 1.1.

Table 1.1

	Mode	Median	Mean
Student Weights	155-164	155	154.2
Boston Selling Prices	$25,000-$29,999	$28,000	$32,809

The relationships among the mode, median, and mean indicate that the weight data are fairly symmetric, while the price data are clearly skewed.

These three measures of centrality have different sensitivities to a variable's distribution and units of measurement. The mode is perhaps most subject to the way the data are measured, and therefore least typical. For example, if students' weights are recorded to the nearest five pounds, the mode may be substantially different from the mode we would observe if weights were recorded to the nearest ten pounds; if weights were recorded to the nearest ounce, there might be no mode at all since every student might have a different recorded weight. By contrast, the mean is sensitive to extreme values of a variable. For example, if you were reporting starting salaries of 100 students graduating from college, and one of the students signed as a professional basketball player with a starting salary of $1 million, the mean might be a distorted measure of centrality. For its part, the median is not sensitive to extreme values, but for this reason fails to capture such variation in the data.

The choice between the mean and the median is often a matter of judgment, as the following example illustrates. Two groups of workers with comparable skills and job requirements receive the salaries shown in Table 1.2.

Table 1.2

Group A	Group B
$15,000	$25,000
$16,000	$26,000
$30,000	$30,000
$31,000	$45,000
$33,000	$60,000

Both groups have the same median salary of $30,000. However, a quick glance at the data reveals that Group A receives substantially lower salaries than Group B: the means of $25,000 and $37,200, respectively, more accurately convey the differences between the two groups.

---

[7] If the values of a variable are listed in a column of a spreadsheet, you can find the median by sorting the values and finding the value in the middle of the sorted list. The =MEDIAN() function in Excel finds the median value for you.

[8] The mean is easily computed in Excel using the =AVERAGE() function.

The Appendix to this chapter shows how the mode, median, and mean can each be thought of as a measure that is the best solution for a particular decision problem.

▼ Measures of Location

Fractiles serve as a measure of location. They represent the lowest value of a variable below which some specified fraction of the observations lie.[9] Any fractile can be found from a cumugram or a sorted list of the values. For example, you can use a cumugram to find the 0.75 fractile by locating 0.75 on the vertical axis, and then reading over to the curve and down to the horizontal axis. You should verify from the weight data in Figure 1.4 that the 0.75 fractile is approximately 170 pounds. Alternatively, you can locate the 0.75 fractile by sorting the values of the variable and then finding a value below which three-quarters of the observations lie. The median is the same as the 0.50 fractile.

▼ Measures of Spread

In analyzing a set of observations, we are often interested in knowing how the values of the variable spread out. Do they cluster tightly around the central value, or do their values vary widely? Several measures report the **spread**, or dispersion, of a variable's values.

Standard Deviation. The standard deviation is the most widely used measure of spread. It measures the dispersion of values around the mean of a variable. It is computed by first determining the "deviations," or differences, between each value of a variable and the mean of the variable, then squaring these deviations, finding their average, and taking the square root of this average. More succinctly, the standard deviation is the square root of the average of the squared deviations.[10]

In general, the closer the observations are to the mean, the smaller is the standard deviation. The process of squaring the deviations accentuates values that are far from the mean.

Range. The range also reports the spread of a variable. It simply measures the difference between the highest and lowest values of the variable. It should be used with caution, however, because it is strongly affected by outliers or, if the data consist of a sample, by the chances that a particular observation will be included in or excluded from the sample.

Interquartile Range. Still another useful measure of spread is the interquartile range. This measures the difference between the 0.75 and the 0.25 fractiles, a range that contains half the observations.

▼ Types of Variables

Good data analysis depends on recognizing that the measures of distribution, centrality, location, and spread that can be used in summarizing a variable depend on that variable's type, or scale of measurement.

---

[9] Fractiles and percentiles essentially measure the same thing: the 0.75 fractile is the 75th percentile, for example. Occasionally, the definition of percentile is reversed: depending on context, an exam grade in the 5th percentile may be in the upper 5% or lower 5% of all grades given.

[10] In Excel the =STDEVP() function computes the standard deviation.

Ratio-scale variables. A variable is measured on a **ratio scale** if its values have no natural upper bound and cannot be negative. Dollar or unit sales, dividends, interest rates, inventories, number of employees, gross national product, salaries, prices, and time needed to perform a task are examples of ratio-scale variables. All the measures introduced above can be used to describe ratio-scale variables.

Difference-scale variables. A variable is measured on a **difference scale** if its values can be either positive or negative, with no natural upper or lower bound. Profits, inflation rates, a budget surplus or deficit, trade balances, and growth rates are all difference-scale variables. All the measures of distribution, centrality, location, and spread can also be used to describe difference-scale variables. However, they differ from ratio-scale variables in one important way. Since the variables may take on positive or negative values, it is sometimes inappropriate to use ratios and percentage changes to discuss difference-scale variables. Even though many people would characterize an increase in profits from $10 to $15 million as a 50% increase, that kind of characterization works only when both the starting and ending values are positive, and even so may be a poor descriptive measure. Is a 200% increase in profit from $100 thousand to $300 thousand better or worse than the aforementioned increase?

Ordinal variables. **Ordinal** variables are also numeric, but while "greater than" and "less than" relations are meaningful, differences between values are not. Questionnaires sometimes ask people whether they "strongly disagree, disagree, neither agree nor disagree, agree, or strongly agree" with a statement. If the five possible responses are coded 1 through 5, then the higher a person's numeric value, the more he or she agrees with the question. Nevertheless, the difference between a 3 and a 4 is not necessarily the same as the difference between a 4 and a 5 for a given individual, nor is the difference between a 3 and a 5 necessarily twice as great as the difference between a 4 and a 5.

Because the numerical differences between ordinal numbers are not equal, the mean and standard deviation are, strictly speaking, inappropriate measures. Nonetheless, the mean may convey more about the data than the median. If a group of five respondents replied to one question with coded responses of 1, 1, 3, 4, 4, and to another question with 2, 2, 3, 5, 5, the mean probably tells you more about the difference in the group's reactions to the two questions than the median. Similarly, the standard deviation may provide a better description of spread than strictly correct measures.

Categorical variables. **Categorical** variables consist of both qualitative variables—such as religion and marital status—and numeric labels—such as SIC (standard industrial classification) codes, Social Security numbers, part numbers, and zip codes. Differences, relations of "less than" or "greater than," and ratios have no meaning for categorical variables. For example, there is no meaningful interpretation to the statement that my zip code is 47% higher than yours, or 1,750 higher than yours, or even higher than yours. For this reason, cumugrams, means, medians, and fractiles are not suitable summary measures for categorical variables, although histograms are appropriate.

Dummy variables. A special case occurs when we are considering categorical variables with just two values, such as male vs. female, or married vs. single, or bankrupt vs. solvent, or pass vs. fail. By convention, these variables are coded so that one of the numerical values is a 0 and the other is a 1; the numerical value assigned to each category is completely arbitrary. Since "less than" and "greater than" have no meaningful interpretations, fractiles and medians are not applicable measures. However, the mode and mean are meaningful: the mode reveals which category occurs with higher frequency; and the mean is the fraction of observations in the data that are coded as 1. A dummy variable's standard deviation can be calculated in the usual way, but it also can be expressed as a formula, $\sqrt{f*(1-f)}$, where f is the fraction of 1s, i.e., the mean of the dummy variable. No matter what the value of f, the standard deviation of a dummy variable cannot exceed 0.5.

Table 1.3 lists the key characteristics of, and measures that are applicable to and inappropriate for, each type of variable.

Table 1.3

Key Characteristics of Variable Types

TYPE OF VARIABLE	CHARACTERISTICS	EXAMPLES	MEASURES THAT DON'T APPLY
Ratio-scale	Positive, no upper bound	Prices, GNP, # of employees	
Difference-scale	Positive or negative, no upper or lower bound	Profits, growth rates	Percentage change*
Ordinal	Numeric variable, "ordered"	Class rank, responses on scale of 1 to 5	Mean,* standard deviation,* percentage change
Categorical	Qualitative variables and numeric labels	Marital status, religion, SS#, zip codes	Fractiles, cumugram, median, mean, percentage change
Dummy	Two-valued qualitative variables whose values are coded 0 or 1	Political party affiliation, gender, yes/no	Fractiles, median, percentage change

* Sometimes, although not strictly correct, these measures may summarize variables more effectively than any others.

DESCRIPTION OF TWO OR MORE VARIABLES

Independent and Dependent Variables and the Question of Causation

In many analyses, we are particularly interested in how the value of one variable changes when the value of one or more other variables changes. How do sales of a product change when advertising and price change? What is the effect of education, experience, seniority, and special skills on employees' salaries in an organization? How are motor-vehicle accident rates in various states affected by alcohol consumption, speed limits, amount of driving, and motor-vehicle inspection requirements? In all of these cases, there is one variable (sales, salaries, accident rates) whose variability we are trying to understand or explain. This is commonly called the **dependent variable**.

At the same time there are a number of other variables, changes in whose values are accompanied by changes in the value of the dependent variable. These variables are usually called **independent**, or **explanatory**, **variables**.

When we observe that the value of the dependent variable changes as the value of an independent variable changes, we sometimes use the language of causation to describe this relationship. We say, for example, that the independent variables (e.g., advertising and price) *affect* the dependent variable (e.g., sales), or that the *effect* of a $1 per unit decrease in price is to increase sales by 1,000 units per month. Because the language is so suggestive, it is easy to fall into the trap of assuming that this observed statistical association between variables is evidence of causation: that if we reduced price by $1, sales would necessarily increase by 1,000 units per month on the average. Some statistics texts caution against ever inferring causation based on the relationships among variables in observational (nonexperimental) data. It is easy to be misled. For example, suppose that the greater the number of police per capita in U.S. cities, the higher the violent crime rate, on average. Should we conclude that police cause crime?

By refusing to make any causal inferences on the basis of observational data you can avoid making blunders, but at a great cost. Managers, policy makers, and individuals do look at relationships in past data to try to learn what variables affect some particular variable of interest, and by how much. Managers of retail stores and fast-food establishments decide where to locate new stores on the basis of statistics showing the relationship between various characteristics of existing sites and the success of the stores located on them. Legislators enact vehicle-inspection laws based on statistics showing a relationship between inspection requirements and motor-vehicle death rates. Informed smokers try to decide whether to give up smoking based on statistics showing a relationship between smoking and death rates. The leap from "effects" observed in data to causal effects is a large one. In the remainder of this chapter and in subsequent chapters we discuss pitfalls to be wary of, and methods of acquiring and analyzing data that enhance the chances of your making reasonable causal inferences from the data.

Scatter Diagrams

One way to see how one variable varies with another is to plot a **scatter diagram**. Possible values of an independent variable (x) are displayed on the horizontal axis, and possible values of the dependent variable (y) on the vertical axis. Each point corresponds to the values of the dependent and independent variables on an observation. Figure 1.5 shows weight as the dependent variable and height as the independent variable for the student data discussed previously. The diagram shows that weight tends to increase as height increases, that the overall relationship between weight and height could be represented by a straight or slightly curved line, and that there would be a great deal of scatter around such a line.

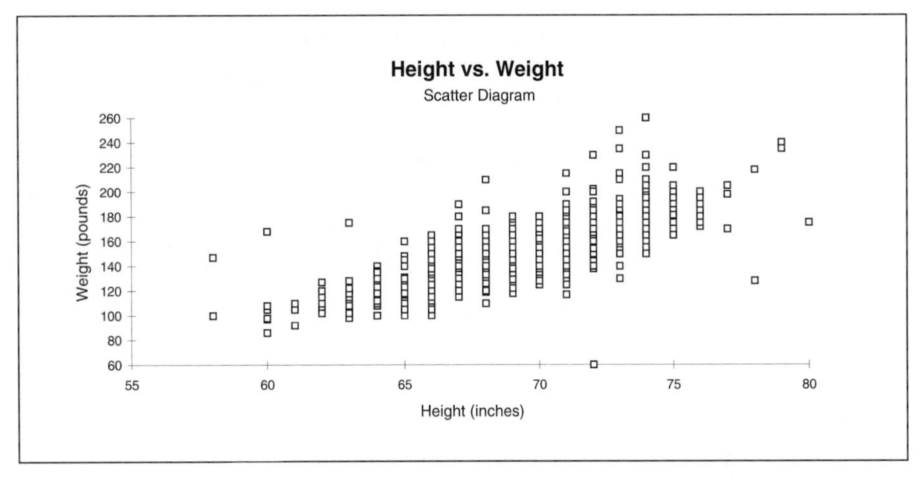

Figure 1.5

One possible explanation for some of the scatter is that the relationship of weight to height may be different for women than for men. Suppose women of a given height weigh less, on the average, than men of the same height. Then the overall relationship between women's weights and heights might be represented by a line or curve below the corresponding curve for men, and this in itself could account for some of the scatter. To explore this hypothesis, we could look at separate scatter diagrams for men and for women, but all of the information can be conveyed on a single diagram in which points for men are represented by a different symbol than points for women. Figure 1.6 shows a scatter diagram identical to that of Figure 1.5, except that each point identifies the person's gender (a circle for women; a square for men). Although there is much overlap, the diagram shows that women tend to cluster to the lower left (shorter and lighter), and for any given height, women tend to weigh less than men. This confirms the hypothesis we started with, and accounts in a small but important way for some of the scatter. Using an **identifier** to represent values, or brackets of values, of a second independent variable in a scatter diagram is often a useful diagnostic tool.

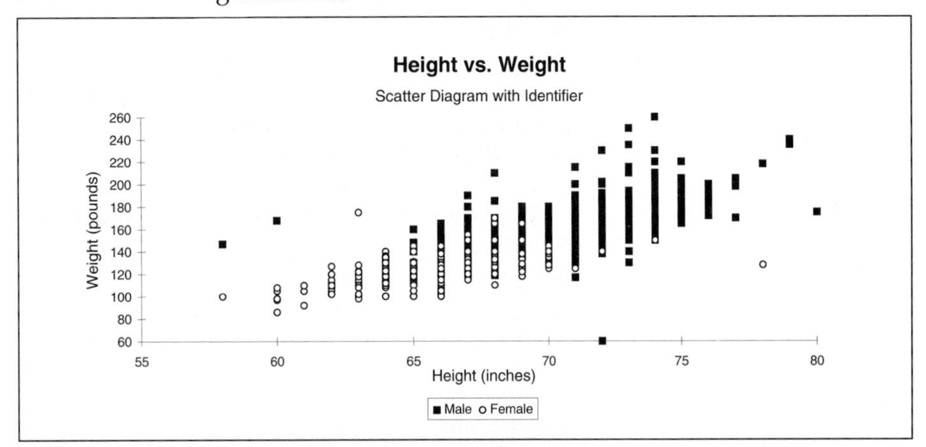

Figure 1.6

Of course, considerable scatter remains. Some of it may be explained in terms of other variables (heights and weights of parents, health, nutrition, ethnicity), and some may be unexplainable or random. Even if we had the data on additional potential explanatory variables, however, we would have trouble graphing their joint effects: we must wait for more powerful analytic tools.

Correlation

When high values of one variable accompany high values of another variable and, similarly, when low values accompany low values, the variables are positively correlated. As an example, companies with a large number of employees also tend to have large asset bases, while those with relatively few employees tend to have smaller asset bases. There are of course exceptions, but if the relationship is true in general, we can say that number of employees and assets are **positively correlated**.

If high values of one variable accompany low values of another (and vice versa), the variables are **negatively correlated**. For example, low inflation tends to accompany high unemployment, and high inflation tends to accompany low unemployment; therefore, inflation and unemployment are negatively correlated.

Two variables that are positively correlated have a scatter diagram whose points cluster around an upward-sloping line that goes from southwest to northeast. Roughly speaking, the closer the points are to an upward-sloping (northeasterly sloping) line, the higher the correlation. The scatter diagram of weight vs. height (Figure 1.5) shows that weight and height are positively correlated. By contrast, a scatter diagram whose points cluster around a downward-sloping (southeasterly sloping) line indicates that the two variables are negatively correlated. In the case of a scatter diagram whose points cluster around a horizontal line, the two variables are uncorrelated with each other.

The **correlation coefficient** is a measure of correlation. To compute it, we first find the **covariance** between two variables, which is a measure of how x and y covary. Suppose we have n observations on each of two variables, x and y. Let m_x be the mean of x on the n observations available, and similarly let m_y be the mean of y. For each observation, compute the product $(x - m_x)*(y - m_y)$, and average these products. This average is called the **covariance**. If, on a particular observation, both x and y are above their respective means, the product for that observation will be positive. The same will be true when x and y are both below their respective means. If most of the observations have x and y values that are jointly above or jointly below their means, the covariance will tend to be positive. If, on the contrary, whenever x is above its mean y tends to be below its mean, and vice versa, the covariance will tend to be negative. The magnitude of the covariance depends on the variability of the x values and the y values, and the degree to which they vary together.

To find the **correlation coefficient,** we next divide the covariance by the product of the standard deviations of the two variables. The resulting number cannot exceed +1, nor be less than –1. A correlation coefficient of +1 represents a case where the points in a scatter diagram all lie exactly on an upward-sloping straight line; a coefficient of –1 represents a case where the points all lie on a downward-sloping straight line. The correlation coefficient for height vs. weight is +0.745.

The correlation coefficient is a useful way of describing the extent to which two variables are linearly related. It is a pairwise measure: if you have three variables, x, y, and z, you can compute the correlation between x and y, between x and z, and between y and z. These pairwise correlations are often presented in a tabular form that is called a **correlation matrix**.[11] Table 1.4 gives a sample correlation matrix of height vs. weight. Since the correlation of x vs. y is the same as y vs. x, a correlation matrix typically gives the correlation between a pair only once.

Table 1.4

	HT	WT
HT	1.000	
WT	0.745	1.000

[11] The =CORREL() function in Excel will calculate the correlation between two variables.

As was true in our discussion of independent and dependent variables, high correlation between x and y does not necessarily mean that there is a causal relationship between x and y. For example, high correlation between stock prices and house sales over time may be due wholly or in part to the fact that both tend to increase as interest rates decline, whether or not they are causally related.

Simple Description of Effects

We observed in Figure 1.6 that height and weight are positively correlated and this relationship holds true for women as well as men. If we want to specify these relationships in a quantitative way, we could list the average weight of men in one-inch height increments, and do the same for women. The result would produce a lot of detail, but little understanding of the way weight is related to height. At the opposite extreme, we could assert that each additional inch of height is accompanied by an average of four additional pounds for both men and women, but that women of any given height weigh 25 pounds less, on the average, than men of the same height. This description conveys a large amount of information succinctly and simply.

If the relationships among height, weight, and gender were adequately summarized by the preceding assertion, we would call the relationship **linear** and **additive**. Because we assert that each one-inch increment in a man's or woman's height is accompanied by the same average increment in weight, whether the increment is from 62 to 63 inches or from 72 to 73 inches, the relationship between height and weight for either sex is graphed as a straight line: it is linear. Furthermore, for a man and a woman of the same height, the man's weight, on the average, is asserted to be 25 pounds greater than the woman's, regardless of whether the man and woman in question are both 62 inches tall or are both 72 inches tall. Under these assumptions, the effect of gender is additive.

While linear and additive descriptions of relationships are simple and intuitive, they are not necessarily right. Looking at the same data, one could instead assert that weight should increase with the cube of height, since height is a linear measure and weight a volume measure. Careful analysis of the data might reveal that this nonlinear relationship provides a better description. Alternatively, if the relationship between height and weight were linear, it might be that an additional inch of height increases weight for men by six pounds, and for women by only three pounds, on the average. In this case, the relationship between height and gender and weight would be linear but not additive.

The tradeoff between simplicity and more accurate description often comes down to a matter of judgment. Sometimes, as explained later in this chapter, we can transform a relationship that is neither linear nor additive into one that is both, and thus achieve a much simpler description.

Time Series

In a **time series**, the observations are ordered chronologically and one of the independent variables may be time itself. When time is the only independent variable, the time series reveals the values of some dependent variable over time. As an illustration, Figure 1.7 on the following page displays monthly sales of all retail stores in the United States from June 1982 through February 1988.

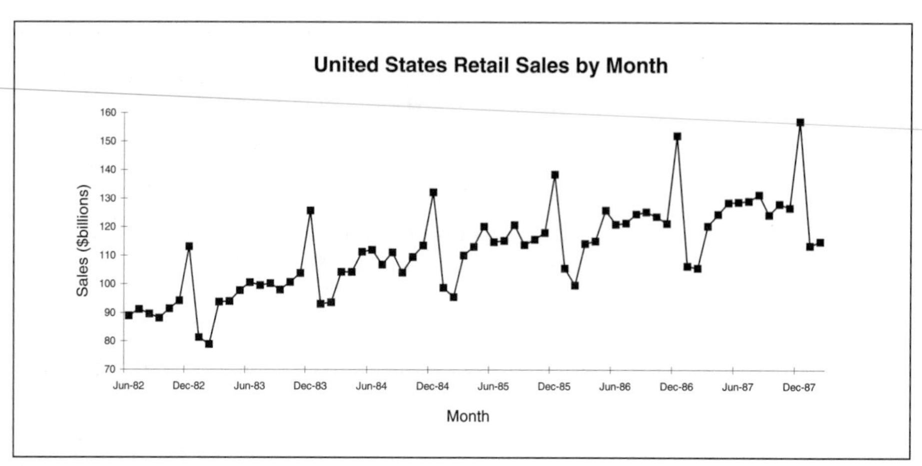

Figure 1.7

The graph reveals two characteristics of this time series:

- There is a pronounced seasonal pattern: sales shoot up in December and then decline precipitously in January and February, before returning to a more normal level for the rest of the year.
- There is a steady upward trend: the December peaks are higher each year, and even the January-February troughs are generally higher; in between, the level also tends to increase with time.

Trends and Seasonals. How can we capture the trend and seasonal effects graphically and display them more vividly? One simple device is to lay out a twelve-month scale on the horizontal axis and display individual line graphs for each year. This is done in Figure 1.8, which shows that each year's data moved higher on the graph, and that there was a pronounced seasonal pattern within each year. Would this display help you forecast retail sales for December 1988? Why or why not?

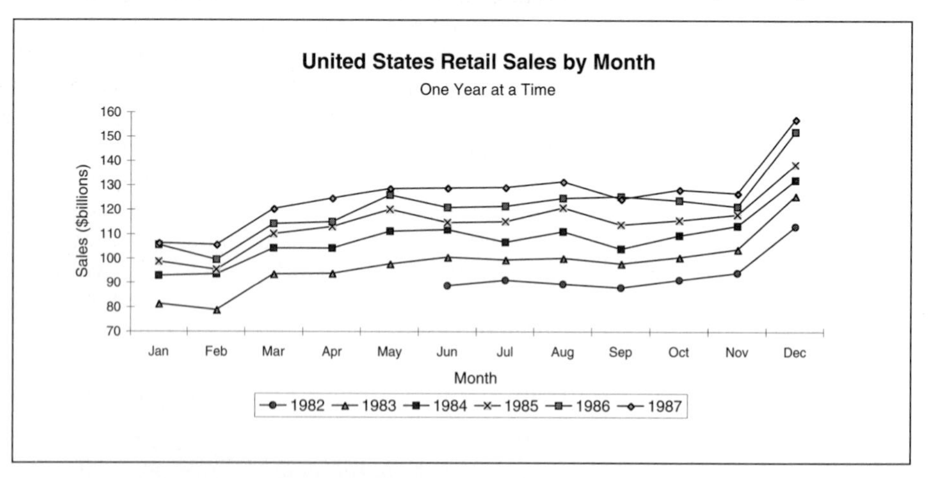

Figure 1.8

Multiple Time Series. Often, when we are trying to understand the behavior of one time series, we introduce another time series to serve as an independent variable. In the retail sales example, we could introduce advertising expenditures over time as an independent variable to help explain how retail sales change over time. We might hypothesize that the stores' advertising expenditures generate sales. Figure 1.9 shows both retail sales and advertising as time series, with different scales for the two series.

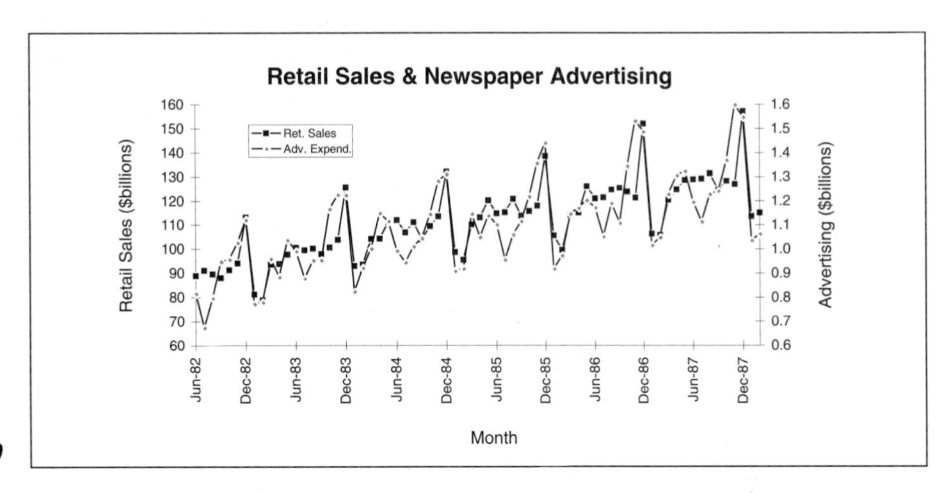

Figure 1.9

This graph shows that:

- Advertising expenditures constituted roughly 1% of retail sales dollars for the entire period.
- The seasonal pattern of advertising expenditures closely mirrors that of retail sales, except that advertising expenditures build up to the December peak more gradually, increasing more in October and November and less in December, than sales do.

Deseasonalization. Because so much of the month-to-month fluctuation in retail sales is due to purely seasonal effects, it is common to report (and think about) sales on a deseasonalized basis. The fact that January sales are below the previous December's is not in itself cause for despair; what you would really like to know is the trend in deseasonalized sales. We shall learn later how to take seasonality into account in forecasting; for now, all you need to know is that retail sales and many other time series that exhibit strong seasonal effects are reported both in natural and in deseasonalized form. Figure 1.10 is a graph of deseasonalized sales. From that graph, you can more easily detect trends in the data.

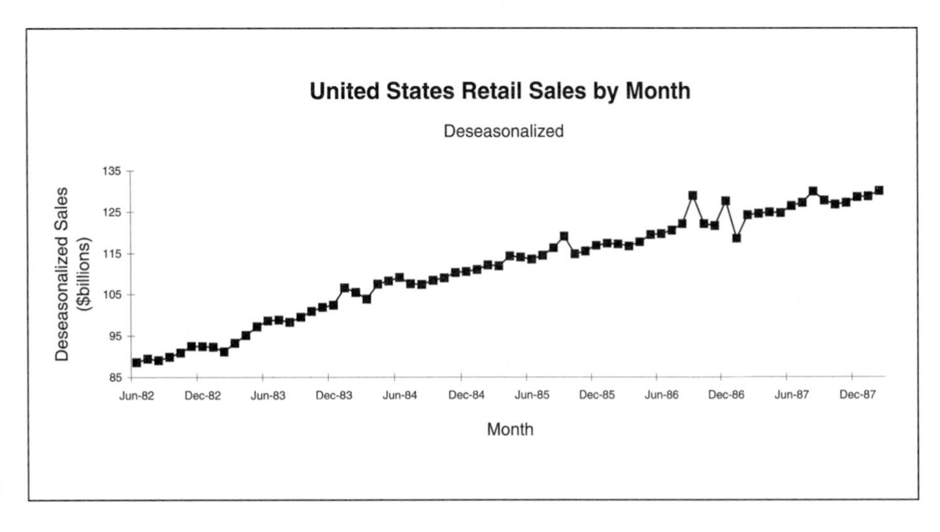

Figure 1.10

Age as an Independent Variable: Life-Cycle vs. Cohort Effects

Data about people (customers, employees, taxpayers, students, etc.) often include age as an independent variable. Markets are often segmented by age: consider sales of popular records, or air travel, or breakfast cereals, for example. Certain behavioral phenomena tend to be thought of as age-related: it is commonly believed that people tend to become more conservative politically as they grow older, for example. Other phenomena seem to depend on a particular generation's exposure to ideas or products: in the 1990s, young people tend to be more adept at using computers than older people. We might predict that a twenty-year-old person is likely to become more conservative politically in thirty years, but that her ability to use a computer will remain high when she turns fifty (and will be greater than that of most people who are currently fifty years old). If the general tendency for people to become more conservative politically as they age will be as clear twenty years from now as it is now, this phenomenon is called a **life-cycle effect**. If the computer ability of people who are currently young persists over their lifetime, this phenomenon is called a **cohort effect**. In interpreting cross-sectional data in which age or some variable related to age (experience, date of birth, year of graduation, seniority) has an important effect on some other variable, it is impossible to determine from the data alone whether the observed effect is a life-cycle or a cohort effect. Because misinterpretation of age-related variables is so common, it is useful to understand how easily one can be misled.

An Example

In 1993 about 500 randomly selected alumni of a university were asked to describe their personal feeling of gratitude toward their school. Figure 1.11 shows the average score on a ten-point scale (with 1 low and 10 high) as a function of year of graduation. The figure shows that older alumni felt more gratitude, on average, than more recent graduates. Is this because whatever the university does to make alumni grateful was less effective in recent years than in the past (a cohort effect), or is it because graduates' feeling of gratitude grows as they age (a life-cycle effect)?

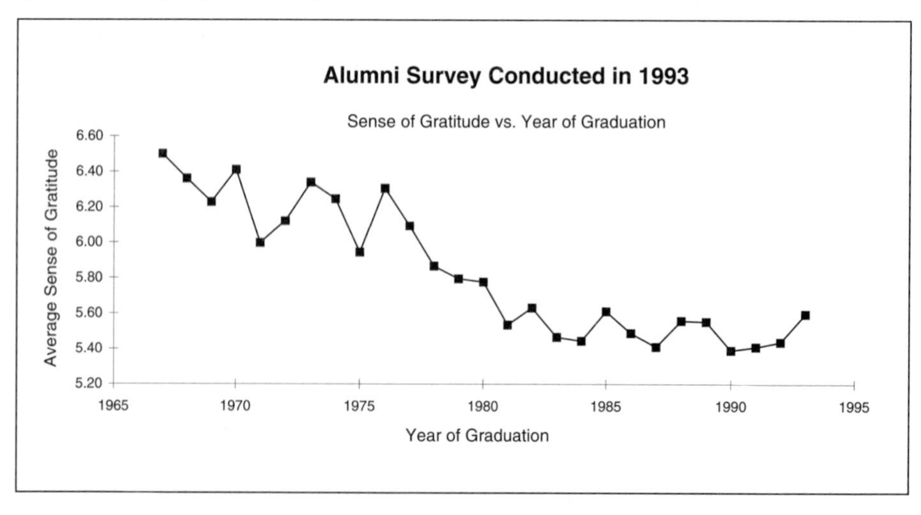

Figure 1.11

If gratitude of alumni is important to the university, then the university's administration should be concerned if the observed trend is a cohort effect, but should not be concerned if it is a life-cycle effect. There is no way of knowing from the data, however, which effect accounts for the trend, or whether it is the result of a combination of effects.

Suppose the identical survey were conducted again five years later, in 1998, and resulted in Figure 1.12, which for classes from 1967 through 1993 looks exactly like Figure 1.11. The evidence suggests that graduates of the class of 1993 remain as ungrateful in 1998 as they were in 1993, while graduates of the late 1960s and early 1970s continue to be as grateful: the effect is a cohort effect. If, on the other hand, the graph for the 1998 survey looked like Figure 1.13, we would attribute the observed trend to a life-cycle effect: in 1998 graduates of the class of 1998 have the same level of gratitude that graduates of the class of 1993 did in 1993; in 1998 graduates of the class of 1980 have the same level of gratitude that graduates of the class of 1975 did in 1993, etc. Of course, the 1998 graph might look different from either of these extremes, but we might find that it looked close enough to one of them to convince us that one of the two effects was more important.

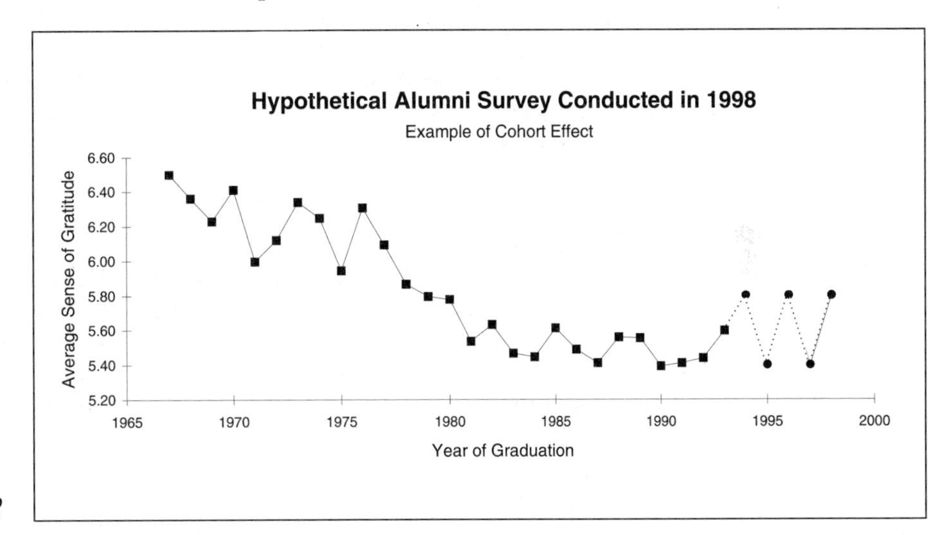

Figure 1.12

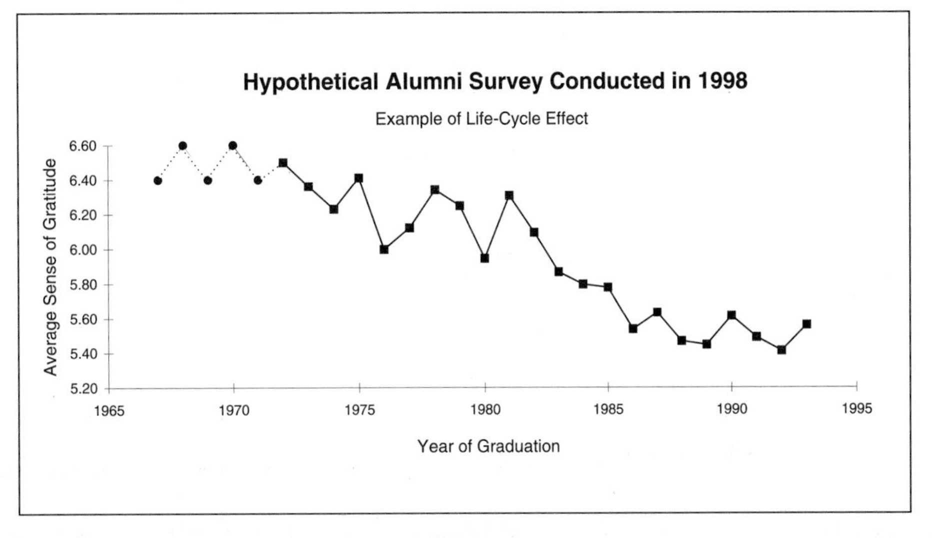

Figure 1.13

The moral of this example is that a single snapshot showing the relationship between age (or an equivalent variable) and some variable of interest is insufficient to sort out life-cycle vs. cohort effects. At least two such snapshots, separated in time, are needed.

Logarithms and Multiplicative Effects

Multiplicative Symmetry: A Second Look

In our discussion of histograms earlier in this chapter we showed that some variables have multiplicatively symmetric distributions. We stated, without any justification, that the logarithmic transformation of some variables has a distribution that exhibits ordinary symmetry. The reason for this is not hard to understand if you recall one key fact about logarithms: the logarithm of the product of two or more numbers is the sum of their logarithms.
Thus, for example,

$$\log(a*b*c) = \log(a) + \log(b) + \log(c) \quad .$$

This is true whether you are using so-called natural logarithms (logarithms to the base e , where e = 2.71828, approximately), or logarithms to the base 10.[12] You should convince yourself, using a spreadsheet or a calculator, that

$$\log(2*3*4) = \log(2) + \log(3) + \log(4)$$

for both natural logarithms and logarithms to the base 10.

Distributions that are multiplicatively symmetric often arise when the value of each observation is the product of a number of small random effects. The logarithm of such a value will then be the sum of a number of small random effects and the distribution of such a set of values is likely to be symmetric. Thus, logarithms convert or **transform** multiplicative effects into additive effects.

Multiplicative Seasonals and Constant Growth Rates

If you turn back to Figure 1.7, the graph of retail sales over time, you will notice that the amplitude of the swings between the December peak and the January-February trough seems to get larger over time. One possible explanation is that the seasonal effect is multiplicative. Suppose, for example, that every December tends to be 20% above normal and every January 15% below normal. Then the difference between December and January will be larger for high levels of sales, and smaller for low levels. Because the level of sales has been increasing over time, the differences will therefore get larger over time. If the seasonal effect of each month is, in fact, a constant multiple of some normal level of sales, then a graph showing a time series of the **logarithm** of sales will depict seasonal effects that add to or subtract from normal the same constant amount for each month: on the logarithmic scale, the seasonal effects will not change with the level of the series. Figure 1.14 shows such a graph; it seems to confirm the multiplicative-seasonal hypothesis.

---

[12] In Excel, =LN() calculates the natural logarithm and =LOG() the base-10 logarithm. On pocket calculators, the LN and LOG keys usually serve the same purposes, respectively.

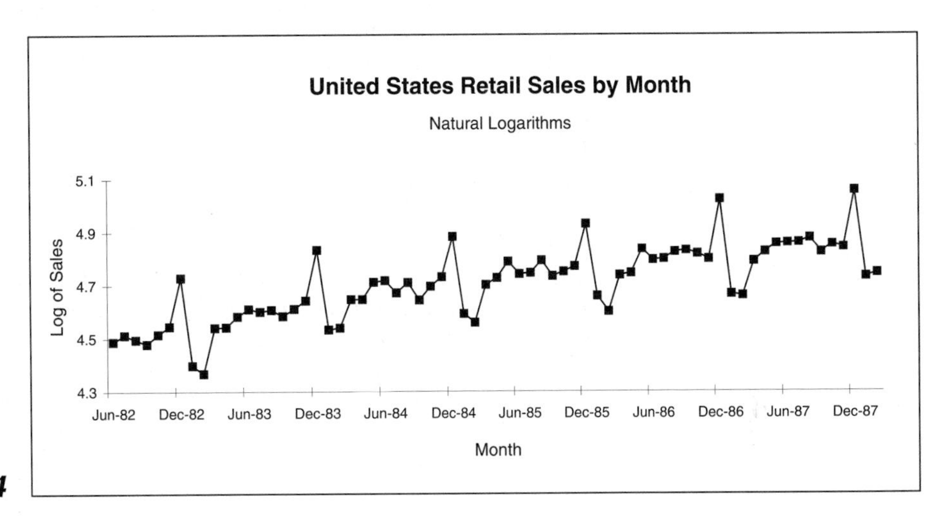

Figure 1.14

A constant growth rate means that successive values in a time series are constant multiples of preceding values. For instance, if a variable increases by 5% per year, each year's value will be 1.05 times the preceding year's. If this is so, the logarithm of the values in the time series will have values that are constant **increments** above preceding values, and therefore a graph of the logarithmic series will have a shape close to a straight line. The U.S. population, the consumer price index, gross national product, and electric power generation all exhibit patterns of multiplicative growth over sufficiently long periods of time. The values increase at an increasing rate, often described as exponential growth. Figure 1.15, a graph of electric energy production in the United States from 1920 to 1983, exhibits this kind of exponential growth, at least through the mid-1970s. Figure 1.16, a graph of the logarithm of the series, is more nearly linear. Notice that by "linearizing" the series we can more easily detect the effects of the 1930s depression, the end of World War II, and the OPEC oil crisis of the 1970s on energy production.

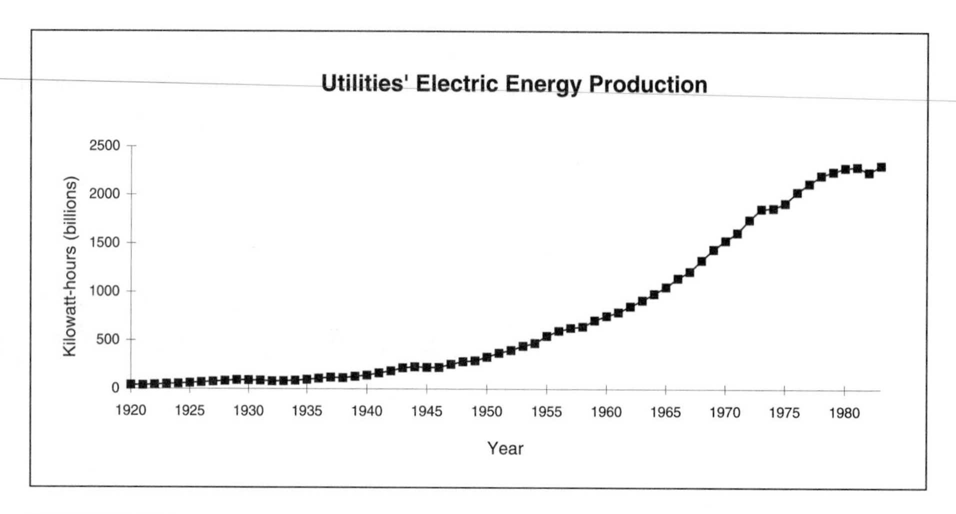

Figure 1.15

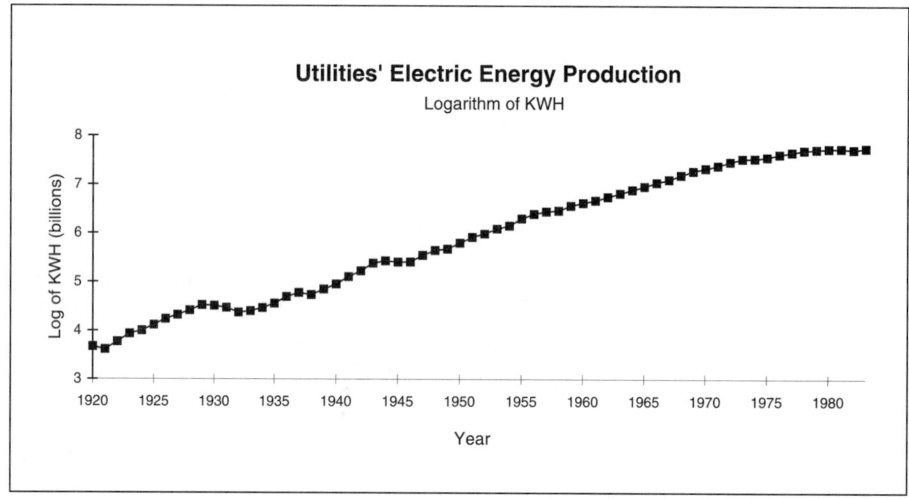

Figure 1.16

Multiplicative Effects in Cross-Sectional Data

Even in cross-sectional data, the values of a variable may depend on the values of other variables in a multiplicative way. A company's salary structure may reward seniority and education, among other things. Suppose that in a given year starting salary for a person without a college education is $20,000, that college-educated employees earn 25% more than employees with comparable seniority, and that each year of seniority confers 5% more salary. Figure 1.17 shows what salaries would be if there were no additional factors affecting salary. Figure 1.18 shows the effects of seniority and education on log(salary). For a given level of seniority, education adds the same amount to log(salary), no matter what the level of seniority; we say that the effect of education on log(salary) is **multiplicative**. When we look at the effect of seniority on log(salary) for either level of education, we observe that it is linear, and this implies that the effect of seniority on salary is multiplicative.

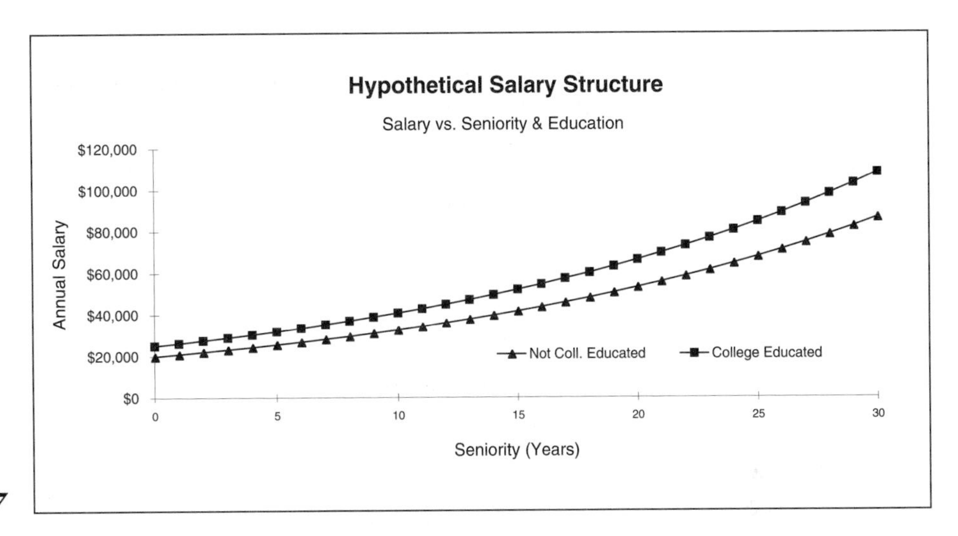

Figure 1.17

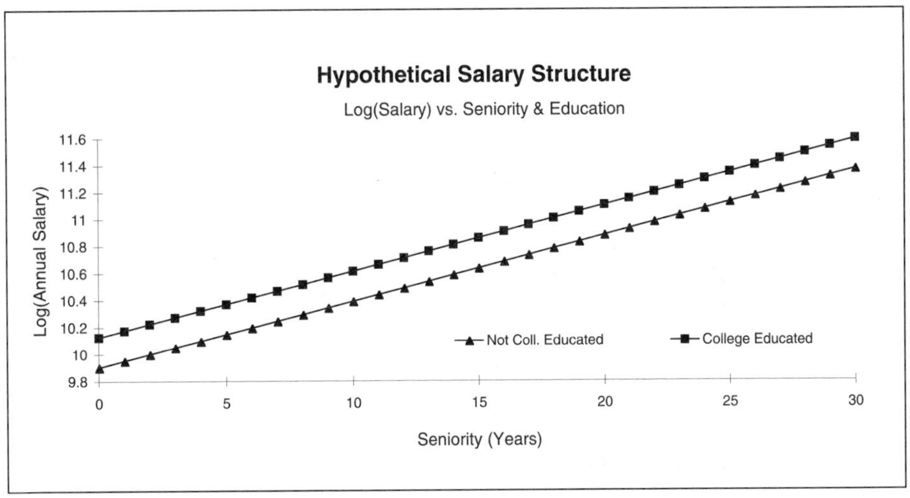

Figure 1.18

In general, when the relationship between two variables is not linear and the dependent variable is measured on a ratio scale, it is a good idea to see whether effects can be more simply explained by looking at graphs of the logarithm of the dependent variable. A logarithmic transformation of a ratio-scale dependent variable may:

- produce a symmetric, rather than a skewed, distribution of the dependent variable.
- produce a linear and additive relationship between the independent variables and the transformed dependent variable, if the effects on the natural dependent variable were multiplicative.

In performing transformations on our data, such as a log transform, we trade off complexity to get simplicity in the form of straight lines and additive effects.

Graphs with Logarithmic Scales

Ordinary graphs are plotted on an arithmetic scale, so that each increment on an axis represents equal distance (i.e., the distance from 1 to 2 is the same as the distance from 2 to 3). Sometimes graphs are plotted on a ratio or logarithmic (log) scale, in which equal distances represent equal percent changes (i.e., the distance from 1 to 2 is the same as the distance from 2 to 4).

Figures 1.19A–C show three different ways of plotting monthly values of the Standard and Poors 500 Stock Index, from January 1968 through January 1993. Part A shows values of the Index on an arithmetic scale, B shows values of log (Index) on an arithmetic scale, and C shows values of the Index on a log scale.[13] Both B and C are essentially the same graphs: when you want to capture multiplicative effects of a ratio-scale variable graphically, graphs of either sort are equally appropriate.

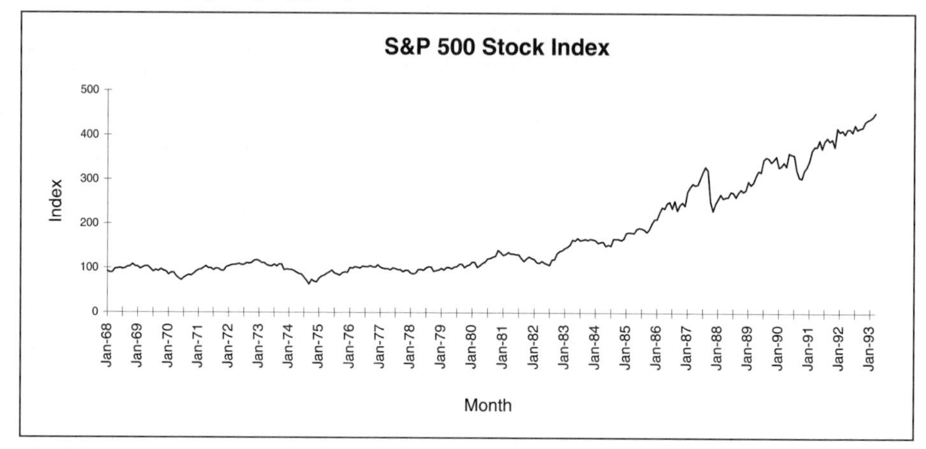

Figure 1.19A

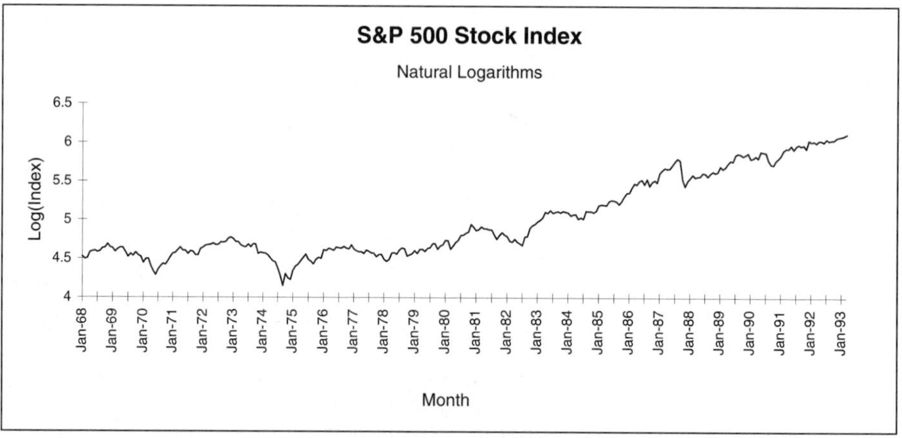

Figure 1.19B

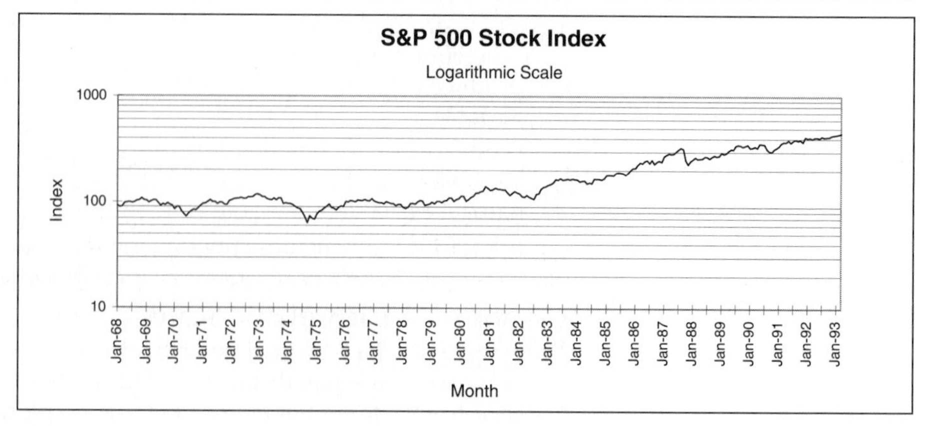

Figure 1.19C

13 Excel permits you to create either line charts or XY (scatter) charts with logarithmic scales.

Notice that both the B and C graphs convey a great deal more information about stock-market fluctuations than does A. Looking at A, for example, it appears that the major stock-market declines occurred during the crash of 1987 and prior to the Gulf War in late 1990. Both B and C show, however, that the percentage decline from 1969 to 1970 was as severe as—though more protracted than—the 1987 crash, while the decline from late 1972 to late 1974 was substantially more severe than the crash.

Appendix: Measures of Centrality as Solutions to Decision Problems

It is instructive (and will prove useful in the context of regression) to think of the mode, median, and mean as solutions to decision problems involving finding the optimal value to represent all the values of a variable.

Consider the following fifteen observations on a variable: 1, 3, 4, 7, 7, 7, 8, 9, 12, 16, 19, 19, 25, 30, 37. Suppose you were asked to pick a particular number (not necessarily one corresponding to one of the observed values of the variable), after which one of these fifteen values would be picked at random. If you were penalized $1 if you were wrong, what number would you pick? You could draw the decision tree, but you can easily see that the number that minimizes your expected penalty is the one that occurred most frequently—the mode, or 7. You would have only 3 chances in 15 (20%) of avoiding the $1 penalty, but there is nothing that you could do that would improve your chances. The expected value of the penalty would be $0.80 if you picked 7, and would be higher if you picked any other number.

Now suppose that you were asked to pick a number, but were penalized $1 times the amount of your error. For instance, if you picked 7 and the number 3 were drawn at random, you would be penalized $4; if, instead, 30 had been picked at random, you would have been penalized $23. What number should you pick? This is a critical-fractile problem: if you think tentatively of picking a number, say 7, and then ask for the incremental gain or loss of picking 8 instead, you would *gain* $1 (relative to the penalty you would have incurred had you picked 7 instead) if the random draw were 8 or higher, but would *lose* $1 if the random draw were less than 8. The same incremental gains and losses would apply for any decision to add one to the number you had tentatively picked. Therefore, to minimize your expected penalty, you should pick the G/(G + L) = 1/(1 + 1) = 0.50 fractile, or the **median**, which in this case is 9. The expected value of the penalty would be 8/15 + 6/15 + 5/15 + 2/15 + 2/15 + 2/15 + 1/15 + 0/15 + 3/15 + 7/15 + 10/15 + 10/15 + 16/15 + 21/15 + 28/15 = 8.07, and would be higher (or at least no lower) if you picked any other number.

Suppose that instead of being penalized $1 times the error, you were penalized $1 times the *square* of the error. If you picked 7 and the random draw were 3, you would be penalized $16, for example. It can be shown that under this penalty structure, you should pick the *mean*, or 13.6. You can compute the minimum expected penalty as 103.97.

The mode is the right number to pick if you avoid a penalty only if the random draw is exactly equal to your pick; the median is correct if the penalty is proportional to the error; and the mean is correct if the penalty is proportional to the square of the error. In this sense, the mean is said to be a **least-squares** estimate.

Figures 1.20,[14] 1.21, and 1.22 show the expected penalty as a function of the number picked for the three problems we have considered. The black dot shows the pick for which the expected penalty is minimized.

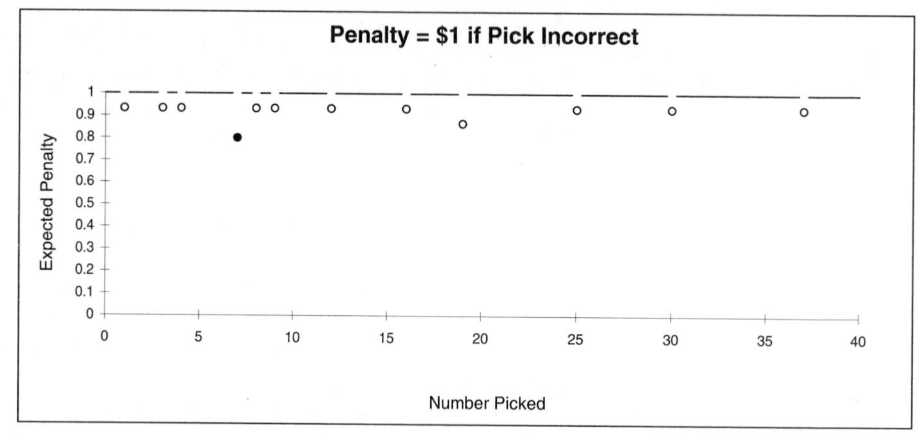

Figure 1.20

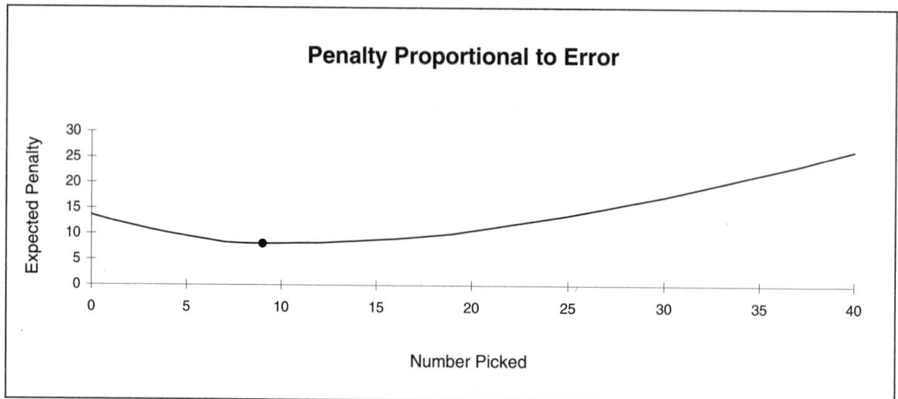

Figure 1.21

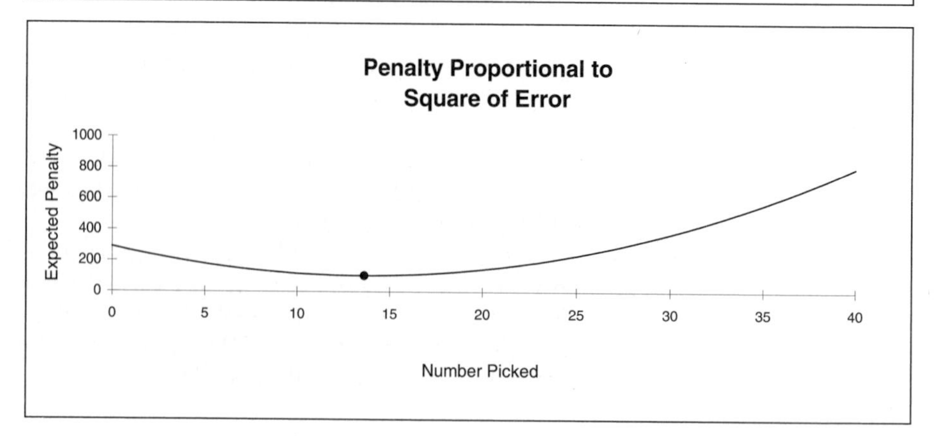

Figure 1.22

14 The broken horizontal line at the value 1 in Figure 1.20 indicates that picking any number that is not one of the twelve values actually observed will result in a certain penalty of $1.

Exercises on Interpreting Data [CHAP1EX.XLS]

Question
Who was the more valuable player, with respect to batting statistics, in 1991? How could Murray beat Merced both as a left-handed and a right-handed batter and still have a lower batting average?

Problem 1: Batting Averages

In 1991, Eddie Murray of the Los Angeles Dodgers and Orlando Merced of the Pittsburgh Pirates both played first base in baseball's National League. Murray and Merced were solid "switch hitters," which means they could bat either left-handed or right-handed.

A player's batting average (BA) is one of the most important statistics in baseball. The BA is calculated by dividing the number of hits by the number of plate appearances—or at bats. So a batter who gets a hit one out of every three times at the plate would have a BA of .333. When contracts are negotiated and player trades are considered, a difference of only a few BA points can be worth hundreds of thousands of dollars.

Switch hitters are valuable players because they can adjust to different types of pitchers. Facing a left-handed pitcher, a switch hitter would usually bat right-handed because the ball is more likely to move in toward the batter, making it easier to hit. Managers need to know a switch hitter's BA from each side of the plate when deciding whether to have that player face a left-handed or right-handed pitcher.

The switch-hitting statistics for Eddie Murray and Orlando Merced looked like those shown in Table 1.5.

Table 1.5

1991 Left- and Right-Handed Batting Statistics

	Eddie Murray Batting Average	Orlando Merced Batting Average
Left-Handed	.295	.285
Right-Handed	.217	.208
Total	.260	.275

Question:
Provide a forecast of the number of replacement tires that will be demanded. (Remember, every passenger mile is driven on four tires.)

Problem 2: Replacement-Tire Forecast

The tire industry manufactures tires for a variety of vehicles and aircraft around the world. Demand forecasts for tires[15] are developed separately for different vehicle types, by geographic region, and by whether the tires are for "original equipment" (i.e., tires that are bundled in the sale of new vehicles) or "replacement."

We shall be concerned here with automobile passenger tires sold in North America in the replacement market. Tires on automobiles are replaced when they wear out or are damaged beyond repair. The annual demand for replacement tires depends, therefore, on their longevity (the number of miles driven before a tire wears out) and the total number of passenger miles driven per year.

[15] This problem is based on data in *Firestone Tire and Rubber Company* (see Chapter 4).

Table 1.6

PROJECTED LONGEVITY (MILES)	PERCENT OF TIRES
18,000	20%
24,000	20%
32,000	20%
42,000	20%
80,000	20%

Forecasts of the latter can be made with considerable accuracy, but tire longevity varies considerably, depending on road conditions encountered, driving style, condition of the vehicle, and type of tire. Data on tire longevity can be acquired by measuring treadwear on original-equipment tires, observing the mileage odometer, and making projections based on knowledge of the rate at which treadwear increases with mileage.

Recently, a study of treadwear on 50,000 vehicles projected longevity[16] as shown in Table 1.6. Passenger miles were forecast to be one trillion in the year in question.

QUESTION: Interpret the data in the table. Do they support or refute the proposition that the rich got richer, the poor got poorer, and the middle class stagnated?

Problem 3: Did the Rich Get Richer?

In the spring of 1992, a debate among economists and demographers played a role in the presidential campaign. Did the "rich get richer and the poor get poorer" during the Reagan-Bush years? Bill Clinton said "yes," and he cited supporting data. President Bush and his supporters were quick with rebuttals.[17]

Two reports, "correcting" errors made by analysts who concluded that the income gap grew, were released in the summer of 1992, and both reports were widely praised and denounced—depending on one's political leanings.[18] One table, in particular, seemed to show that people who were relatively poor in 1977 did much better in the 1980s than people who started out relatively rich. This was precisely the opposite conclusion being preached by candidate Clinton (by then the Democratic nominee). In fact, one might even argue, based on this analysis, that "trickle down" economics actually worked.

The table in question (Table 1.7) is based on responses to a panel survey in which the same people (aged 25 to 54 in 1977) were followed from 1977 through 1986.

Table 1.7

Average Family Income and Change in Average Family Income, 1977–1986, of Families Grouped by Their Income Quintile in 1977[19] (1991 Dollars)

QUINTILE IN 1977	AVERAGE FAMILY INCOME		CHANGE IN AVERAGE FAMILY INCOME	
	1977	1986	AMOUNT	PERCENTAGE
First	$15,853	$27,998	$12,145	77%
Second	31,340	43,041	11,701	37
Third	43,297	51,796	8,499	20
Fourth	57,486	63,314	5,828	10
Fifth	92,531	97,140	4,609	5

16 The data are presented in just five categories for the sake of simplicity.

17 As background on the debate, see Anne B. Fisher, "The New Debate Over the Very Rich," *Fortune,* June 29, 1992, pp. 42-54, and Marvin H. Kosters, "The Rise in Income Inequality," *The American Enterprise,* November/December, 1992, pp. 29-37.

18 Isabel V. Sawhill & Mark Condon, "Is U.S. Income Inequality Really Growing? Sorting Out the Fairness Question," *Policy Bites,* The Urban Institute, June 1992; and "Household Income Changes Over Time: Some Basic Questions and Facts," U.S. Department of the Treasury, Office of Tax Analysis, July 1992.

19 Derived from Isabel V. Sawhill & Mark Condon, *op. cit.* Reproduced with permission.

WORKED EXAMPLES IN DATA ANALYSIS USING SPREADSHEETS

All spreadsheets permit you to produce charts and other analyses like those contained in this chapter. Because summarizing data in the form of charts, tables, or summary statistics gives you insight and lets you communicate your findings to others, this section explains a few tricks that make reasonably sophisticated summaries quite easy to produce. We illustrate those tricks using Excel 5.0. Other spreadsheets and other versions of Excel can do similar analyses, but you must learn different commands.

We assume that you have taken the Excel tutorials and have sufficient knowledge of Excel to perform basic functions without detailed instructions on keystrokes. Where appropriate, we give tips on how to perform certain operations using special tools built into Excel.

In this section, we show how to obtain summary statistics, make histogram and cumugram charts, compute fractiles and correlation coefficients, and produce various forms of scatter diagrams and time-series charts. We illustrate these techniques using two data files that are included in your data diskette: HTWT.XLS, which shows height, weight, and gender of each of 768 students in a recent first-year MBA class at Harvard Business School; and RETAIL.XLS, which shows retail sales and advertising in the United States by month from June 1982 through February 1988.

Analysis of a Single Variable

Summary Statistics. To start the analysis, open the file HTWT.XLS. (To avoid accidentally overwriting the data, start by saving the file as HTWT1.XLS.) Table 1.8 displays the first few rows of data (the actual data extend down to row 774).

Table 1.8

	A	B	C	D	E	F	G
1	**HTWT: Survey Response**						
2	Height, Weight and Gender of a First Year MBA Class						
3							
4			Height	Weight	0=Male		
5			(Inches)	(pounds)	1=Female		
6			**HT**	**WT**	**M/F**		
7			68	140	0		
8			69	155	0		
9			66	120	0		
10			72	180	0		
11			72	165	0		
12			69	175	0		
13			72	165	0		
14			71	130	0		
15			71	175	0		

Table 1.9 shows how to compute summary statistics for the height data that appear in column C in Table 1.8. The first column of Table 1.9 on the following page gives the name of the statistic, the second column gives the Excel function that computes the statistic, and the third column gives the value of the statistic.

Harvard Business School note 2-191-113. Worked Examples in Data Analysis Using Spreadsheets, was prepared by Professor Arthur Schleifer, Jr.

Table 1.9

SUMMARY STATISTIC	EXCEL FUNCTION	VALUE OF STATISTIC
Number of Observations	=COUNT(C7:C774)	768
Smallest Value	=MIN(C7:C774)	58
Largest Value	=MAX(C7:C774)	80
Mean	=AVERAGE(C7:C774)	69.5156
Median*	=MEDIAN(C7:C774)	70
Mode	=MODE(C7:C774)	72
Standard Deviation**	=STDEVP(C7:C774)	3.4422

*If there are an odd number of observations, Excel will give the value of the middle observation; if the number of observations is even, Excel will give the average of the two middle values. Thus it will give 5 as the median of 1, 2, 5, 17, 100, and 3.5 as the median of 1, 2, 2, 5, 17, 100. These results may differ from the results you get using a cumugram (see the discussion on fractiles later in this section). The differences will seldom be material, but in case of doubt, use the cumugram method.

**The function =STDEVP gives the population standard deviation, the value that corresponds to the standard deviation as defined in this chapter. The function =STDEV gives the sample standard deviation, a statistic that need not concern us now, but one that we shall find useful in later chapters.

Histograms. To create a histogram, (1) define "bins" into which the individual observations can be sorted, then (2) use Excel's Histogram routine to count the frequency with which values fall into each bin and to plot the result.

For the height data, we have seen that the smallest value is 58 inches and the largest 80 inches. Therefore we can create bins as a column of integers starting at 58 and ending at 80.

For these exercises you will use the file named HTWT1.XLS, which you previously saved.

To create the bins:

- Type **Bin** in cell F6 and **Frequency** in cell G6.
- Type **58** in cell F7 and **59** in cell F8.
- Highlight cells F7 and F8, release the mouse button, then move the cursor to the lower right-hand corner (the "fill handle" of the cell containing 59). The cursor changes to a black plus sign (+).
- Drag the fill handle down until you reach 80 (row 29). If you fall short of this row, repeat from where you stopped. If you overshoot, delete the entries above 80. When you finish, the range F7:F29 should be highlighted.

Now, to perform the analysis and create the histogram:

- Click the **Tools** menu, then click **Data Analysis.*** A menu of analytic tools displays.
- Highlight **Histogram** and click **OK**. A Histogram dialog box displays.
- In the Histogram dialog box, type **C7:C774** in the box next to Input Range.
- Type **F7:F29** in the Bin Range section of the dialog box.

* If you don't see a Data Analysis option on the Tools menu, click Add-Ins, highlight Analysis ToolPak in the Add-Ins dialog box and click OK. Then click the Tools menu again. Now you should see a Data Analysis option.

- If you like, you may specify an Output Range that puts the histogram on the same worksheet as your original data. However, in most cases you will want to accept the default New Worksheet Ply, which puts the output into a separate worksheet.
- Make sure that the boxes for Pareto and Cumulative Percentage do not contain an x, and that the box for Chart Output is selected (contains an x). (Click on the box to add an x, or click again to remove an x you don't want.)
- When all the dialog box settings are correct, click **OK**.

Excel produces two columns in a new worksheet, as shown in Table 1.10. The first column contains the bin values and the second contains the number of observations in each bin. The number 2 in the Frequency column to the right of 58 in the Bin column means that there were two observations with values of 58 or more but lower than 59 (the next value in the Bin column). (In this sample, all values are integers, so you can interpret this as meaning that two heights were reported as 58 inches exactly. Had heights been reported to one-tenth of an inch, the Frequency column would give the number of observations for which reported heights were 58.0, 58.1, ... 58.9 inches, but not for 59.0 inches.) Notice that there is one height of exactly 80 inches, but no heights opposite "More" (meaning more than 80 inches) in the Bin column.

Table 1.10

	A	B	C	D	E	F	G	H	I
1	*Bin*	*Frequency*							
2	58	2							
3	59	0							
4	60	6							
5	61	3							
6	62	10							
7	63	17							
8	64	26							
9	65	36							
10	66	51							
11	67	51							
12	68	76							
13	69	82							
14	70	98							
15	71	60							
16	72	105							
17	73	57							
18	74	50							
19	75	21							
20	76	9							
21	77	3							
22	78	2							
23	79	2							
24	80	1							
25	More	0							
26									

Next to the two columns of numbers, Excel produces a histogram chart, as shown in Figure 1.23. Notice that the bars are labeled with the bin value. In this case, the labeling conveys exactly the right information: there were 51 heights of 66 inches, for example, as the chart reports.

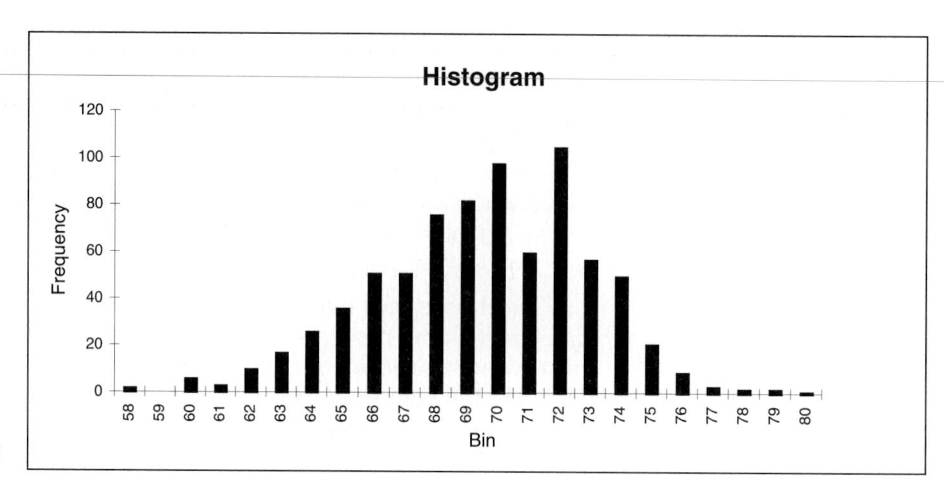

Figure 1.23

Let's plot the data again using bin values increasing by two—58, 60, 62 ..., 80. Copy the height, weight, and gender data to a new worksheet, using the Copy and Paste commands, create the bins and then the histogram, as in Figure 1.24.

The label for the first bar, 58, represents the number of heights that were either 58 or 59 inches; similarly, the label 70 represents the number that were either 70 or 71, etc. You have to be careful when you read histograms produced by Excel.

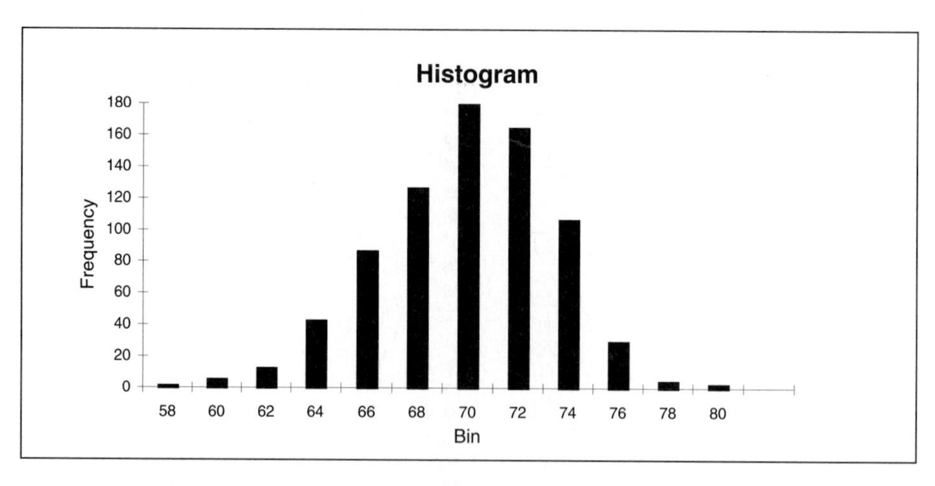

Figure 1.24

Histogram of Relative Frequencies. Taking the results in Figure 1.24, suppose we wanted the vertical axis to represent *relative* frequencies, i.e., the fraction of observations that fell into each bin. For instance, 180 observations fell into the 70 bin, representing the number of students with heights of 70 or 71 inches. This is a fraction (180/768 = 0.234, or 23.4%) of all students. Let's construct a graph to show frequencies in terms of such fractions.

Start with the two-column output generated by the second histogram example (see Table 1.11). At the bottom of the Frequency column (cell B15), have Excel compute the total number of observations (we know that the total is 768, but suppose we didn't).

Table 1.11

	A	B	C	D	E	F	G	H	I
1	*Bin*	*Frequency*							
2	58	2							
3	60	6							
4	62	13							
5	64	43							
6	66	87							
7	68	127							
8	70	180							
9	72	165							
10	74	107							
11	76	30							
12	78	5							
13	80	3							
14	More	0							
15									

TIP:

You can type = , then simply put the cursor on cell B2, type /, then place the cursor on cell B15 and press the F4 key to add $. Because you want to divide each value in the B column by 768 (the value in B15), you want Excel to recognize B15 as an "absolute" address. The format $B$15 signifies an absolute address.

To sum the column:

- Click cell **B15**, click the **AutoSum** (Σ) **button** on the toolbar, then press **[Enter]**.
- In cell C2, type the formula **=B2/$B$15** and press **[Enter]**.
- Now drag the fill handle in cell C2 down through cell C15. The results are shown in Table 1.12.

Table 1.12

	A	B	C	D	E	F	G	H	I
1	*Bin*	*Frequency*	*Rel Freq*						
2	58	2	0.002604						
3	60	6	0.007813						
4	62	13	0.016927						
5	64	43	0.05599						
6	66	87	0.113281						
7	68	127	0.165365						
8	70	180	0.234375						
9	72	165	0.214844						
10	74	107	0.139323						
11	76	30	0.039063						
12	78	5	0.00651						
13	80	3	0.003906						
14	More	0	0						
15		768	1						

You can use Excel's ChartWizard to draw the histogram of relative frequencies. To begin, highlight the Bin column and the Relative Frequency column:

- Hold down **[Ctrl]** (if you're using Windows) or **[Command]** (on the Macintosh), click in cell **A2**, and drag from A2 to A13.
- Release the mouse button (but continue to hold down [Ctrl] or [Command]), click in cell **C2**, and drag from C2 to C13.

- Click the **ChartWizard icon** on the toolbar. The cursor becomes a plus sign (+) with a bar chart at the lower right.
- Click and drag the cursor on your worksheet to form a rectangle (if you prefer, you can use a new worksheet).

Now the ChartWizard takes over.

- It asks whether the range you have selected ($A$2:$A$13,$C$2:$C$13) is correct. Since it is correct, click **Next**.
- Now it asks you to select a chart type. Double-click **Column**.
- Select a chart format by double-clicking on "**1**."
- In the Control box that appears, specify that the data are in Columns, that you are using the first 1 column for category (X) axis labels, and that you are using the first 0 rows for legend text. Then click **Next**.
- ChartWizard then invites you to provide various titles. We do not need a legend, so click **No** under Add a Legend? You may, at your option, add chart and axis titles. Click **Finish** to tell the ChartWizard to create the chart. After it appears, you can use the resize arrow at the lower-right corner of the chart to change its width and height until it looks something like Figure 1.25.

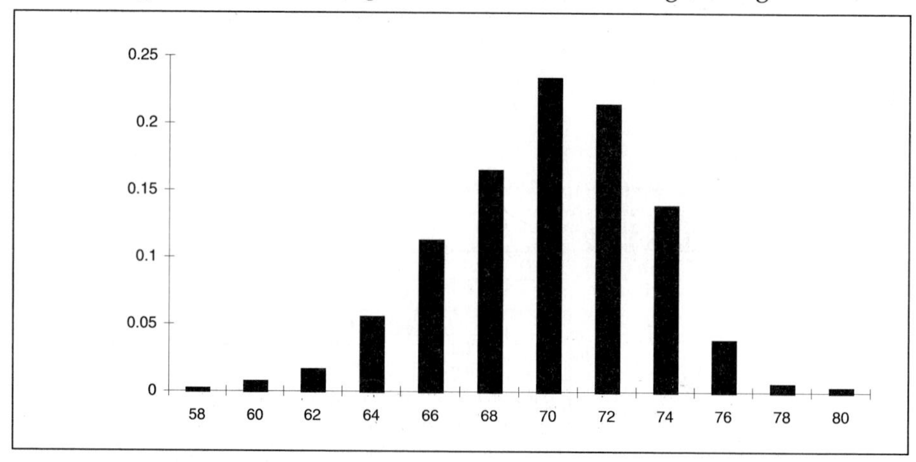

Figure 1.25

Cumugrams. Excel has various routines for drawing cumulative distributions, but none of them produces a stairstep cumugram that correctly displays the fraction of observations that lie at or below any specified value.

A cumugram add-in (CUMUGRAM.XLA) is provided on your data diskette. Copy it to a directory or folder. When you want to use it:

- Open the file that contains the add-in, click the **Window** menu, and switch back to your original workbook.
- Highlight the column of data that contains the variable whose cumugram you want, and simultaneously press **[Ctrl]** (if you're in Windows) or **[Command]** (on the Macintosh), **[Shift]**, and **[C]**. A cumugram chart appears on a worksheet called Cumugram. See Figure 1.26 for a cumugram of the height variable.

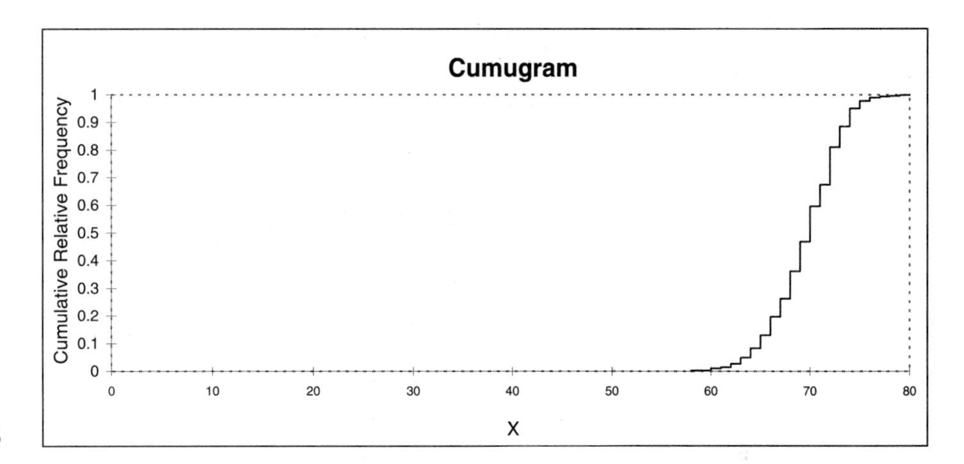

Figure 1.26

TIP:
The cumugram add-in doesn't always do a good job of setting the horizontal scale values. You can get rid of all the white space to the left of the cumugram by changing the scale. To do so, double-click the horizontal axis. When a Format Axis menu pops up, choose the Scale tab, then change the minimum value to 55 and the maximum value to 85. Click OK to get a cumugram like that shown in Figure 1.27.

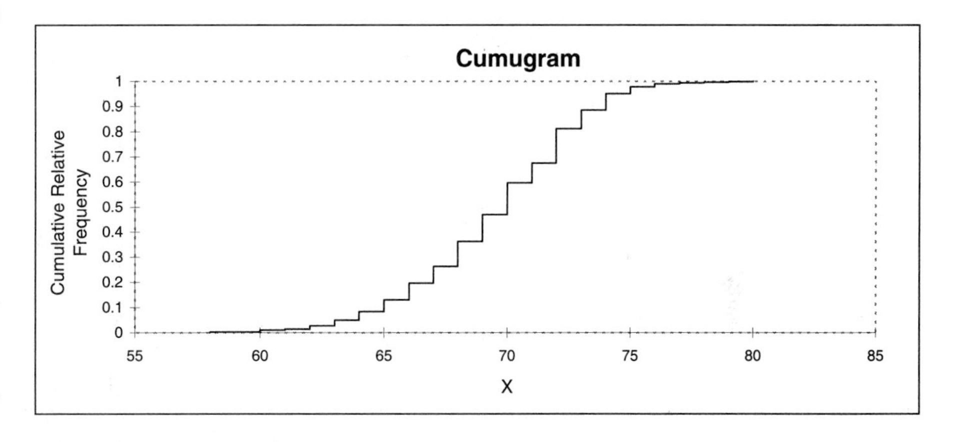

Figure 1.27

Fractiles. Excel has various routines that are supposed to give fractile-like results, but they do not conform to the standard (and economically useful) definition of a fractile. The easiest way to get a fractile is by way of the cumugram. Figure 1.28, which has a more detailed scale, shows that the three quartiles (the 0.25, 0.50, and 0.75 fractiles) are respectively 67, 70, and 72 for the height data.

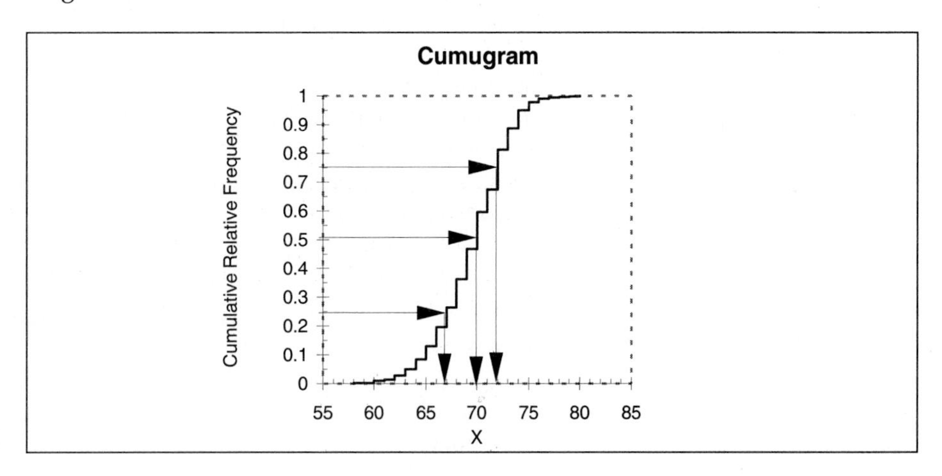

Figure 1.28

Analysis of Two Variables

Open the HTWT1.XLS data file again, and move the data to another worksheet in the same location. You'll use this file for your work on scatter diagrams.

Scatter Diagrams. You can use Excel's ChartWizard to plot a scatter diagram of height against weight. To start:

- Highlight the range **C7:D774** (the fastest way to do this is to highlight cells C7 and D7, then press **[Ctrl]** **[Shift]**), and the [↓], then click the ChartWizard button. Finally, use the mouse to draw a rectangle on your spreadsheet. The ChartWizard takes over.
- Look at the Range text box, and verify that you have chosen C7:D774. Click **Next**.
- Choose **XY (Scatter)** for the chart type, and click **Next**.
- Choose chart format **1**, and click **Next**.
- In the control boxes to the right of the sample chart, indicate that the Data Series is in **Columns**, that the first **1** column is for category (X) data, and the first **0** row is for legend text. Click **Next** when you have checked your selections.
- Select **No** under Add a Legend? In the Chart Title text box, type **Scatter Diagram**, then type **Height** next to category (X) and **Weight** next to Value (Y) and click **Finish**. When your diagram appears on the spreadsheet, adjust its shape with the resize arrow until it looks something like the one shown in Figure 1.29.

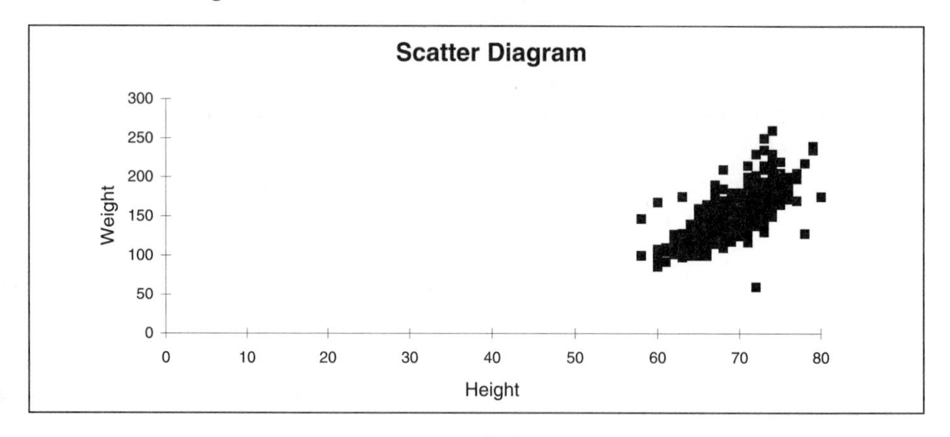

Figure 1.29

Now you may want to rescale the horizontal axis. To make this change:

- Double-click the chart to activate it, then double-click on the **horizontal scale markers**. A Format Axis dialog box displays. Click the **Scale** tab, then change the Minimum value to **55** and change the Major Unit to **5** to spread out the data points. Click **OK**.
- You could double-click the vertical scale and repeat the process if you wanted to expand the vertical scale.

To reformat the data points:

- Double-click the chart if you need to activate it, then click on any data point in the chart. A Format Data Series dialog box displays (you may have to click on several different data points to obtain the correct dialog box).
- In the dialog box, choose the **Patterns** tab, which offers you a choice of marker shapes as well as background and foreground colors. Try different combinations.

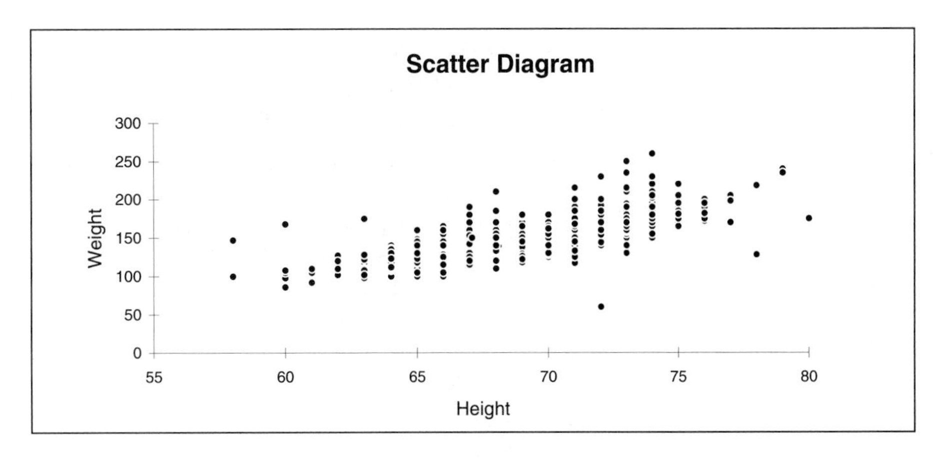

Figure 1.30

In the scatter diagram shown in Figure 1.30, the marker has been changed from a solid square to a circle with a white foreground and a black background.* To do that:

- On the Patterns tab, click the scroll arrow opposite Style to open the drop-down menu and click the circle.
- Click the scroll arrow opposite Foreground and choose the white square.
- Click the scroll arrow opposite Background, choose the black square, and click **OK**.

Scatter Diagrams with Identifiers. Scatter diagrams with identifiers convey a great deal of information, but are difficult to produce. In this section, we show how to produce a diagram like Figure 1.6 of this chapter, which shows height vs. weight, with different markers for men and women.

Start by copying the three columns in HTWT.XLS (height, weight, and gender) to a new worksheet in the same location. Then sort the data so that all the men (Gender = 0) appear first, followed by all the women. To do that:

- Highlight the range **C7:E774** and click **Data**, then click **Sort**. The Sort dialog box opens.
- In the dialog box, click the scroll arrow in the Sort By box and select M/F. The Ascending option should be selected. Click **OK**.

Now :

- Move the height column (C4:C774) to **B4:B774**. (Adjust column width if necessary.)
- Move the weight data for men (D4:D603) to **C4:C603**.

You now have one column of height data and two columns of weight data: column C has 171 blank rows at the bottom, and column D has 597 blank rows at the top. Save your work so far.

Highlight the range B7:D774, click the **ChartWizard** button, and drag a rectangle on the spreadsheet. To plot the scatter diagram with the ChartWizard:

- Look at the Range text box, and verify that you have chosen B7:D774. Click **Next**.
- Choose **XY (Scatter)** for the chart type, then click **Next**.

---

* This may be easier to see on-screen if you choose bright colors such as reds, blues, greens, etc.

- Choose chart format **1**, then click **Next**.
- In the control boxes to the right of the chart, select **Data Series** in Columns, first 1 column for category (X) data and **first 0 row** for legend text.
- For the Chart Title, type **Scatter Diagram with Identifier**, with Height for Category (X) and Weight for Value (Y).

After rescaling the axes and changing markers, you will obtain a chart like Figure 1.31, where Series 1 represents men and Series 2 represents women.

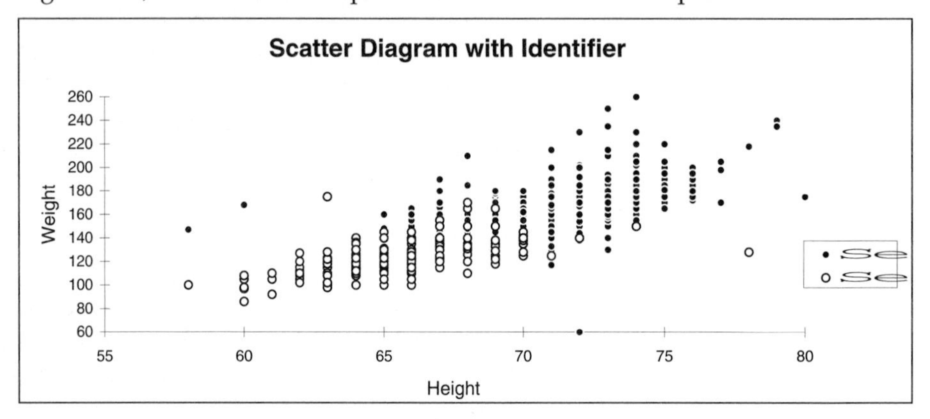

Figure 1.31

Correlation. You can obtain the correlation coefficient using Excel's Correl function. To obtain the correlation between height and weight, click any empty cell on the worksheet and type: **=Correl(C7:C774,D7:D774)**, then press [**Enter**].

You will get a correlation coefficient of 0.745255.

Time Series

Finally, we show how to obtain the simple time series of Retail Sales shown in Figure 1.7 of this chapter, as well as the stack graph showing sales one year at a time (Figure 1.8).

Open RETAIL.XLS and save it as RETAIL1.XLS. Highlight cells **C9:D77** (the column of Dates and the column of Actual Retail Sales). Click the ChartWizard and draw a rectangle with the mouse. This time, choose Line as the chart type (sales corresponding to successive months are plotted at equally spaced intervals in a Line chart) and choose 2 as the format. After following the procedures described in the ChartWizard, then rescaling the axes, you should obtain a chart that looks something like Figure 1.32.

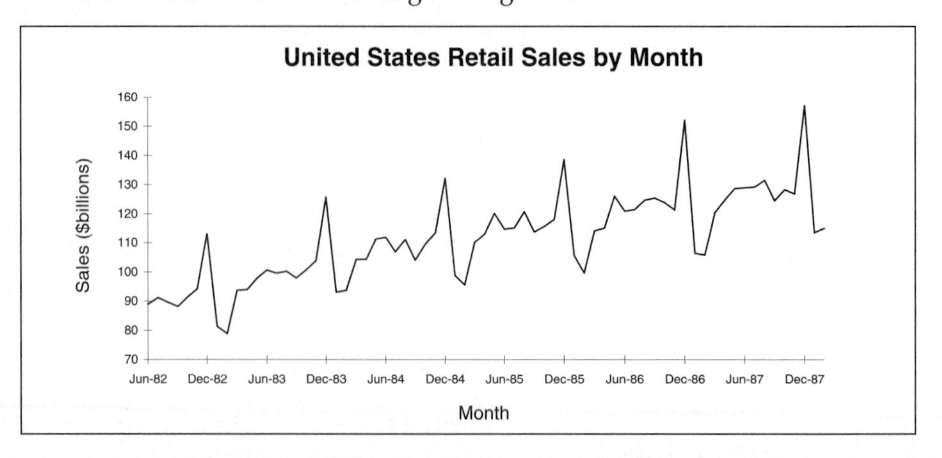

Figure 1.32

To create the stacked graph:

- First find an empty grid 13 rows high by 8 columns wide (start at cell H6, for example). In cell H7, type **Jan**, then drag the fill handle down the column, which will automatically fill in three-letter abbreviations for the other months. Don't go too far; the yearly cycle will repeat if you drag further than 12 months.
- In cell I6, type **1982** and in J6 type **1983.** Highlight these two cells and then drag the fill handle to the right, to cell O6.
- Now copy the column of actual sales (D9:D77) to column I, starting with row 12 opposite Jun, since the data series starts with June 1982. Then copy the actual sales data starting in January 1983 (cell I19) into Column J, row 7 (opposite Jan) and so forth (you may need to widen the columns to make the data fit). When you finish, save your work; the grid should look like Table 1.13.
- Next, highlight the grid you have just created, invoke the ChartWizard, specify Line as the chart type and 1 as the format. In Step 4 of the ChartWizard procedure, indicate that the Data Series is in Columns, that the first **1** column will be used for category (X) axis labels, and that the first **1** row will be used for legend text. After scaling the axes, you should obtain a chart that looks something like Figure 1.33.

Table 1.13

	H	I	J	K	L	M	N	O
6		1982	1983	1984	1985	1986	1987	1988
7	Jan		81.34	93.09	98.82	105.64	106.39	113.64
8	Feb		78.88	93.69	95.59	99.66	105.80	115.10
9	Mar		93.76	104.29	110.17	114.24	120.44	
10	Apr		93.97	104.34	113.11	115.13	124.74	
11	May		97.84	111.31	120.19	126.09	128.69	
12	Jun	88.97	100.61	111.98	114.78	120.98	128.99	
13	Jul	91.21	99.56	106.88	115.23	121.47	129.26	
14	Aug	89.64	100.23	111.16	120.77	124.72	131.54	
15	Sep	88.16	97.97	104.03	113.84	125.44	124.52	
16	Oct	91.42	100.67	109.55	115.75	123.84	128.30	
17	Nov	94.20	103.87	113.54	118.06	121.37	126.90	
18	Dec	113.19	125.76	132.26	138.65	152.11	157.19	

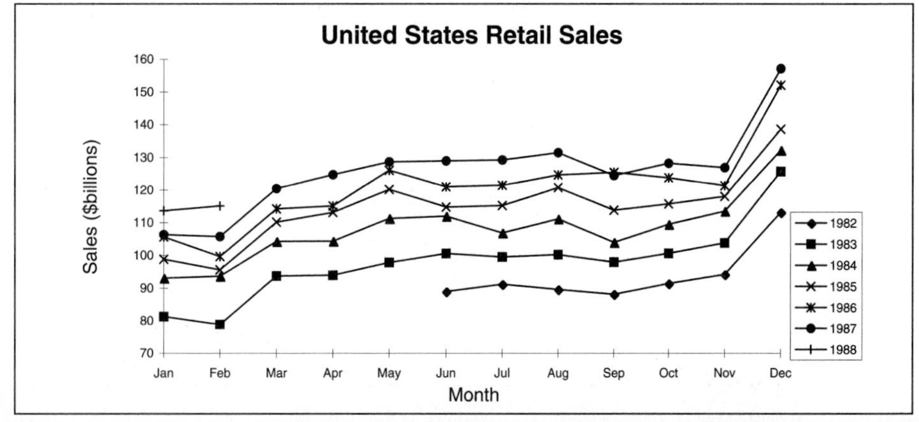

Figure 1.33

BOSTON EDISON VS. CITY OF BOSTON

In March, 1979, the Supreme Judicial Court of the Commonwealth of Massachusetts handed down the so-called Tregor decision, upholding the decision of the Appelate Tax Board that "a taxpayer has a right to have his assessment reduced so that it is proportional to the assessments of the class of property valued at the lowest percentage of fair cash value." This decision unleashed a flood of lawsuits across the Commonwealth by companies seeking tax abatements. By 1985, the City of Boston alone had returned $115 million in abatements for overassessed commercial properties.

Boston Edison was the city's largest taxpayer, with annual real estate and personal property taxes amounting to $40 million. Roughly one-third of these taxes were assessed on the L Street and New Boston generating plants. In 1984 Boston Edison entered into a protracted negotiation with the city to abate taxes paid from 1977 through 1982. These negotiations ultimately broke down, and in 1985 Boston Edison brought suit before the Appelate Tax Board.

The Tax Assessment Process

Each of Massachusetts' 351 cities and towns assesses taxes on real property by applying a fixed tax rate to the assessed value of a particular property (or "parcel"). If a parcel has an assessed value of $20,000, and the tax rate in a given year is $100 per $1000, then the tax on that parcel is $2000. Tax rates may differ from community to community, and may change from year to year, but, prior to 1983,[20] the rate in a given community in a given year had to be the same for all parcels. The income which a community receives from property taxes is the product of the tax rate times the sum of assessed values of all parcels in the town. The rate is set so as to generate the income needed to pay for the services which the community provides.

In theory the assessed value of a parcel is supposed to reflect its "fair cash value." One way in which assessors value residential parcels is to look at recent sales of comparable parcels, implicitly equating fair cash value to market value. While it may be quite difficult to establish groups of comparable residential parcels, it is next to impossible to do so for commercial parcels. As a consequence, the latter are assessed in a variety of ways: in accordance with book value, replacement cost, or income generation.

No matter what method of assessment is used, the process is time consuming, judgmental, error prone, requires frequent updating, and is subject to dispute by property owners who deem their assessment too high. In Massachusetts (and other states) in the 1960s and 1970s, assessment practice started to deviate substantially from theory.

Harvard Business School case 9-189-047. This case was prepared by Professor Arthur Schleifer, Jr.

[20] In 1978 the Massachusetts Constitution was amended. The Classification Amendment permitted cities and towns to levy property tax rates that discriminated among different classes of property. Implementing legislation enacted by the General Court led to the so-called Certification Program, which forbade any city or town from thus discriminating until the Commissioner of Revenues certified that it has assessed all parcels at "full and fair market value." The City of Boston was not so certified until 1983.

As strong demand and inflation caused real estate values to increase, assessors tended not to adjust a parcel's assessed value to reflect these changes. The ratio of assessed value to market value (the A/M ratio), which in theory should have been 1, thus became a fraction less than 1 whose value gradually became smaller and smaller; typical values of the A/M ratio for residential properties in the late 1970s were 15% to 25%. This in itself caused no serious problems: communities simply increased their tax rates to offset the effects of inflation and to add to the services provided. But the failure to reassess properties over a period of time resulted in two more serious problems. First, within given classes of parcels such as residential properties, some parcels appreciated more quickly than others (and thus their A/M ratios fell more quickly); if taxes were supposed to be proportional to market value, those parcels that appreciated rapidly weren't paying their "fair share." Second, while nearly all parcels were assessed below market value, commercial parcels were assessed closer to market value than residential. This resulted in *de facto* discriminatory taxation, based on the theory that owners of commercial properties had a greater "ability to pay" than owners of residential properties, but this was in clear violation of the State Constitution prior to enactment of the Classification Amendment, and this violation is what led to the Tregor decision.

Implementation of Tregor

The Commonwealth of Massachusetts recognized nine classes of property. For the purposes of this discussion, the only classes of importance are single-family residences (R1) and commercial. It was generally acknowledged that, throughout the Commonwealth, the A/M ratio for R1 parcels was lower than for any other class. Thus the implementation of the Tregor decision immediately focused on the A/M ratio for R1 parcels. Commercial properties, whose A/M ratio was much higher, could seek a revision of their assessed value by multiplying fair cash value by the R1 A/M ratio. All abatement hearings on commercial properties thus went through the following five-step procedure.

1. Establish the fair cash value of the property in question.
2. Determine the A/M ratio for R1 parcels in the community.
3. Compute the revised assessed value as the product of the fair cash value computed in step 1 times the A/M ratio for R1 parcels determined in step 2.
4. Compute the revised tax as the product of the revised assessed value computed in step 3 times the tax rate.
5. If the revised tax is less than the tax paid, the difference is abated.

Steps 3, 4, and 5 were purely mechanical, but steps 1 and 2 were highly controversial, and typically were resolved only through protracted negotiation or litigation.

The Boston Edison Case

At the end of the first stage of what was to become the longest trial in the history of the Appellate Tax Board, the Board ruled on the fair cash value of the Boston Edison property under dispute. For the purpose of this discussion, we shall concentrate on the year 1979, for which the fair cash value was ruled to be $126,198,318.

The trial was not yet over. In the second stage of the trial, the Board had to rule on the appropriate R1 A/M ratio, which would be multiplied by the fair cash value to obtain a revised assessed value. While there was no question that this ratio was somewhere between 15% and 25%, it was important to pin the number down precisely: a swing of just one percentage point implied a swing of over $1 million in taxes for 1979 alone; with six years in dispute, a systematic deviation of one percentage point would have an effect in excess of $6 million.

Equalization Studies

Before discussing the controversies surrounding computation of the A/M ratio for R1 properties, we must introduce what at first seems like a totally irrelevant statutory process. Its relevance to the Boston Edison case will be made clear in a moment.

The Commonwealth of Massachusetts distributes aid to cities and towns. The general basis of apportionment is that "rich" communities receive less aid per capita than "poor" communities. Operationally, the wealth of a community is construed as the market value per capita of that community's real estate. The number of people in each community is known, but how does the state determine the market value of Boston's, or Worcester's, or Braintree's real estate?

If the assessed value were a reasonable approximation of market value, the answer is easy: the assessed value of every parcel in every community is known, because it is the determinant of real-estate taxes. But, as we have seen, in the late 1970s assessed value was far out of line with market value. If the ratio of assessed-to-market value were roughly the same across communities, this would pose no problem: assessed value per capita would be proportional to market value per capita, and would serve to distinguish relative wealth of communities. But there is no evidence that this ratio is the same across communities.

To obtain a measure of market value, therefore, the Massachusetts Department of Corporations and Taxation was authorized by statute to conduct so-called "equalization studies" each year for all Massachusetts communities. These studies compared the selling price of those properties that actually exchanged hands with the assessed value of those properties. Efforts were made to cull out those sales which were not "at arm's length," for instance, sales to family members, to charitable institutions, foreclosures, etc.

The exact methodology of these equalization studies need not concern us; it is sufficient to note that a by-product of these studies was a data base consisting of all sales of property in the state, including the selling price, the assessed value, whether the sale was "at arm's length," and various other characteristics of the property and the sale. It was this data base which had been used to compute A/M ratios in other tax abatement proceedings.

The Dana Data Base

Because so much was riding on the computations of the A/M ratio, the Boston Edison lawyers did not want to rely on the data bases on which the equalization studies were based. There were two principal reasons for their concern. First, because of the time required for collecting and processing data, the transactions on which any year's equalization study were based consisted of sales six to eighteen months prior to the date as of which market value was to be determined. And second, the state's criteria for distinguishing arm's length transactions from others were not objectively stated.

Slower selling items would be replaced with newer designs and these might be updated during the course of a year if indicated by sales performance. Some chains centralized their purchasing so that Hygiene would receive electronically transmitted orders for their warehouses as well as individual stores every two weeks. Others ordered through the mail for individual stores or consolidated distribution.

Many chains cooperated by sharing their monthly point-of-sale information, which reflected consumer sales. Of course Hygiene could always calculate sales with reasonable accuracy from their own record of shipments. Perhaps more importantly, chains that ran promotions (newspaper inserts or aisle-end displays) usually notified Hygiene many months in advance. Some would also furnish estimates of the additional purchases they believed would be needed. Exhibits 1, 2 and 3 describe some of the data available from two representative customers.

Production

Hygiene was a converter of all roll stock used in the manufacturing of their end product. The raw materials used to manufacture fabric shower curtains were known as greige goods. Greige goods were prepared for either screen printing or dying and finishing. Certain fabrics also received a Teflon finish for water repellency. Vinyl shower curtain materials were purchased with embossed designs to be sold as solid colors or used for printing. Hygiene owned its own vinyl printing facility, and materials were stored there awaiting processing. The processed materials, as well as accessory products made elsewhere, were then shipped to Hygiene's manufacturing facilities in New York, Mississippi, and Los Angeles.

Substantial inventory of roll goods had to be on hand for the manufacturing process at all times. The roll goods were manufactured into the multiple components of finished products which made up more than 2,000 stock keeping units (SKUs). Short delivery cycles and the broad range of products made it impractical to manufacture against specific orders. Merchandise had to be prepared in advance of sales and held in finished goods inventory until packaged. Hygiene utilized two systems for packaging. The finished goods for major customers were pre-packaged for immediate order picking and greater efficiency. A lot system was used to accumulate units by packaging type for other customers. In order to meet individual customer requirements for UPC coding, labeling, and pre-pricing, 32 distinct types of packaging were prepared each week representing over 10,000 different packaged SKUs.

Exhibit 1

HYGIENE INDUSTRIES

Monthly Sales by Bradlees

Pattern	Color	Size	Cost	Feb-87	Mar-87	Apr-87	May-87	Jun-87	Jul-87	Aug-87	Sep-87	Oct-87	Nov-87	Dec-87	Jan-88	Feb-88
Candystripe	Navy	6 X 6	$3.62	349	289	279	304	324	374	361	381	369	445	431	340	313
Capistrano	Blue	6 X 6	$4.85	670	670	819	814	798	916	901	875	871	930	980	715	716
Chorus Line	Black	6 X 6	$5.87	251	223	257	302	278	358	372	308	264	303	371	267	243
Do-Si-Do	Blue	6 X 6	$9.20	177	206	231	235	210	240	249	254	196	141	120	122	150
Elite	Ecru	6 X 6	$3.75	539	435	492	519	616	615	633	614	684	638	748	466	419
Flamingo Rd	Pink	6 X 6	$5.87	22	42	72	187	316	374	465	487	455	470	569	441	356
Hms	Clear	6 X 6	$3.25	1,039	1,009	991	1,121	1,195	1,282	1,209	1,264	1,055	1,149	1,048	951	911
Maytime	Green	6 X 6	$3.62	380	368	418	350	324	496	431	399	464	518	520	305	309
Pandora	White	6 X 6	$5.87	372	321	355	358	425	519	487	419	533	615	819	529	653
Sanibel	Peach	6 X 6	$11.33	112	143	155	147	171	202	179	80	83	125	157	87	95
Siri	White	6 X 6	$11.85	189	223	255	232	197	194	145	101	228	305	295	196	165
Theodore	Brown	6 X 6	$5.87	459	226	252	255	243	361	369	300	281	406	443	281	167
Tuliptime	Blue	6 X 6	$4.28	278	236	258	287	319	368	363	341	359	325	379	311	279
Victoria	Blue	6 X 6	$11.57	123	95	126	156	122	153	145	151	162	160	122	63	87
Wildlife	Brown	6 X 6	$9.20	258	222	387	268	258	318	309	306	252	318	337	210	194
Sutton	Rasp	Deluxswg	$15.00	501	437	371	384	392	591	254	494	536	330	688	203	367
Sutton	Ginger	Deluxswg	$15.00	192	352	283	294	271	323	171	365	345	226	501	133	298
Sutton	Navy	Deluxswg	$15.00	277	290	240	231	241	324	155	273	303	129	275	88	231
Sutton	White	Deluxswg	$15.00	156	298	183	199	185	205	93	147	237	92	278	67	138
Sutton	Jade	Deluxswg	$15.00	180	242	221	197	185	264	131	299	377	172	299	52	207
Sutton	Ecru	Deluxswg	$15.00	224	196	216	178	197	232	100	197	220	103	347	69	178
Sutton	Blue	Deluxswg	$15.00	304	297	340	352	318	502	204	394	484	240	738	145	380

Pattern	Color	Size	Cost	Mar-88	Apr-88	May-88	Jun-88	Jul-88	Aug-88	Sep-88	Oct-88	Nov-88	Dec-88	Jan-89	Feb-89	Mar-89
Candystripe	Navy	6 X 6	$3.62	311	434	338	397	420	476	533	400	372	452	154	91	177
Capistrano	Blue	6 X 6	$4.85	790	1,005	868	946	967	1,091	1,176	926	1,104	952	629	600	1252
Chorus Line	Black	6 X 6	$5.87	189	174	117	123	111	103	59	38	74	127	65	66	106
Do-Si-Do	Blue	6 X 6	$9.20	197	278	287	233	271	259	233	230	274	220	132	100	178
Elite	Ecru	6 X 6	$3.75	469	682	612	530	623	720	750	511	579	502	309	301	587
Flamingo Rd	Pink	6 X 6	$5.87	420	579	478	477	648	664	627	433	515	442	380	334	674
Hms	Clear	6 X 6	$3.25	978	1,318	1,092	1,134	1,270	1,471	1,338	1,161	1,238	1,317	809	777	1567
Maytime	Green	6 X 6	$3.62	412	489	404	484	525	490	437	459	413	428	157	178	324
Pandora	White	6 X 6	$5.87	610	591	536	498	637	722	567	439	485	397	274	294	558
Sanibel	Peach	6 X 6	$11.33	99	149	121	105	121	118	100	76	71	63	46	35	89
Siri	White	6 X 6	$11.85	185	253	201	174	248	187	227	212	238	169	126	134	266
Theodore	Brown	6 X 6	$5.87	185	216	185	169	254	276	215	204	219	210	105	81	135
Tuliptime	Blue	6 X 6	$4.28	280	424	361	410	437	392	376	292	346	308	194	212	466
Victoria	Blue	6 X 6	$11.57	109	199	195	176	375	247	437	184	207	148	66	94	188
Wildlife	Brown	6 X 6	$9.20	187	314	241	245	352	304	311	307	361	312	173	214	412
Sutton	Rasp	Deluxswg	$15.00	541	285	460	379	413	432	473	408	562	718	142	400	682
Sutton	Ginger	Deluxswg	$15.00	432	351	381	360	315	166	210	213	348	507	116	292	649
Sutton	Navy	Deluxswg	$15.00	341	319	287	234	266	275	292	196	264	356	79	196	397
Sutton	White	Deluxswg	$15.00	211	171	189	162	163	203	129	122	165	378	95	214	415
Sutton	Jade	Deluxswg	$15.00	373	308	243	258	273	269	221	231	268	500	82	245	566
Sutton	Ecru	Deluxswg	$15.00	226	203	195	208	243	237	129	140	156	297	28	76	199
Sutton	Blue	Deluxswg	$15.00	443	367	405	320	380	331	326	308	359	668	78	326	692

Note: These data reflect actual sales by stock keeping unit as reported to Hygiene Industries by Bradlees. The SKUs were chosen from among the shower curtains carried continuously during the period Feb–87 through Mar–89.

Exhibit 2

HYGIENE INDUSTRIES

Monthly Deliveries by Hygiene to Kmart

Pattern	Size	Cost	Color	Jan-87	Feb-87	Mar-87	Apr-87	May-87	Jun-87	Jul-87	Aug-87	Sep-87	Oct-87	Nov-87	Dec-87	Jan-88	Feb-88
Capistrano	6 X 6	$4.50	Peach	2,788	3,336	5,932	2,805	3,869	4,009	4,317	4,855	2,801	4,835	1,986	1,886	1,669	2,138
Chorus Line	6 X 6	$5.34	Black	9,426	2,545	3,035	2,016	2,416	2,726	2,958	3,365	2,783	4,224	1,737	1,665	1,466	1,940
Empress	6 X 6	$12.00	Ecru	706	882	1,327	365	403	573	322	538	890	1,502	593	452	367	558
Flowerbox	6 X 6	$10.14	Yellow	1,185	1,583	2,602	1,344	1,683	1,662	1,900	2,112	1,360	2,256	1,008	891	707	903
Mums	6 X 6	$2.29	Brown	5,876	7,818	11,199	5,887	7,234	7,683	19,818	11,493	4,199	8,076	3,662	4,426	3,869	4,976
Mums	6 X 6	$2.29	Royal	7,618	9,469	13,811	7,291	9,649	10,084	22,327	14,269	6,164	11,378	5,448	6,700	5,981	6,862
Mums	6 X 6	$2.29	Yellow	5,614	7,679	9,653	6,269	7,300	8,377	20,497	11,699	4,424	7,817	3,440	4,471	3,430	4,488
Mums	6 X 6	$2.29	White	7,142	9,531	12,976	7,270	8,848	10,485	24,637	15,489	6,255	10,710	4,977	5,944	5,168	6,363
Paris	6 X 6	$5.87	Ecru	1,587	2,070	2,809	1,448	2,219	2,203	2,369	2,605	1,305	2,569	1,274	1,207	973	1,181
Phoenix	6 X 6	$3.35	Brown	4,211	5,282	7,904	4,319	15,739	8,135	5,208	5,575	3,707	7,015	3,302	3,480	4,016	3,816
Rainbow	6 X 6	$5.34	Red	9,117	3,146	4,107	2,358	2,970	3,143	3,857	4,639	2,774	4,833	1,685	1,766	1,455	2,177
Renowned	6 X 6	$3.00	Rasbry	2,933	5,771	7,995	3,148	8,359	3,466	4,479	9,914	3,180	4,831	2,355	8,431	4,637	6,584
Renowned	6 X 6	$3.00	Jade	1,826	3,102	3,906	2,363	2,714	3,894	4,163	4,370	3,016	5,383	2,284	3,104	2,595	4,032
Renowned	6 X 6	$3.00	Peach	3,142	6,607	8,538	3,290	9,461	5,000	5,685	11,307	4,284	6,572	3,177	9,270	5,317	7,604
Renowned	6 X 6	$3.00	White	4,225	9,212	11,696	4,417	12,660	7,274	9,502	16,764	5,607	6,340	4,445	13,092	7,154	10,652
Renowned	6 X 6	$3.00	Yellow	1,871	3,459	4,170	2,257	3,320	3,540	4,222	4,500	2,550	4,235	1,553	2,348	1,702	2,902
Renowned	6 X 6	$3.00	Blue	5,351	8,829	11,895	5,296	11,390	7,005	8,832	14,551	5,844	8,805	4,413	11,274	6,770	9,348
Renowned	6 X 6	$3.00	Ecru	5,791	11,112	14,400	6,022	14,093	8,315	9,921	17,938	7,103	9,810	5,316	14,201	8,152	11,378
Renownedc	6 X 6	$2.75	Clear	5,929	15,527	21,873	5,535	22,331	11,195	13,663	28,574	9,416	12,184	6,958	24,649	15,757	20,238
Wild Life	6 X 6	$8.54	Brown	1,490	1,835	2,973	1,436	1,904	1,830	2,014	2,440	1,477	2,866	1,287	1,289	1,155	1,527
Cortina	Deluxe45	$12.00	Ecru	172	248	411	1,410	374	316	254	334	184	340	1,961	219	195	202
Cortina	Deluxe45	$12.00	Evergr	342	403	678	1,361	480	363	316	489	243	414	1,967	387	336	387
Cortina	Deluxe45	$12.00	Gray	126	227	309	1,129	260	181	195	300	155	306	1,834	147	144	133
Cortina	Deluxe45	$12.00	Blue	303	294	603	1,629	505	386	368	451	244	463	2,904	245	199	206
Dobby	Deluxe45	$11.47	Rasbry	565	817	1,693	686	636	560	617	775	1,745	972	483	527	380	496
Dobby	Deluxe45	$11.47	Ecru	189	368	820	272	334	287	291	342	835	570	248	240	166	226
Dobby	Deluxe45	$11.47	Brown	153	215	660	239	242	188	268	316	871	505	188	208	148	175
Dobby	Deluxe45	$11.47	Blue	268	734	1,607	652	750	557	662	777	1,679	872	477	420	456	441
Cortina	Deluxswg	$14.25	Evergr	726	932	1,291	2,441	1,292	1,204	1,073	1,269	717	1,115	2,643	1,007	864	1,001
Cortina	Deluxswg	$14.25	Gray	344	553	764	1,875	772	620	589	694	337	716	2,464	427	340	408
Cortina	Deluxswg	$14.25	Ecru	506	717	1,320	2,584	1,121	971	829	887	431	1,010	2,773	629	538	526
Cortina	Deluxswg	$14.25	Blue	652	831	1,388	2,974	1,350	1,181	1,158	1,198	745	1,174	3,882	643	543	608
Dobby	Deluxswg	$13.87	Rasbry	1,199	1,853	3,809	1,584	1,683	1,409	1,747	1,917	3,959	2,238	862	911	622	1,129
Dobby	Deluxswg	$13.87	Ecru	477	910	1,667	905	911	772	972	1,211	1,579	1,444	646	536	551	692
Dobby	Deluxswg	$13.87	Brown	527	394	1,419	699	437	612	1,041	1,228	1,640	1,588	693	609	527	657
Dobby	Deluxswg	$13.87	Blue	836	1,815	3,974	1,641	1,717	1,659	1,783	2,086	4,036	2,173	1,202	1,160	1,088	1,302

Note: These data reflect shipments from Hygiene to Kmart. The SKUs shown are the shower curtains that were carried continuously during the period Jan–87 through Feb–88.

Exhibit 3

HYGIENE INDUSTRIES

Promotion Forecasts by Kmart

Pattern	Size	Cost	Color	Jan-87	Feb-87	Mar-87	Apr-87	May-87	Jun-87	Jul-87	Aug-87	Sep-87	Oct-87	Nov-87	Dec-87	Jan-88	Feb-88	Mar-88	Apr-88
Capistrano	6 X 6	$4.50	Peach	0	0	0	0	0	0	0	0	0	0	0	0	0	0	0	0
Chorus Line	6 X 6	$5.34	Black	7,288	0	0	0	0	0	0	0	0	0	0	0	0	0	0	0
Empress	6 X 6	$12.00	Ecru	0	0	0	0	0	0	0	0	0	0	0	0	0	0	0	0
Flowerbox	6 X 6	$10.14	Yellow	0	0	0	0	0	0	0	0	0	0	0	0	0	0	0	0
Mums	6 X 6	$2.29	Brown	0	0	0	0	0	0	8,083	0	0	0	0	0	0	0	0	0
Mums	6 X 6	$2.29	Royal	0	0	0	0	0	0	8,083	0	0	0	0	0	0	0	0	0
Mums	6 X 6	$2.29	Yellow	0	0	0	0	0	0	8,083	0	0	0	0	0	0	0	0	0
Mums	6 X 6	$2.29	White	0	0	0	0	0	0	8,083	0	0	0	0	0	0	0	0	0
Paris	6 X 6	$5.87	Ecru	0	0	0	0	0	0	0	0	0	0	0	0	0	0	0	0
Phoenix	6 X 6	$3.35	Brown	0	0	0	0	0	22,000	0	0	0	0	0	0	1,200	0	0	0
Rainbow	6 X 6	$5.34	Red	6,953	0	0	0	0	0	0	0	0	0	0	0	0	0	0	0
Renowned	6 X 6	$3.00	Rasbry	5,372	0	5,537	0	5,000	0	0	3,642	0	0	0	3,356	0	0	0	0
Renowned	6 X 6	$3.00	Jade	0	0	0	0	0	0	0	0	0	0	0	0	0	0	0	0
Renowned	6 X 6	$3.00	Peach	5,372	0	5,610	0	5,000	0	0	3,642	0	0	0	3,356	0	0	0	0
Renowned	6 X 6	$3.00	White	8,053	0	8,240	0	7,500	0	0	5,463	0	0	0	5,034	0	0	0	0
Renowned	6 X 6	$3.00	Yellow	0	0	0	0	0	0	0	0	0	0	0	0	0	0	0	0
Renowned	6 X 6	$3.00	Blue	5,372	0	6,338	0	5,000	0	0	3,642	0	0	0	3,356	0	0	0	0
Renowned	6 X 6	$3.00	Ecru	8,053	0	8,748	0	7,500	0	0	5,463	0	0	0	5,034	0	0	0	0
Renownedc	6 X 6	$2.75	Clear	22,080	0	16,059	0	17,000	0	0	11,496	0	0	0	10,240	0	0	0	0
Wild Life	6 X 6	$8.54	Brown	0	0	0	0	0	0	0	0	0	0	0	0	0	0	0	0
Cortina	Deluxe45	$12.00	Ecru	0	0	0	1,187	0	0	0	0	0	0	1,744	0	0	0	0	0
Cortina	Deluxe45	$12.00	Evergr	0	0	0	1,053	0	0	0	0	0	0	1,714	0	0	0	0	0
Cortina	Deluxe45	$12.00	Gray	0	0	0	1,000	0	0	0	0	0	0	1,679	0	0	0	0	0
Cortina	Deluxe45	$12.00	Blue	0	0	0	1,300	0	0	0	0	0	0	1,268	0	0	0	0	0
Dobby	Deluxe45	$11.47	Rasbry	0	0	1,000	0	0	0	0	0	1,081	0	0	0	0	0	0	0
Dobby	Deluxe45	$11.47	Ecru	0	0	500	0	0	0	0	0	509	0	0	0	0	0	0	0
Dobby	Deluxe45	$11.47	Brown	0	0	500	0	0	0	0	0	509	0	0	0	0	0	0	0
Dobby	Deluxe45	$11.47	Blue	0	0	1,000	0	0	0	0	0	1,081	0	0	0	0	0	0	0
Cortina	Deluxswg	$14.25	Evergr	0	0	0	1,692	0	0	0	0	0	0	2,131	0	0	0	0	0
Cortina	Deluxswg	$14.25	Gray	0	0	0	1,536	0	0	0	0	0	0	2,046	0	0	0	0	0
Cortina	Deluxswg	$14.25	Ecru	0	0	0	1,966	0	0	0	0	0	0	2,205	0	0	0	0	0
Cortina	Deluxswg	$14.25	Blue	0	0	0	2,213	0	0	0	0	0	0	3,289	0	0	0	0	0
Dobby	Deluxswg	$13.87	Rasbry	0	0	2,250	0	0	0	0	0	2,331	0	0	0	0	0	0	0
Dobby	Deluxswg	$13.87	Ecru	0	0	750	0	0	0	0	0	777	0	0	0	0	0	0	0
Dobby	Deluxswg	$13.87	Brown	0	0	750	0	0	0	0	0	777	0	0	0	0	0	0	0
Dobby	Deluxswg	$13.87	Blue	0	0	2,250	0	0	0	0	0	2,331	0	0	0	0	0	0	0

Note: Kmart runs special promotions from time to time on various shower curtain lines. The above data, supplied to Hygiene many months in advance, are Kmart's *estimates* of the additional shower curtains that will be needed in a given month. Actual orders may differ. The data under Feb–88 through Apr–88 are zero because there were no scheduled promotions, and not because the data are missing.

The Stride Rite Corporation (A)

Demand Forecasting Process

"This is going to surprise you, Mark," said Florence Anderson to Stride Rite's director of merchandising, Mark Cocozza. "We're selling a lot more Growing Girls shoes than we expected. I'm going to have to adjust the forecast: I have the figures here for you to check."

"You're right—I didn't expect that. But it's not too late in the season to make some more. Have you talked with Production Control?"

"Yes, we're all set there. But I'm having the usual problems deciding how much more to order. Demand might be starting high, and then continuing at the rate we originally expected. Or it may continue to grow at the higher rate of the past few weeks. I guess it might even drop off, and bring sales back to our original estimate. And, of course, I have to figure out how to distribute any increase among the various styles. I'd like to go over my data with you to see where we stand."

Anderson spread out several sheets of paper containing various statistics and comparisons, some tabulated by computer, and others by hand. She proceeded to explain how, 11 weeks into the fall 1980 season, she had become convinced that the forecast for Growing Girls shoes would have to be substantially amended.

Stride Rite Manufacturing Corporation

The Stride Rite Manufacturing Corporation, the original and largest component of the Stride Rite Corporation, made and sold high-quality shoes for boys and girls up to the age of about 12. Their traditional line began with baby shoes, and included leather oxford styles, dress shoes, and casual wear. Stride Rite had recently expanded Zips, its sneaker line, in response to the greatly increased canvas-shoe market, and had also introduced a number of sandal models. Their shoes were carried by independent retail dealers all over the country, as well as by a number of "captive" stores owned and managed by the Stride Rite Retail Corporation, a wholly-owned subsidiary of the Stride Rite Corporation. These captive stores accounted for about 20% of Stride Rite's total retail sales. Stride Rite also sold 3%-4% of its shoes to the military, for retail sale in the PXs.

Shoes were planned and produced for two 26-week selling seasons: the fall season, June through November, with its retail peak just before children returned to school, and the spring season, which peaked just before Easter. Some styles were new for a given season, while others were continuing shoes that had been available during the past season. In the latter case, most orders were reorders, called "at-once" orders because they were to be filled immediately.

Harvard Business School case 9-181-122. This case was prepared by Alice B. Morgan, Research Associate, under the supervision of Professor Arthur Schleifer, Jr.

If, however, the style was new rather than continuing, the sales staff of about 35 obtained advance orders for the shoes from retail dealers. These dealers and Stride Rite's own retailers did most of their new-shoe buying on a future-order basis. Orders placed January through May would be delivered during the summer. To smooth production, reduce finished goods inventory, and induce retailers to anticipate their demand, such future orders did not have to be paid for until late September. For the spring line, future orders were placed August to November, and delivered December to February. Because the spring shoe line accounted for only about one-third of annual retail sales, there was no special credit arrangement for spring shoes.

While future orders were standard for the shoe industry, Stride Rite was exceptional in using the at-once system as well. Many shoe manufacturers accepted orders on a futures basis only, since that made production scheduling simpler and less subject to alterations. A retailer who ran out of stock in a particular style and size would be unable to reorder from such a manufacturer during the season. Stride Rite's dealers, however, could renew their stock of nearly all styles and sizes with at-once orders.

Making the Initial Forecast

Stride Rite used a top-down forecasting scheme, starting with the budget goal set by the company's management, for total shoe sales (in number of pairs). Mark Cocozza broke down this aggregated forecast with the help of the four merchandisers. As a first step, Cocozza established estimates for each general category: baby shoes, for example, might be forecasted at 25% of total sales; Girls Service shoes at 30%.[24]

The merchandisers had the tasks of forecasting the season's future and at-once orders for every style in each of five size ranges, and placing the appropriate orders with the factories. They divided the line among themselves, with one person responsible for all the sneakers, and another for a single size range, the baby shoes. One of the remaining two merchandisers had the full Girls line, the other the full Boys line. Florence Anderson dealt with the 38 styles in the Girls line each season. Most of these styles came in more than one color, and all came in at least two of the Girls size ranges: Infants, Childs, Misses, and Growing Girls. For each style in a particular color and size range, there was a unique stock number.[25] Anderson created a forecast of the season's sales for each of the stock numbers in the Girls line. She then placed a factory order for the production of the necessary shoes by an appropriate date. Shoes were made primarily by Stride Rite's own factories, with a small percentage bought from independent suppliers. The manufacturers started to make the shoes long before the retail season began, and forecast accuracy was critical in striking a reasonable balance between over- and understocking.

The merchandisers worked on forecasts for at least two seasons simultaneously. For the approaching season, they were estimating the future orders, and tracking these as they arrived to be sure the production schedule was a reasonable one. The process was more complicated for the current season and involved greater time pressure. Once a selling season began, there were virtually no future orders; instead, at-once orders began to arrive.

---

[24] Girls shoes were divided into three "sales groups": Service, Dress, and Welt Sandals. Stride Rite's merchandising categories are displayed in Exhibit 1.

[25] Each style had a basic number: for example, the Varsity was 2617. To this, a color code was added: 26175 was a black-and-white Varsity. A size code appeared as a prefix: 926175, a black-and-white Growing Girls Varsity. A final digit served as a "check digit." Thus, the black-and-white Growing Girls Varsity had the stock number 9261751.

These had to be met immediately, or the sales might be lost. Back ordering was not standard industry practice, as retail sellers usually wanted the shoes right away or not at all. If Stride Rite was out of stock, the order would be automatically canceled. A retailer did, however, have the option of resubmitting a canceled order. Using the forecast as a guide, the merchandisers strove to keep stockouts low while avoiding overproduction. They kept close watch on factory production and shipping, and on the demand for each of their shoes. When necessary, they altered their forecasts and negotiated a revised manufacturing rate with Production Control. As the season advanced, such revisions were less and less useful, because about six weeks were needed to produce a given shoe. When Florence Anderson had her conversation with Mark Cocozza, in mid-August, she was preparing to alter her at-once fall 1980 forecast.

Growing Girls Service Shoes, Forecast for Fall 1980

To determine the forecast for Growing Girls Service shoes, Anderson had begun by looking at the historical data.

Most of Stride Rite's shoes were similar in style to those made in previous years, so there was a great deal of relevant information available about past demand. Size distribution was nearly constant from one year to the next, and the overall children's shoe market was fairly stable. (See Exhibit 2.) Exhibit 3 compares Growing Girls yearly sales (all sales groups) with Stride Rite total shoe sales. (Figures for 1980 are early estimates.) Since 1971, the trend for Growing Girls was clearly down, from 468,000 pairs to 215,000 in 1979, a 54% drop. Stride Rite's total sales had also dropped, though much less sharply, so as a percentage of total sales, Growing Girls shoes had gone from 9.2% in 1971 to 4.8% in 1979, with slightly less estimated for 1980. Anderson was not expecting any reversal of this downward trend.

In addition to historical information, Anderson had access to a good deal of current data provided by the selling staff. Before they started showing the next season's shoes, salespeople estimated how many of each of the new styles they expected to sell; they made new estimates after they had been traveling for a few weeks. Exhibit 4 shows the first and second fall 1980 estimates of future orders by 31 of the sales staff for Girls Service shoes. (Additional information for the forecast appears in columns 3 and 4, which include anticipated Retail and Military orders.) The sales staff also passed on retail dealer response to shoes being sold in the current season, reporting special problems with fit and color, or customer interest in particular styles. Stride Rite's own Retail Corporation provided further current information through monthly reports that showed total sales by style and size classification.

The sales force's information bore out Anderson's general feeling about the Growing Girls size range. Older girls and pre-teens liked to buy adult shoes if they could: they would sooner buy a Bass style that had been sized down, than a Stride Rite that had been sized up. The high-fashion shoes sought by many in this group were not even represented in the Stride Rite line. By the end of February, she had her own initial estimate and both of the sales staff's forecasts, and was already ordering shoes to fill future orders, guided by her current information.

Anderson learned of the season's future orders as soon as these were tabulated. Because a high proportion of sales for a new style came through futures (as much as 80%), they provided a strong foundation for the forecast. A continuing shoe, on the contrary, was rarely ordered in advance; in this case, about 80% of the orders were at-once orders. Here, however, there was always a recent history on which to base the forecast. Past experience indicated that by early May, 75% of the season's future orders would be in hand. Thus, Anderson was able to refine the forecast at this time, developing the at-once component. Again, nothing in her information seemed out of line with her earlier conclusions. By mid-May she had established a Growing Girls Service shoe forecast of 96,100 pairs, 11.3% of the Girls Service total of 850,000 pairs. Of these, 72,771 were future sales (as compared with 566,000 for the entire Girls Service Group) and 23,329 were at-once sales (as compared with 284,000 for Girls Service). (In 1979, future sales had been 85,700; at-once gross demand 29,400; and at-once sales 23,100.)[26]

She then assigned this total of 96,100 pairs to the 22 separate stock numbers in the Growing Girls category (representing 13 styles, some in more than one color). To make the stock number forecasts she again used historical data, the sales staff's information, her own sense of the trends, and the future orders. She continued to track the forecasts, making any changes necessary as, for example, the full complement of future orders arrived. Her major source of information was the Merchandising Summary.

The Merchandising Summary

The Merchandising Summary appeared every Monday, detailing the previous week's orders, shipments, stockouts, and factory production for every stock number. The total sales to date, season forecast, and percentage of forecast sold to date were also listed. (See Exhibit 5.) Of the 16 columns of data offered by the Merchandising Summary, the first nine provided information about demand, sales, and the current forecast. The final seven presented supply data, showing shoes ordered, available for unfilled future orders, finished during the preceding week, on hand, in process, and scheduled. Thus the merchandisers could review very quickly the current supply and demand for each stock number—that is, for each style of shoe in the particular size ranges and colors in which it was made.

The merchandisers used various techniques to track and evaluate their demand forecasts. They like not only to compare the incoming orders with the season prediction, but to make comparisons among different styles in a given size range, and among the various size ranges in which a particular style was offered.

To facilitate such cross comparisons, two versions of the Merchandising Summary were provided. Both were sorted by sales group. The first had stock numbers, sorted by style, within style by color, and within color by size range. The second was sorted by size range, then style, and then color. Exhibit 5 is a page of the latter, showing two size ranges in the Girls Service Group: Misses and Growing Girls.

The top-down process of allocating forecasts to stock numbers is illustrated in Exhibit 6. The top half (Exhibit 6A) shows forecasts for Girls Service shoes within the framework of Stride Rite's shoe business. Exhibit 6B, based on data in the Merchandising Summary, shows details of forecasts and actuals for the Varsity style.

---

[26] At-once gross demand includes both sales (shoes actually shipped) and orders which were canceled because the shoes were not available ("out-of-stocks").

Another way of using the information in the Merchandising Summary was to make comparisons with previous years. Merchandisers could retrieve data that had appeared in the Summary by computer, and they routinely used such data to provide a perspective on the current information. The concept of "relative rate" of demand at a particular week in the season played an important role in the analysis. This year's relative rate was the ratio of season-to-date at-once demand to the season's at-once forecast; last year's was the season-to-date at-once demand (for last year) to last year's total at-once demand. If the relative rate of a particular shoe was about 40% both for this year and for last year, then the at-once forecast was probably on target: the shoe was selling this year at about the same rate as last year. Merchandisers looked for exceptional cases where relative rates were out of line. Sometimes, exceptions could be explained by problems on the supply side: if the factory were seriously behind in production, the rate of sales would be low, and the gross demand figures might well include repeat orders that had been canceled when they first arrived. Sometimes a demand problem was an obvious consequence of an external situation: if the winter weather were unusually mild, the demand for boots was likely to be lower than anticipated.

The At-Once Orders and Adjusting the Forecast

When the Fall 1980 season started in June, Anderson had received almost all the future orders, and the at-once orders were just starting to arrive. To permit easy comparison with the overall patterns for preceding years, Anderson developed graphs of weekly and cumulative at-once demand, as well as relative rates, for the whole Girls Service group and for Growing Girls Service shoes. Exhibit 7 shows graphs of Girls Service shoes[27] demand for the 1975 through 1979 fall seasons and for the first 11 weeks of 1980; Exhibit 8 shows similar graphs for Growing Girls Service shoes for 1978, 1979, and through Week 11 of 1980. Looking at the cumulative graph in Exhibit 8, Anderson was impressed by the fact that 1980 sales had been consistently ahead of the two previous years, although she noted with concern that 1979 had been consistently ahead of 1978 through Week 14, but had nevertheless ended the year 12% below 1978.

Looking at the corresponding relative-rate graph, she noted that 1979 substantially exceeded 1978 in Week 11, and this meant that the remainder of the 1979 season had to account for a smaller fraction of total sales than the comparable period in 1978: if the 1980 forecast were to hold up, this would indicate that weeks 12 through 26 would produce sales only slightly more than total sales to date. She could see no reason for the 1978-1979 reversal to recur, however, and was convinced that she had underforecast 1980 demand.

She began her revisions from the top down, starting with the information she had about all the Girls shoes, comparing the aggregated demand to that of previous years, and to her forecasted demand. She looked carefully at orders that had been canceled because the required shoes were out of stock. The information as of Week 11 was necessarily slight: further changes in the forecast would be possible as the weeks passed and fuller information became available, an argument for keeping the early changes small. On the other hand, early changes disrupted the factory schedule least, and provided more time to produce the required stock.

---

27 Although Anderson forecasted and tracked all three Girls sales groups, references from this point on are to just one of these, the Girls Service category.

After considering the situation in detail, Anderson settled on a revised season forecast for Girls Service shoes. She then reviewed the current ratio of demand among the four size ranges that comprised the Girls Service program. She assumed that demand for the rest of the season would continue in about the same proportions—that is, if Misses shoes were selling twice as well as Growing Girls shoes, but only half as well as Childs shoes, that those relationships would be stable as the season progressed. To generate the revised forecasts for each size group, she determined how many reforecasted Girls Service shoes remained to be sold, and allocated that number proportionally among the four size ranges. In the case of Growing Girls shoes, she increased the forecast by 5,000 pairs. It was this substantial upward revision that she wanted to discuss with Mark Cocozza.

"It looks as if we've got some good items in the line for this age group, Mark," Florence noted. "The boat shoe style and the saddle shoes are selling especially well." (See Exhibit 9 for pictures of the Mariner and the Varsity, respectively the "boat shoe" and "saddle shoe" styles.)

"Well, great. Our salespeople tried to tell their accounts that those would be big sellers, but some of the retailers were hard to convince. Now I guess they wish they had a lot more in stock."

Cocozza went over the figures with Anderson and accepted her revisions. "Better settle on some figures for those individual stock numbers," he remarked.

This was the third and final step in the forecast adjustment process. Anderson had first altered the overall Girls shoe forecast and had then allocated the remaining anticipated demand among the four size ranges. She still had to revise the forecast for each of the stock numbers within the size categories. Her procedure can be illustrated with the Varsity 9261751, a black-and-white saddle shoe.

Revising the 9261751 Forecast

The order for this shoe during Week 11 appears in Exhibit 5, as 1,205 pairs. Anderson considered demand at this rate to be surprisingly high. Growing Girls shoes typically sold one-third to one-half as many as the same style in the next size range down (Misses)—yet, in Week 11, the two size ranges of the Varsity were selling at the same rate. (See Exhibit 5, the comparable Misses shoe is 7261753.) The week's order of 1,205 pairs accounted for most of the season-to-date at-once sales of 1,264 pairs, and meant that Anderson's original total forecast (futures and at-once) of 3,200 pairs was already oversold. She felt she had to explore more fully the information about the Varsity.

On August 20, she called Retail Sales to inquire about the sales information she had received. She spoke to Liz Buckley, area merchandiser for the region where a large Varsity order had originated.

"Any idea what's going on with the Varsity?" Anderson asked. "We certainly didn't expect it to sell at this rate."

"Oh, sure. Don't you know what happened with that shoe? Bamberger's is selling it like crazy. All the cheerleaders in New Jersey are wearing them," answered Buckley.

Bamberger's, a department store chain in the New Jersey area, had Stride Rite booteries in 18 of its stores, and the large Bamberger order accounted for much of the demand for the black-and-white Varsity. It was hard to know whether the information about New Jersey cheerleaders had predictive value.

Buckley thought that all the cheerleaders had now made their shoe purchases and that demand would drop off. On the other hand, perhaps the rest of the student population would be influenced by the cheerleaders' choice of footwear. In that case, demand would continue high. And what, if anything, could be concluded about demand elsewhere from the Varsity's popularity with New Jersey cheerleaders? It was most extraordinary for a Stride Rite shoe to be bought in any significant quantity by girls in junior and senior high school. Anderson accepted Buckley's assessment, and did not anticipate repeat orders on the scale of Bamberger's recent one.

That order accounted for about 600 of the at-once Varsity demand in Week 11, so Anderson decided to base her calculations on an adjusted order level of 605 shoes for that week, thus eliminating the exceptional Bamberger's order from her figures. The Merchandising Summary indicated that total at-once orders for the week had been 5,339 in the Growing Girls category (Exhibit 5), from which she deducted the exceptional retail order of 600 pairs, leaving an adjusted total of 4,739. The black-and-white Varsity, without the exceptional order, had accounted for 605/4,739 = 12.8% of the week's adjusted demand for Growing Girls shoes. Anderson felt that this adjusted demand percentage, though based on a single week's data, was a reasonable guide to the Varsity's proportional demand (the demand for the 9261751 as compared to overall Growing Girls demand) for the rest of the season. She therefore wanted to determine what 12.8% of the remaining available Growing Girls shoes would be.

The aggregated Growing Girls demand had been forecasted at 96,100 pairs (Exhibit 5) and was now reforecasted to be 5,000 pairs more, or 101,100. From this number she deducted the pairs already shipped by the end of Week 11, plus the futures unreleased. She was now ready to allocate the remaining 18,519 pairs among the stock numbers in the Growing Girls line. Her allocation was based on each stock number's percentage of the overall Week 11 demand, with any adjustments, such as those described for the 9261751, taken into consideration.[28] As noted above, the week's adjusted demand for that shoe was 12.8% of the week's Growing Girls demand. On the assumption that the demand would continue in about the same proportions, the remaining demand (for the rest of the fall season) for the 9261751 would be 12.8% of 18,519 or 2,370 pairs. (The calculation is presented in detail in Exhibit 10.)

Once she had made a direct calculation of the likely demand for the Varsity, Anderson subjected it to further judgmental analysis. Her original forecast of 3,200 had been exceeded already. An additional order of about 2,500 pairs might therefore be reasonable. But the 9261751 would be out of stock for perhaps six of the 14 weeks remaining in the fall season. She could see from the factory figures in Exhibit 5 that the total position was less than 900, and that only 335 had been added during Week 11. Some sales would surely be canceled and thus lost, unless they were resubmitted much later in the season. She next conferred with Production Control, and learned that the factory could produce as many as 2,000 additional pairs of the black-and-white Varsity. She also knew that the shoe would continue into the coming seasons. After reviewing the apparent demand, the inevitable out-of-stocks, and the factory requirements, Anderson decided that a reasonable sales estimate for the rest of the season was another 1,600 pairs. Since 3,773 pairs had been sold (Exhibit 5), she reforecasted the 9261751 at 5,400 pairs (3,773 + 1,600 + 28 futures unreleased = 5,401).

She made a comparable calculation and analysis for the rest of the stock numbers in the Growing Girls category, and then for each stock number in the other size categories of the Girls line.

28 Without this adjustment, the 9261751 would be expected to account for 22.6% of the Growing Girls fall demand.

Exhibit 1

Stride Rite's Merchandising Categories

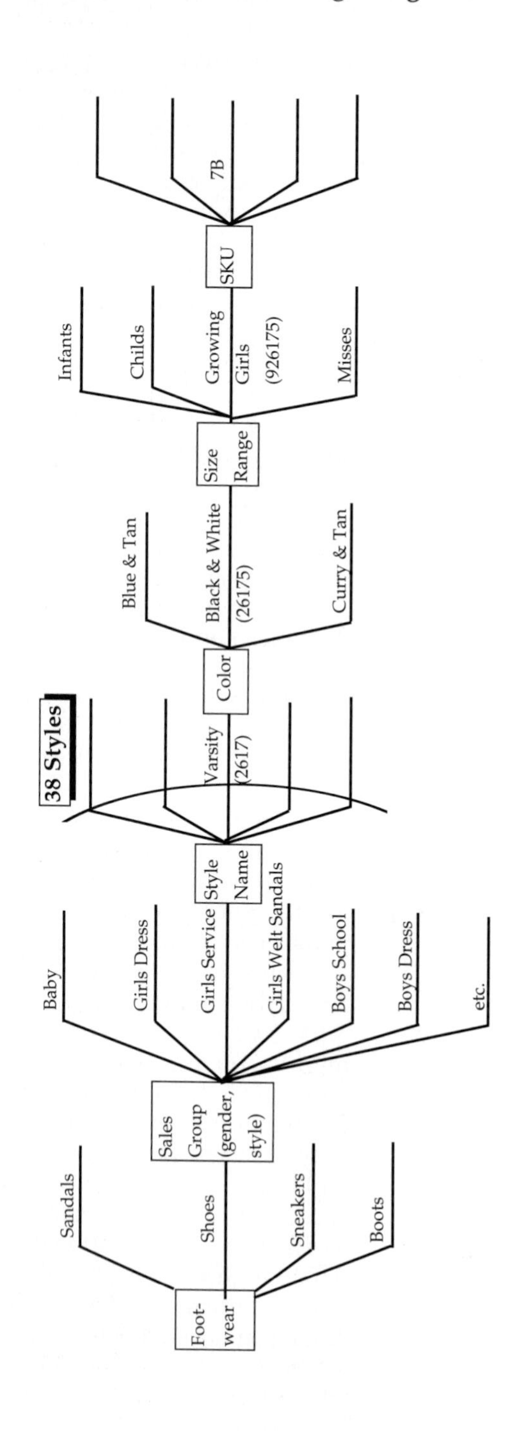

Exhibit 2

Domestic Production and Imports: Non-Rubber Footwear (millions of pairs)

YEAR	TOTAL	% CHANGE OVER PREVIOUS YEAR	CHILDREN'S	CHILDREN'S % CHANGE
1975	669.5	-7%	107.9	-4%
1976	786.5	+17%	140.8	+30%
1977	781.8	-1%	120.8	-14%
1978	785.5	+0%	109.7	-9%
1979	776.7	-1%	102.9	-6%

Source: Stride Rite Shoe Company

Exhibit 3

Growing Girls and Total Stride Rite Shoe Sales, in Pairs

YEAR	GROWING GIRLS SALES	TOTAL SALES	GG SALES AS PERCENT OF TOTAL SALES
1971	468,000	5,081,000	9.2%
1972	423,000	5,251,000	8.1%
1973	373,000	5,056,000	7.4%
1974	326,000	4,981,000	6.5%
1975	381,000	5,438,000	7.0%
1976	305,000	5,490,000	5.6%
1977	206,000	4,491,000	4.5%
1978	212,000	4,597,000	4.6%
1979	215,000	4,437,000	4.8%
1980 (est)	201,900	4,549,500	4.4%

Exhibit 4

Girls Service Shoes Fall 1980

STYLE	31 SALESPEOPLE 1ST Est	31 SALESPEOPLE 2ND Est	+RETAIL	+MILITARY
Mariner	118,900	116,880	136,580	143,000
Misty	53,500	47,840	49,490	55,400
Harbor	37,260	37,790	43,390	43,390
Varsity	50,800	29,700	33,200	34,300
Jessie	9,355	8,720	8,720	8,720
Heather	39,700	39,200	46,400	46,400
Tammy	68,650	64,800	77,300	79,200
Tracy	23,100	19,940	19,940	19,940
Joy	67,850	46,700	50,700	52,700
Tweety	46,400	53,700	61,100	61,100
Flair	37,800	48,000	48,000	48,000
Camper	7,950	4,615	4,615	4,615
Total	**561,265**	**517,885**	**579,435**	**596,765**

Exhibit 5

Merchandising Summary Report

Week Ending 16-Aug-1980

		At-Once		Total	Total	Out-of-Stk		% of	This	Next			FG	Finish	Work		
			Season		Season		Seas		Season	Season	Total	Future	Added This	Gds-On	In		Total
Style No.	Style Name	This Week	Total	Future Ship	Sales	This Week	To-Dt	F-Cast Sold	F-Cast	F-Cast	Ordered	Unrlsd	Week	Hand	Process	Sched	Pos
7214646	Heather	384	949	8,795	9,744	10	28	86%	11,300		12,011	139	516	2,147	120		2,267
7214836	Misty	163	74	12,638	12,712	19	21	83%	15,400		15,108	63	582	764	1,632		2,396
7214869	Misty	136	116	13,719	13,835	8	25	84%	16,400		16,150	130	496	755	1,152	408	2,315
7225238	Harbor	290	385	13,531	13,916	2	22	84%	16,500		18,027	85	359	751	1,752	1,608	4,111
7225261	Harbor	334	475	11,439	11,914		20	88%	13,600		13,118	143	72	268	1,128	-192	1,204
7239536	Kim	56	127		127			42%	300		2,803			2,676			2,676
7239544	Kim	18	71		71			71%	100		2,414			2,343			2,343
7261746	Varsity			1,488	1,488	6	24	99%	1,500		1,544	38	36	44	12		56
7261753*	Varsity	1,232	1,315	4,542	5,857		5	87%	6,700	1,900	8,071	15	308	1,422	792		2,214
7261787	Varsity	176	187	3,150	3,337		11	68%	4,900	600	5,295	24	144	914	732	312	1,958
7261795	Varsity	811	937	15,767	16,704		28	71%	23,400	1,100	22,752	172	1,311	3,600	1,752	696	6,048
7286131	Tracy	277	583	8,706	9,289	30	49	87%	10,700		11,335	171	1,016	1,302	744		2,046
7286149	Tracy	53	316	4,797	5,113	54	85	90%	5,700		5,848	143	84	39	696		735
7417504	Jessie	235	877	180	1,057	111	209	96%	1,100		1,224	-1		167			167
7417801	Jessie	131	406	47	453	4	34	57%	800		887	16		434			434
7464597	Fairway	212	405	243	648		37	29%	2,200		5,771			5,123			5,123
7573934	Penny	1,086	2,146	1,629	3,775	352	777	60%	6,300	3,300	7,896	126		2,213		1,908	4,121
7960032	Greta	27	96	11	107			6%	1,800	1,000	9,429	28		9,322			9,322
Misses Total		9,968	17,892	191,282	209,174	596	1,381	79%	265,800	33,300	299,100	3,281	12,751	64,150	17,052	8,724	89,926
9201732	Kate	38	128	24	152			15%	1,000	2,900	6,458	107		6,306			6,306
9210832	Mariner	556	1,291	14,951	16,242	156	213	86%	18,900	7,300	18,983	192	356	101	1,704	936	2,741
9210865	Mariner	1,256	2,179	14,718	16,897	106	132	92%	18,400	2,700	19,515	291	96	506	2,112		2,618
9214636	Heather	197	495	6,104	6,599	43	81	87%	7,600	500	8,424	102	756	1,273	456	96	1,825
9214644	Heather	76	220	1,415	1,635	26	54	96%	1,700		2,239	18	24	580	24		604
9214834	Misty	19	-1	3,033	3,032	14	110	87%	3,500		3,510	35	83	214	216	48	478
9225236	Harbor	201	109	7,263	7,372	102	154	88%	8,400		9,056	166	72	-8	1,188	504	1,684
9225269	Harbor	249	277	6,773	7,050	8	13	94%	7,500		7,567	114	96	301	192	24	517
9239534	Kim	13	31	14	45			90%	50		1,203			1,158			1,158
9239542	Kim	4	8		8			16%	50		724			716			716
9240532	Camper	196	936	455	1,391			41%	3,400	2,500	8,204	-27	215	1,920		4,893	6,813
9261744	Varsity			650	650		7	93%	700		720	1		70			70
9261751*	Varsity	1,205	1,263	2,510	3,773	27	62	118%	3,200	700	4,657	28	335	-148	624	408	884
9261785	Varsity	27	27	855	882		6	74%	1,200	200	1,653			471		300	771
9261793	Varsity	522	539	7,699	8,328		119	85%	9,700	600	10,301	38	618	1,055	1,008		2,063
9286139	Tracy	12	134	2,591	2,725	9	45	88%	3,100		3,486	53	491	557	204		761
9286147	Tracy	36	28	1,772	1,800	15	70	82%	2,200		2,342	82	96	-106	648		542
9417502	Jessie	65	370	100	470	71	293	78%	600		696			226			226
9417809	Jessie	31	143	7	150	5	49	75%	200		606			456			456
9464595	Fairway	102	172	8	160		10	36%	500		2,357			2,177			2,177
9573932	Penny	505	1,313	542	1,855	71	162	53%	3,500	1,700	4,324	12		1,497		972	2,469
9960030	Greta	29	78	19	97			14%	700	500	7,123	26		7,026			7,026
Growing Total		5,339	9,740	71,503	81,313	653	1,580	85%	96,100	19,600	124,148	1,268	3,238	26,348	8,376	8,181	42,905

* Black-and-white Varsity "saddle" shoes

Exhibit 6

Fall 1980 Forecasts

A. Season Forecasts of Girls Service Group within overall Forecasting Framework

Sales Group

BABY

GIRLS

		Size Range				
		Infant	*Child*	*Growing Girls*	*Misses*	*Total*
Dress		. . .	. . .	. . .	. . .	. . .
Service	Futures	. . .	. . .	73K	195K	566K
	At-Once	. . .	. . .	23K	71K	284K
	Total	. . .	. . .	96K	266K	850K
Welt Sandals		. . .	. . .	. . .	. . .	. . .

BOYS

School
Dress
etc.

B. Forecasts and Actuals for Varsity and Related Styles

Week Ending 16-Aug-1980

		Growing Girls				Misses			
		Forecast			Actual	Forecast			Actual
Style	*Color*	*Futures*	*At-Once*	*Total*		*Futures*	*At-Once*	*Total*	
Kate	. . .								
. . .	. . .								
. . .	. . .								
Varsity	. . .								
Varsity	B+W	2,538	662	3,200	3,773	4,557	2,143	6,700	5,857
Varsity	. . .								
Varsity	. . .								
Total Varsity		11,811	2,989	14,800	13,543	8,865	27,635	36,500	27,386
Penny	. . .								
Greta	. . .								
Total		72,770	23,330	96,100	81,243	194,563	71,237	265,800	209,174

Exhibit 7

Fall Gross Demand, Girls Service Shoes

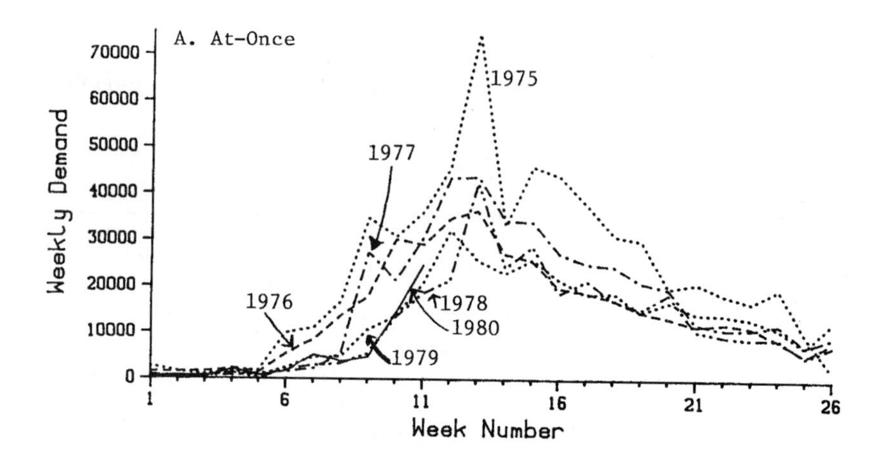

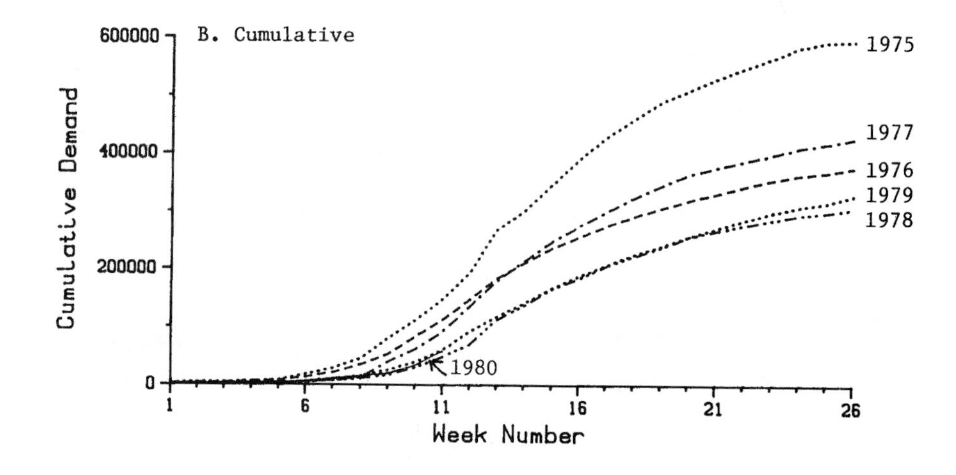

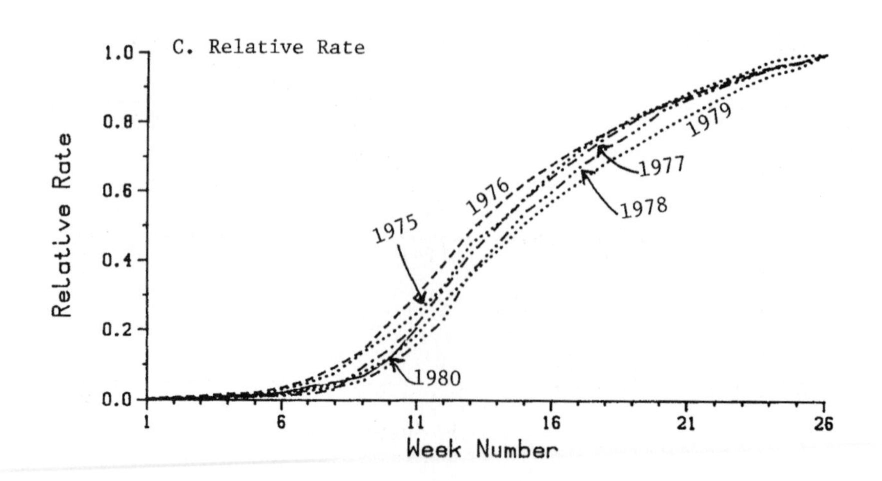

Exhibit 8

Fall Gross Demand, Growing Girls Service Shoes

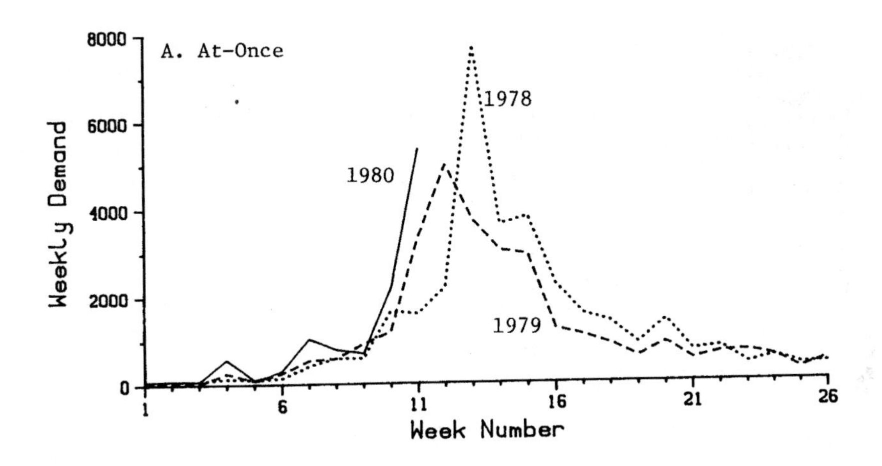

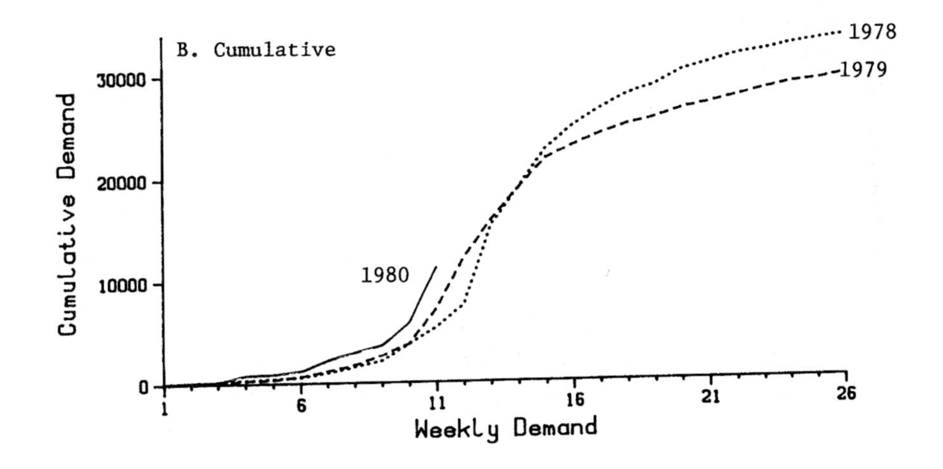

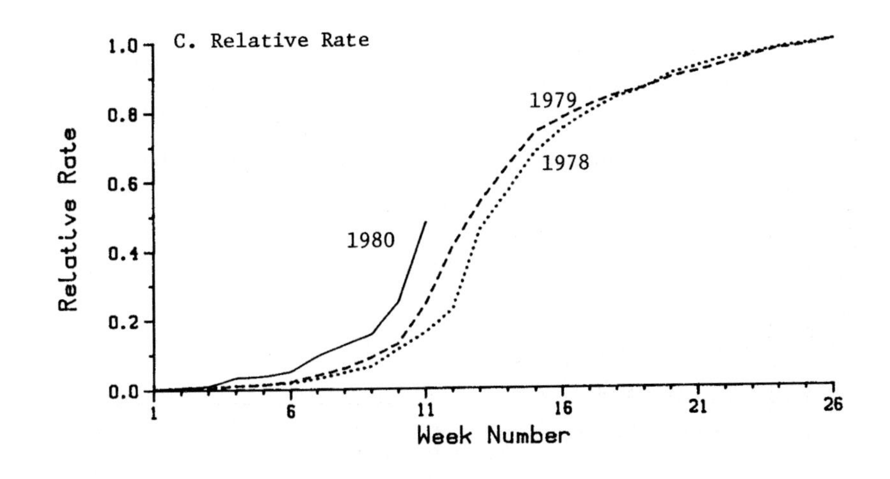

Exhibit 9

VARSITY	72617 53	Black/White	12½ to 3	B to EE	14.50	29.00
	92617 51		5 to 9	AA,B to D	16.00	32.00
	72617 87	Blue/Tan	12½ to 3	C to E	14.50	29.00
	92617 85		5 to 9	AA to C	16.00	32.00
	72617 95	Curry/Tan	12½ to 3	B to EE	14.50	29.00
	92617 93		5 to 9	AA,B to D	16.00	32.00
MARINER	52108 36	Tan	8½ to 12	B to EEE	14.00	28.00
	72108 34		12½ to 3	B to EEE	15.50	31.00
	92108 32		5 to 9	AA,B to E	17.00	34.00
	52108 69	Burgundy	8½ to 12	C to EE	14.00	28.00
	72108 67		12½ to 3	C to EE	15.50	31.00
	92108 65		5 to 9	AA,B to D	17.00	34.00

Exhibit 10

Calculation of Revised Demand Forecast for No. 9261751

Data are taken from the Merchandising Summary for Week 11 (Exhibit 5)

Actual Growing Girls demand, Week 11	5,339 (final entry, column 3, Exhibit 5)
Exceptional demand for week 11, No. 9261751	− 600
Adjusted Growing Girls demand, Week 11	4,739
Old forecast, Growing Girls, season	96,100 (final entry, column 10, Exhibit 5)
Adjustment	+ 5,000
New Growing Girls forecast season	101,100
Demand for No. 9261751 as percent of demand for Growing Girls, Week 11	12.8% (= 605/4,739)
Growing Girls shoes already shipped	81,313 (Exhibit 5, col. 6)
Futures unreleased	+ 1,268 (Exhibit 5, col. 13)
Total already sold	82,581
Total forecasted	101,100
Total already sold	− 82,581
To be sold during rest of season	18,519

Revision of 9261751 forecast

2,370: (12.8% of 18,519): rest-of-season demand

1,600: rest-of-season forecast of sales (reduced from 2,370 because of possible stockouts)

5,400: revised forecast (= 3,773 sold to date +28 futures unreleased +rest-of-season forecast, rounded)

SAMPLING AND STATISTICAL INFERENCE

INTRODUCTION

This chapter describes what you can infer about a population from a sample of observations drawn from that population. It focuses on inferences about population means and percents.[1] Examples of population means include the average annual per capita purchases by a population of customers, and average waiting time for customers trying to contact an airline reservation system. An example of a population percent is the percent of the United States workforce that is unemployed.[2]

The key concept of statistical inference is that although a sample provides useful information about a population mean or percent, that information is imperfect. We are left with some uncertainty about its true value. Statistical inference provides ways of estimating the true value and of quantifying our uncertainty about that estimate.

Samples are taken when it is impossible, impractical, or too expensive to obtain complete data on a relevant population. For example, you ask a sample of people drawn from a target population how much each intends to spend on a product, and from their responses you make an inference about actual sales per capita for the whole target population. As another example, you take a sample of telephone calls coming into an airline reservation system, measure the amount of time (if any) each call is kept waiting, and from this make an inference about the average waiting time for all calls.

This chapter gives you a conceptual overview of issues in sampling and inference. Some of the basic ideas are subtle. The best way to test your understanding of these ideas is to work through the examples and the exercises. When you get a numerical result, ask yourself what that result means. The formulas that are important in sampling are quite simple; they are collected together in the Appendix to this chapter so that you can readily access them.

Harvard Business School note 9-191-092. This note was prepared by Professor Arthur Schleifer, Jr.

1 Sometimes we are interested in population totals rather than in means or percents. For example, we might be more interested in total purchases made by a population of customers than in average per capita purchases. Whatever we know about a population mean or percent can readily be translated into knowledge about a population total by multiplying the mean or percent by the size of the population.

2 Inferences can also be made about processes. For example, we might want to make inferences about the long-run average rate of defects of a production process. Everything we say about populations is applicable to processes as well, but we shall focus our discussion on populations only.

For those of you who have previously studied sampling theory, it is important to understand that there are different levels of refinement in approaching the subject. You may have learned more accurate or more powerful methods than those presented here. These come at a cost of increasing complexity. In many managerial situations, it is not critical to know that a 95% confidence interval really extends from 200 to 250, when a rough-and-ready calculation might indicate that it extends from 190 to 260; either result will often lead to the same conclusion. In situations where very precise results are critical, you should seek expert advice.

Almost all the concepts discussed in this chapter come up again in the context of regression. Some of the concepts included here (for instance, the t-statistic or degrees of freedom) are not crucial for your understanding of sampling and inference, but introduce topics that are important in regression. This and subsequent chapters on regression should make you aware of the close linkages among sampling, inference, and regression, and give you two contexts in which these concepts play an important role.

SAMPLING ERROR

A sample may fail to tell you the exact value of a population mean or percent[3] due to sampling error. Sampling error arises from the fact that the sample mean may differ from its counterpart in the population due to the "luck of the draw." It is one of several sources of error in making inferences about a population from a sample.

▼ Inferences from a Sample

Even if a sample is "representative"[4] of the population from which it is drawn, it provides only imperfect information about that population, owing in part to sampling error. As an illustration, suppose you ask 100 potential customers how much they will spend on a proposed new product next year. The first says $10, the second $92, the third "nothing," and so on. You add up the 100 responses, divide by 100, and obtain $32.51 as your **sample average**. At this point, you want to make an inference from these responses about how much will be spent by the average potential customer in your entire market (the target population) next year. You could make the following inferences:

a) "My best estimate of average sales per potential customer is $32.51."

b) "Average sales per potential customer will be between $27.37 and $37.65 with 95% confidence."

c) "Average sales per potential customer will be greater than the breakeven amount of $27 at a 2½% level of significance."

An inference like (a) is called **statistical estimation**, (b) is a **confidence interval**, and (c) is a **test of** (statistical) **significance**.

[3] For most of this section we shall discuss population means only; at the end, we shall introduce the very minor modifications that apply for population percents.

[4] The only way to assure that the sample is representative is via some form of random sampling. We discuss this concept more fully later in this chapter.

Estimation and Confidence Intervals

Statistical estimation is a relatively straightforward procedure: when the distribution of values in the population is fairly symmetric and there are no extreme outliers, the **sample mean** (m) serves as a good estimate of the population mean. In the sales per customer example, $32.51 serves as a good estimate of the population mean.

Confidence intervals, on the other hand, are somewhat more subtle. Confidence in how close a sample estimate is to the true population mean depends on the **sample size** (n), and on the dispersion of the sample observations, as measured by the **sample standard deviation** (s).[5] Everything else being equal, your uncertainty about the value of a population mean will decrease as the sample size increases and the dispersion in the sample values decreases. To illustrate, in the example above, you would feel more confident that the population mean is close to $32.51 if your sample mean of $32.51 came from a sample of 400 respondents, instead of only 100. Similarly, you would feel more confident that the population mean is close to the sample mean if each of your 100 respondents expected to spend between $32 and $33 on the product next year, and less confident if their responses ranged from "nothing" to $500.

The level of confidence about the value of a population mean can be expressed in terms of a **confidence distribution.** In making inferences about a population mean, the mean of the confidence distribution is equal to the sample mean m, and its standard deviation (called the **standard error**) is equal to the sample standard deviation s divided by the square root of the sample size, n:

$$\text{standard error} = s/\sqrt{n}$$

The standard error decreases as the dispersion in the sample decreases and as the sample size increases.

The shape of a confidence distribution is essentially normal or bell-shaped, regardless of the shape of the distribution of values within the population itself. As is the case for *any* normal distribution, a value within one standard deviation above or below the mean occurs with 68% confidence; a value within two standard deviations above or below the mean has 95% confidence; and a value within three standard deviations has 99.7% confidence. Figure 2.1 portrays these relationships for a confidence distribution.

From the confidence distribution, you can construct a confidence interval for the population mean. A confidence interval states both a range within which the true value of the population mean may lie, as well as a level of confidence that the population mean does, in fact, lie within that interval. Confidence intervals of different lengths, and correspondingly different levels of confidence, can be constructed from a given sample. In general, we would prefer confidence intervals to be narrow rather than wide, since narrow intervals imply greater certainty about the population mean, and we would like the confidence level to be as high as possible. Unfortunately, for a sample that has already been taken, these preferences constitute a tradeoff: the narrower the confidence interval, the lower the confidence that the true value of the population mean lies within the interval. Conversely, the higher you want your confidence to be that the population mean lies within a given interval, the wider the interval must be. The only way you can achieve both objectives—to have a narrow interval with high confidence—is to take a large sample.

---

[5] See the opening section of the Appendix to this chapter for a discussion of the sample standard deviation.

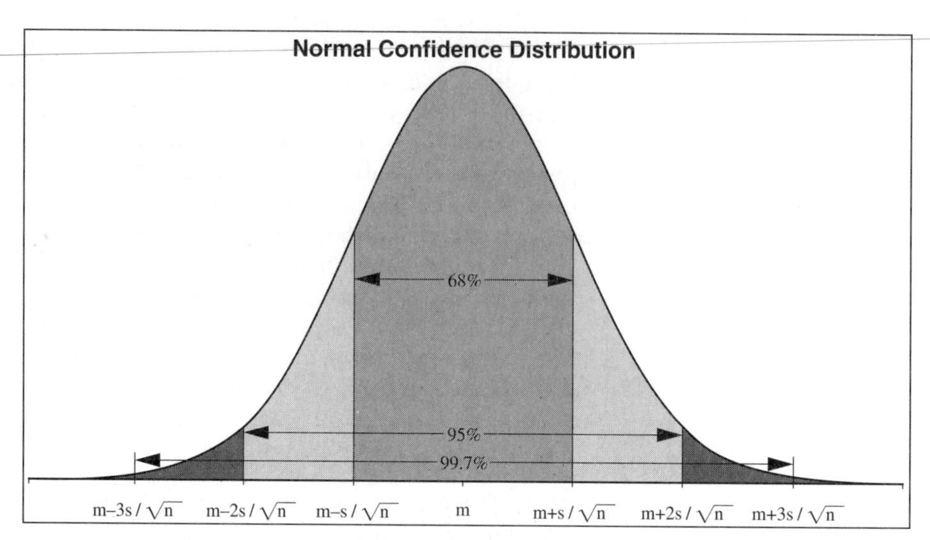

Figure 2.1

From the confidence distribution, you can construct a confidence interval for the population mean. A confidence interval is based entirely on three characteristics of a sample: the sample size, n; the sample mean, m; and the sample standard deviation, s. The population size is irrelevant. A confidence interval based on a sample with given values of n, m, and s will be the same, whether the sample in question is drawn from a population of 1,000 or a population of 1,000,000. It is a very common error for people to believe that an adequate sample should be some fixed percentage of the population, say 10%. Quite the contrary. It is the absolute size of the sample that determines accuracy.[6]

The following example illustrates how a confidence interval is constructed. A market researcher asks a sample of 100 potential customers how much they plan to spend on a product next year. The mean of this sample turns out to be $32.51 and the standard deviation is 25.7. The best estimate of what the average potential customer will spend is, therefore, $32.51. The standard error is $25.7/\sqrt{100} = 25.7/10 = 2.57$. Thus, with 68% confidence,[7] we can say that the average potential customer will spend between $32.51 – $2.57 and $32.51 + $2.57 (or between $29.94 and $35.08). Similarly, with 95% confidence, average expenditure will be between $27.37 and $37.65, and with 99.7% confidence, it will be between $24.80 and $40.22.

Statistical Significance

A third type of inference you can make from a sample is a **test of significance**. Rather than estimating a point at which, or an interval within which, the population mean lies, you can indicate if the population mean is likely to fall above or below a critical value of interest, called c. For example, you may want to know whether average sales per capita is likely to exceed a breakeven level of $27. If the sample mean is $32.51 and thus lies above the critical value of $27, how likely is it that the population mean is on the same side of the critical value?

---

[6] For a slight qualification of this statement, see the section called Finite–Population Correction in the Appendix to this chapter.

[7] The 68% confidence can be interpreted as follows: if you sampled many different populations and constructed a 68% confidence interval for each sample, then 68% of the time the true population mean would lie within the corresponding interval.

To answer this question, you can first construct a confidence interval and then determine whether the interval overlaps the critical value. If it does not, the sample outcome is said to be **statistically significant**. If it does overlap the critical value, the sample outcome is said to be **not significant**. These tests have a corresponding **level of significance**, and the two should be reported together. A sample result is statistically significant at a level of 2½% when a critical value falls outside a 95% confidence interval and therefore lies in a region in which we have only 2½% confidence[8] that the population mean lies. Similarly, a test is statistically significant at a level of 0.15% when a critical value falls outside a 99.7% confidence interval.[9]

The case of launching a new product serves to illustrate tests and levels of significance. Suppose $n = 100$, $m = 32.51$, and $s = 25.7$. Then, as we saw before, a 95% confidence interval extends from 27.37 to 37.65, and does not cover the critical value of $c = 27$. Therefore, relative to 27, the sample outcome of 32.51 is statistically significant at the 2½% level.

The t–statistic. Rather than determining statistical significance by constructing a confidence interval, the following shortcut procedure produces the same results. If c represents the critical value, compute the statistic

$$t = (m - c)/\text{standard error} = (m - c)/(s/\sqrt{n})\ .$$

Then if $t > 2$ or $t < -2$, the sample outcome is significant at the 2½% level. (Try the computation on the two examples above, and convince yourself that it yields the same results.) If $t > 3$ or $t < -3$, the sample outcome is significant at the 0.15% level.

Significance vs. Importance. It is very important to realize that a result can be statistically significant but unimportant, or vice versa. Because it is extremely unlikely that the population mean or percent will be precisely equal to the critical value, a sufficiently large sample usually shows statistical significance, whereas smaller samples may not. For this reason, a statistically significant outcome may not be economically better than an outcome that is not statistically significant. For example, suppose a manager is trying to decide which of two new products, A or B, to introduce. Breakeven sales per capita are $27 for both A and B. The manager obtains a sample of 10,000 potential customers for product A, but only 100 for product B. The sample results are in Table 2.1.

Table 2.1

	Product A	Product B
n	10,000	100
m	27.3	46.0
s	10	100

[8] We have a 95% level of confidence that the population mean lies within the interval, and hence a 5% level of confidence that it lies outside. By virtue of the symmetry of the confidence distribution, we have just a 2½% level of confidence that the population mean lies below the lower limit of the confidence interval.

[9] These are examples of one-sided tests of significance. In the statistical literature, some tests are two-sided (essentially doubling the level of significance), but such tests are mentioned only for completeness, and need not concern us here.

For product A, a 95% confidence interval extends from 27.1 to 27.5, a result that is statistically significant relative to $c = 27$ at the 2½% level. For product B, a 95% confidence interval extends from 26 to 66, and hence the result is not significant at the 2½% level. If the manager must now decide which product to introduce, and can acquire no more data, what should she do?

Product A has higher prospects of breaking even, but not much potential for large profit. Product B has lower prospects of breaking even (more downside risk) but much more upside potential. Unless the manager were very risk-averse, she should prefer B to A.

▼ Sample Size

Taking a sample to obtain information about a population mean requires a decision about the sample size. How large a sample should you take? Since sampling is usually expensive, and the cost of sampling often increases with sample size, there is a tradeoff between more precision at higher cost and less precision at lower cost. To determine the appropriate sample size, you can first specify the length of a confidence interval and an associated confidence level that you would consider satisfactory, and then compute a sample size that will deliver such a confidence level and interval.

Suppose, for example, you want to say with 95% confidence that average annual expenditure for a target population is within an interval of length L. That is, if the sample mean turns out to be \$32.51, you would like the sample size to be such that with 95% confidence you would know the population mean is between 32.51–0.5L and 32.51 + 0.5L. How large a sample must you take to achieve this level of precision? If we knew the sample standard deviation, s, the answer would be simple. You would like the interval from $m - 2s/\sqrt{n}$ to $m + 2s/\sqrt{n}$ to have length L. That means that $4s/\sqrt{n} = L$, or $\sqrt{n} = 4s/L$, or $n = 16s^2/L^2$. If $s = 25.7$ and $L = 2$, then the required sample size would be $n = 2{,}642$. (By similar logic, the sample size for a 68% confidence interval is $n = 4s^2/L^2$, and for a 99.7% interval is $n = 36s^2/L^2$.)

A sample of this size might be prohibitively expensive. If you were willing to double the length of the interval L from 2 to 4, you could reduce the size of the sample by a factor of *four*, from 2,642 to 661. That is because the length of the interval depends on $1/\sqrt{n}$: if you quadruple n, you halve the length; if you reduce n by a factor of 4, you double the length. In general, the process of choosing a sample size is an iterative one, in which you compare the tradeoffs between precision and cost.

The analysis above assumes that you know the value of the sample standard deviation s before the sample is even drawn. Unfortunately, s is a statistic whose value is computed *after* the sample is drawn. The only thing you can do to escape this logical trap is to *guess*[10] the value of s, and hope that your results are not too sensitive to a substantial misestimate.

▼ Confidence vs. Probability

When reporting that a 95% confidence interval for a population mean extends from \$27.37 to \$37.65, it is tempting to slip into the language of probability, and say that there is only a 5% probability that the true value of the population mean is outside this interval. Similarly, if the sample mean is \$32.51, then, by virtue of the symmetry of the confidence distribution and the fact that its mean is equal to the sample mean, it is tempting to say that there is a 50-50 chance that the true population mean is greater than \$32.51.

[10] Sometimes you can get a fair estimate of s from a small pilot sample.

Such probabilistic interpretations are much more natural and appealing than the rather convoluted interpretation given in footnote 7. But are they legitimate?

Suppose, as before, a sample of 100 potential customers reveals how much each will spend on your product next year. In the past several years, average expenditure per customer has ranged from $10 to $15, and no product or competitive changes are likely to support a jump to the $27 – $38 range. For this reason, you must conclude that the 95% confidence interval extending from roughly $27 to $38 overstates the probability that demand per customer will fall in that range. You can only conclude that the luck of the draw led to a sample mean of $32.51 that, in this case, was too high.

If, on the other hand, you have no additional information to suggest that this sample result is substantially more or less likely than other possible results, then you can interpret the confidence distribution derived from the sample as a probability distribution, and you can treat a significance level as the probability that the population mean falls above or below a critical value.

▼ Population Percents

As indicated previously, almost everything that we have said regarding inferences for population means applies to population percents. Although it is natural to think in terms of percents (the percent of voters who will vote for candidate X; the percent of people who will buy our product, etc.), it is important for computational purposes to express percents in fractional form, e.g., 20% = 0.20. To emphasize this, we will use the language of fractions and not percents in this section.

When we are dealing with fractions, individual values can be coded as 0s and 1s. A person who will vote for candidate X will be coded 1; a person voting for a different candidate will be coded 0, etc. The mean of this dummy variable is then the fraction of 1s. Suppose we take a sample of n observations, and that a fraction f of those observations are 1s. Then the sample fraction f is the sample mean. Accordingly:

- The sample fraction f is an estimate of the population fraction.
- A confidence distribution for the population fraction is normal with mean f and standard error $s/\sqrt{n}$.
- The 68%, 95%, and 99.7% confidence intervals are defined exactly as before, as are t values.

While the value of s can be computed for a dummy variable in exactly the same way as for any other variable, there is a shortcut formula available. For a dummy variable,

$$s = \sqrt{f * (1-f) * n/(n-1)} \quad ;$$

unless n is quite small (10 or less), the term $n/(n–1)$ is close enough to 1 so that it can be safely ignored. Thus, to a very reasonable approximation,

$$s = \sqrt{f * (1-f)} \quad .$$

Example. A sample of 100 voters was asked for which of two candidates, X or Y, they would vote. Fifty-two said they would vote for X. Then f = 0.52, $s = \sqrt{0.52 * (1-0.52)} = 0.4996$, and the standard error is $0.4996/\sqrt{100}$ = 0.04996. A 95% confidence interval would extend from 0.52 – 2*0.04996 = 0.4201 to 0.52 + 2*0.04996 = 0.6199. The sample outcome is not significant at the 2½% level relative to a critical value of c = 0.50. (Had the sample size been 10,000, with 5,200 in favor of X, the sample outcome would have been statistically significant, as you should verify.)

Figure 2.2 shows how the standard deviation s for a dummy variable depends on the sample fraction f. If you are quite sure, before you take a sample, that the sample fraction will not be less than 0.2 nor greater than 0.8, you can be equally sure that s will be between 0.4 and 0.5.

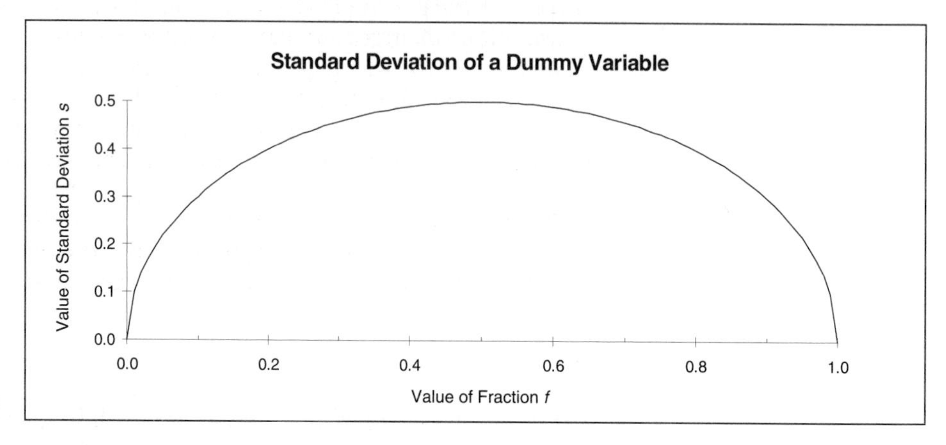

Figure 2.2

This is very helpful in sample–size problems. To illustrate, suppose you are conducting a national poll to determine the percent of voters who will vote for the Republican candidate in the next presidential election. Suppose that you want to state with 95% confidence that your results are within an interval of L= 5 percentage points. We can be quite sure that f will be between 0.2 and 0.8, and hence that s will be between 0.4 and 0.5. Let's assume the worst case—that s = 0.5. Then our sample size n should be such that

$$n = 16s^2/L^2 = 16*0.5^2/0.05^2 = 1{,}600 \quad .$$

In fact, many national polls involve around 1,600 respondents, and are commonly reported as having a margin of error of 2½% (half the length of the confidence interval); the confidence level of 95% is implicit.

Sampling in the Real World

Apart from random sampling error, there are two other sources of error in making inferences about a population from a sample. The first involves response bias and the second concerns the representativeness of the sample.

Response Bias

Almost all questions about people's opinions, attitudes, expectations, or preferences involve response biases of various kinds. When 44% of the voters sampled in a public-opinion poll in June say they will vote for the Republican presidential candidate, and when the actual vote in November turns out to be 53% for the Republican, the polls are often said to have been "wrong." However, the discrepancy between 44% and 53% can seldom be accounted for by sampling error. Instead, it is mostly due to the fact that respondents did not do in November what they said they would do in June.

Such response bias stems from a number of factors. First, opinions change over time. Second, people do not always say what they will do or have done. Third, people cannot answer certain questions realistically; for example, they seldom know how much they will spend on a new product in the coming year. Fourth, the way in which a question is asked sometimes influences the response.

For example, of 764 people who were asked whether they agreed or disagreed with the statement "Advertising often persuades people to buy things they shouldn't buy," 76% agreed, 20% disagreed, and 4% had no opinion, while of 772 people who were asked a similar question about the statement "Advertising seldom persuades people to buy things they shouldn't buy," 40% agreed, 56% disagreed, and 4% had no opinion.[11]

Although the way a question is asked may influence the answer, asking the same question repeatedly can reveal changes in respondents' attitudes over time. The president's "report card" is generated by asking respondents periodically to assess the president's performance. As long as the question remains the same, a sharp drop in performance rating (if it cannot be explained as mere sampling error) is a good indication that the population at large has become less satisfied with performance. But a sharp drop that coincides with a rewording of the question may be nothing more than an indication of response bias.

Representativeness

A sample should be representative of the population from which it is drawn. The easiest way to ensure representativeness is to take a sample in which every member of the target population has an equal chance of being included. Such a sample is called a **random sample**. To obtain a truly random sample, you must have a complete list of every member of the target population and you must select members from the list by a process that gives each member the same chance of being included in the sample. The random process may entail drawing names from a hat, using a table of random numbers, or telling a computer to select observations at random. If your target population is a group of people, collecting a sample entails one more step: you must track down the selected members and convince them to answer your questions. Obtaining a truly random sample is a time-consuming and difficult process. Often much more informal procedures—stopping people on the street, or in shopping malls, or in airports—are used.

Even when a serious attempt is made to obtain a random sample, the people who end up in the sample may not have been chosen in the prescribed manner for a variety of reasons. For example, in the United States, target populations are often based on the decennial census, which gradually becomes outdated, and which falls short of complete accuracy even when current. Door-to-door sampling is plagued by not–at–homes and refusals to be interviewed.[12] Telephone surveys are limited to families with listed telephones, and even many of these cannot be contacted and of those who are reached, many refuse to reply. Nonresponse in mail surveys typically runs around 80%. Thus, even with the best of intentions, it is generally not possible to obtain a random sample from the population.[13]

---

[11] Raymond A. Bauer and Stephen A. Greyser, *Advertising in America: The Consumer View*, Division of Research, Harvard Business School, 1968.

[12] According to an article in *The New York Times*, members of the Council of American Survey Research Organizations, an industry association, reported that 38% of consumers turned down requests for interviews in 1988 ("Surveys Proliferate, but Answers Dwindle," *The New York Times*, October 5, 1990).

[13] One exception, where true random sampling can be carried out, is in sampling a mailing list. In this case, the names on the list constitute the population, and from the responses of a sample of names on the list, one can make legitimate inferences about how the entire mailing list would respond.

Many polling organizations and market research agencies, nevertheless, report sample results in the form of estimates, confidence intervals, and significance tests, as if the samples that provided the data were truly random. Are such results seriously misleading? Sometimes they are.[14] But before throwing the inferential baby out with the nonrandom bathwater, it is important to understand what may go wrong when one tries to make inferences from such a nonrandom sample.

The implications of a nonrandom sample depend on the reason for the lack of randomness and the nature of the study. If you are asking about evening television viewing habits, people who are not at home in the evenings are likely to be very different from people found at home at that time. But if you are interested in which of two spaghetti sauces they prefer, it may be more reasonable to believe that at–homes can be treated as nearly, if not exactly, representative of not-at-homes. It is sometimes possible to test such a proposition by making special efforts to track down a sample of not-at-homes, but one is always left with the lingering doubt that the trackable not-at-homes are different from those who could not be tracked. And it is even harder to test whether respondents who willingly answer an interviewer's questions are representative of those who slam their doors in the interviewer's face, or hang up the telephone.

One corrective action commonly taken by sampling organizations is to replace randomly selected nonrespondents with respondents who have identical, or nearly identical, demographic profiles, such as age, gender, ethnicity, education, and income. To the extent that these demographic characteristics help to distinguish the responses of different segments of the population, this replacement technique makes sense: for example, if the young and the old have markedly different attitudes towards rock music, then replacing a young nonrespondent with an old substitute would distort your sample estimates about the population's musical tastes, while replacing the young nonrespondent with a young substitute would avoid that particular kind of distortion. Nevertheless, the replacement necessarily differs from the nonrespondent with respect to willingness to respond, and to the extent that one can hypothesize a link between willingness to respond and musical taste, a problem remains. If no such link seems plausible, then this corrective action is probably sufficient to render inferences about the population reliable.

APPENDIX: ELEMENTS OF SAMPLING THEORY

Sample Standard Deviation

Spreadsheet Calculation. The sample standard deviation *s* is a measure of how spread out the values in a sample are. If you have a column in an Excel spreadsheet consisting of the value of each sample observation, then the =STDEV function, applied to the column of sample values, gives you the sample standard deviation.

---

[14] In the famous *Literary Digest* poll of 1936 to predict the outcome of the Landon–Roosevelt presidential election, over 10 million ballots were mailed to people whose names were selected from lists of owners of telephones and automobiles. Of the 2,376,523 ballots returned, 54.5% were for Landon, the Republican candidate, who received only 36.7% of the popular vote. Subsequent analysis revealed that most of the huge error was due to the fact that people on the lists were not representative of nonowners of telephones or cars, who were much more likely to vote for the Democratic candidate.

If you want to compute the sample standard deviation without relying on the =STDEV function, go through the following steps:

- Suppose the sample values are in column A. Compute the average of the values in column A using the =AVERAGE function.
- In column B compute the difference between the first number in column A and the average of the values. This is called a **deviation**.
- In column C compute the square of the first deviation in column B.
- Now copy the computations in the first row of columns B and C to all the other sample values.
- Compute the sum of the squared deviations (column C), using the =SUM function.
- Divide this sum by $n - 1$.
- Compute the square root, using the =SQRT function. This is the sample standard deviation.

In words, the sample standard deviation s is computed by taking the sum of the squared deviations, dividing by n–1, and taking the square root.[15]

Standard Deviation of a Dummy Variable. If the values in column A are zeros and ones only, you can verify that the above procedure results in a value equal to

$$s = \sqrt{f * (1 - f) * n / (n - 1)}$$

where f is the fraction of 1s (or the average of the values in column A).

Degrees of Freedom. You may have noticed that the sample standard deviation uses a divisor of n–1 before you take the square root, while the standard deviation defined in Chapter 1 uses a divisor of n. Or, the Excel function for the standard deviation in Chapter 1 was =STDEVP, while here we are using =STDEV. Why the difference?

In Chapter 1 we were computing the mean and standard deviation of a *population*. Here, we are computing the corresponding statistics for a *sample*. The same computations are used for the population and the sample mean, but there is a slight difference (divisor of $n - 1$ vs. n) for the sample standard deviation.

Just as the sample mean is an estimate of the population mean, the sample standard deviation is an estimate of the population standard deviation. We would like the sample mean and the sample standard deviation to be "unbiased" estimators. By this we mean that if you took repeated random samples of fixed size (say $n = 10$) with replacement (i.e., with the possibility of drawing the same element more than once in a given sample), the sample estimate would sometimes be too high and sometimes be too low, but in the long run would "average out" to the corresponding population value. The sample mean is an unbiased estimate of the population mean, but the sample standard deviation using a divisor of n is not an unbiased estimate of the population standard deviation. It will "average out" too low,[16] because in order to compute the sample standard deviation, you first must compute the sample mean from the same set of data.

In statistical terminology, every one of the n sample values provides a **degree of freedom** for estimating sample statistics. One degree of freedom is used to estimate the sample mean, leaving just $n - 1$ for estimating the sample standard deviation.

---

[15] Because the standard error (the standard deviation of the confidence distribution) also involves a division by $\sqrt{n}$, it is easy to become confused. Here we are talking about how to compute s, the sample standard deviation, which is an estimate of the population standard deviation. The **standard error**, as previously stated, is $s / \sqrt{n}$.

[16] In the extreme, if you used a sample of one height to estimate the standard deviation of the heights of first–year MBA students at Harvard Business School, your sample standard deviation, using a divisor of n, would be 0, which clearly underestimates the true value.

In order for the average value of the sample standard deviation to be approximately equal to the value of the population standard deviation, you should, instead of dividing the sum of squared deviations by n (the sample size), divide by $n - 1$ (the degrees of freedom), and then take the square root. (In Excel, the =STDEV function does this for you automatically.)

Except when the sample size is very small (10 or less), this refinement is of no practical consequence, and this discussion is included merely to introduce the concept, not to plague you with additional computational burdens. However, when we come to regression, more degrees of freedom are used up in estimating additional statistics, and the degrees-of-freedom issue will become consequential.

Finite–Population Correction. In this chapter we noted that the accuracy of a sample depends on its absolute size, not on its size relative to the size of the population from which it was drawn: a sample of 100 gives rise to a confidence interval of the same length, whether it was drawn from a population of 1,000 or a population of 1,000,000. That's not quite right, as you can see if you imagine a population size of 101. In that case, a sample of 100 will surely leave you far less uncertain about the value of the population mean than you would be if the population were 1,000,000.

For samples of size n drawn without replacement (i.e., so that the same individual or item cannot appear more than once in the sample) from a population of size N, the standard error is not $s/\sqrt{n}$, but rather

$$s * \sqrt{(N-n)/(N-1)} / \sqrt{n} \quad .$$

The factor $\sqrt{(N-n)/(N-1)}$ is called the **finite–population correction**, or FPC. If N = 1,000,000 and n = 100, the FPC = 0.99995, and can be ignored. If N = 1,000 and n = 100, the FPC = 0.949, so that ignoring the FPC will result in a standard error that is only around 5% too large, and no great harm is done by ignoring it. If N = 101 and n = 100, however, the FPC = 0.1, and ignoring it would result in a standard error that is ten times too large.

In almost all practical sampling situations, you seldom sample more than 10% of the population, and ignoring the FPC does no great harm.

Summary of Formulas

Notation. n: sample size
m: sample mean
f: sample fraction
s: sample standard deviation
c: critical value, relative to which the significance of a sample outcome is measured

Estimate of Population Average. Sample mean or average (m), or sample fraction (f).

Confidence Intervals. $m - s/\sqrt{n}$ to $m + s/\sqrt{n}$: 68% confidence ,

$m - 2s/\sqrt{n}$ to $m + 2s/\sqrt{n}$: 95% confidence ,

$m - 3s/\sqrt{n}$ to $m + 3s/\sqrt{n}$: 99.7% confidence .

Statistical Significance. If $t = (m - c)/(s/\sqrt{n}\,) < -2$ or > 2, the sample mean differs significantly from the critical value at the 2½% level; if $t < -3$ or > 3, the outcome is significant at the 0.15% level.

Sample Size. If you want a confidence interval of length L with:

68% certainty, then $n = 4s^2/L^2$;

95% certainty, then $n = 16s^2/L^2$;

99.7% certainty, then $n = 36s^2/L^2$.

Computations. Define x_i to be the value of the "i–th" observation, $i = 1, ..., n$. Then the following "formulas" can be used. For the Excel formulas (in italics), assume the data is in the range A1:A100.

Sample mean (m) or fraction (f) of a dummy variable. Add the sample values and divide by the sample size n.

$$m = f = \frac{1}{n}\sum_{i=1}^{n} x_i$$

=AVERAGE(A1:A100) .

Sample standard deviation (s). Subtract from each sample value its mean (m), and square the result. Add these squared values, divide by $n - 1$, and take the square root.

$$s = \sqrt{\frac{1}{n-1}\sum_{i=1}^{n} (x_i - m)^2}$$

=STDEV(A1:A100) .

If the population values are all 0s or 1s

$$s = \sqrt{f * (1-f) * n / (n-1)}$$

Standard error.[17] Divide the sample standard deviation (s) by the square root of the sample size.

$$\text{Standard error} = s/\sqrt{n}$$

=STDEV(A1:A100)/SQRT(100) .

Exercises on Sampling and Statistical Inference

1. Several years ago, in conjunction with a computer exercise, all first-year students at Harvard Business School were asked to report their height (in inches), weight (in pounds), and gender. Data on 768 students were acquired in this fashion. The following sample of 20 weights was selected at random from the 768. (See data file HTWTSAMP.XLS on your data diskette.)

160	113	140	148	185
130	185	155	166	161
158	200	144	180	210
170	175	108	155	163

   a) Compute the sample mean and sample standard deviation.

[17] There is no direct formula in Excel for calculating the standard error. However, under the Options menu, Analysis Tools command, Excel has an Add-In called Descriptive Statistics, which calculates, among other things, the mean, standard deviation, and standard error.

b) Compute a confidence interval that has a 95% probability of covering the true average weight of the 768–person population.

c) The average weight of the 768–person population was 154.15 pounds. Did your confidence interval cover the true average?

d) Two people in the sample of 20 were women: one weighing 113 pounds and the other 108 pounds. Of the 768 students on which data were available, 171 were women. On the basis of this information, what is your best estimate, based on the sample, of the average population weight?

2. The presidents at two large midwestern state universities, State University and State College, are bickering over the academic quality of their respective incoming freshman classes. Unbeknownst to either, the combined SAT scores at the two schools are as follows:

	STATE UNIVERSITY	STATE COLLEGE
AVERAGE	950	930
STD. DEVIATION	160	160

a) Assume that both presidents will sample the same number of incoming freshmen at their respective schools. How large a sample do the presidents need in order to establish, with 95% confidence, that the average SAT score for State University freshmen is higher than the average SAT scores for State College freshmen?

b) The sample has been taken, and the president of State University claims: "My students are better. The difference in average scores is statistically significant." The president of State College argues: "On average, your students may be better, but many of my students are better than many of your students." Be prepared to discuss the statistical merit of these two statements. Estimate, without calculating, the probability that a randomly selected student from State University will have a higher SAT score than a randomly selected student from State College.

3. National Cookie Company produces premium chocolate chip cookies for sale in up–scale food markets. Cookies are sold in 250 gram packages of 10 cookies. Unfortunately, the manufacturing process has some inherent variation. After years of improving the process, National has reduced the standard deviation of the process to 0.6 grams per cookie. In the short term, the variation of the production process is outside National's control. However, National is able to adjust the mean weight level for each cookie. For various economic reasons, setting the mean level properly is important. If it is too high, then the cookies are heavier than required and National must incur unnecessarily high raw material costs; if the mean level is too low, then the cookies must be sold as thrift.

a) Assume that National sets the maximum number of underweight packages at 2.5%. Where should National set the mean production level per cookie?

b) Without carrying out any calculations, describe how National should determine the optimal number of underweight packages. It might be helpful to assume the following concrete numbers: raw material costs are \$0.12 cents per gram, and cookies sell to distributors for \$1.75 per package (regular) and \$0.75 per package (thrift).

Note: solving the problem in general requires calculus.

4. The Water's Edge Company sold a variety of products related to water sports (sailing, swimming, water polo, snorkeling, etc.) by mail-order catalog. The Company's entire mailing list contained over 20,000,000 customer names and addresses. In the summer of 1990, they decided to run an experiment to ascertain which of two catalog formats would draw more customers. Accordingly, they sent out a sample of 500,000 issues in their standard 7½-by-8½-inch format, and another sample of 500,000 in an experimental 8½-by-11-inch format. The two customer groups of 500,000 were chosen so as to match one another as closely as possible with respect to past purchasing behavior and geographical location. The standard catalog was sent to the 19,000,000 other customers, but responses from the experimental mailings could be identified by the item numbers that customers used to specify merchandise ordered, which were unique to those catalogs. Orders generated by the catalogs were as follows:

STANDARD:	8,450	or 1.69%
EXPERIMENTAL:	11,472	or 2.29%

a) Construct 99.7% confidence intervals for the percent in their entire mailing list who would respond to each of the two catalog formats. Do the confidence intervals overlap? What can you infer about the difference in response rates for the two formats in the entire population?

b) Would you use the experimental 8½-by-11–inch format in your new catalogs? Why or why not?

c) Before running the experiment, there had been some discussion of simply sending the experimental version of the catalog to all 20,000,000 customers, and comparing the 1990 response rate with that of 1989. List the pros and cons of this method versus the experimental method actually used.

5. The following exercise is designed to demonstrate that confidence intervals can be interpreted as probability statements about a population mean. You will be asked to "sample" three variables, for each of which there are 768 observations in data file HTWT.XLS, which was distributed to you on your data diskette. From the samples, you will compute confidence intervals and determine whether or not those intervals cover the true mean of the population.

The three variables in question are height, weight, and gender. As you can easily verify, the average height of all 768 students is 69.52 inches, the average weight is 154.15 pounds, and the "average gender" is 0.223 (male students were coded 0 and females 1, but the average is simply the fraction of female students in the population).

You will be asked to take a sample of 20 students' heights, a sample of another 20 students' weights, and a sample of still another 20 genders, and compute confidence intervals from those three samples. You could do the sampling by putting 768 slips of paper, numbered 1 through 768, into a hat, drawing 20 such slips, for each slip finding the student in the list corresponding to that number, writing down the value of the appropriate variable, and computing a confidence interval based on the sample you had drawn.

Fortunately, Excel has a much less tedious mechanism for accomplishing the same task. Here's what you should do:

- Open the file HTWT.XLS.
- Click the **Tools** menu, then **Data Analysis**, then **Random Number Generation.**
- When the dialog box appears, click on **3** for Number of Variables, **20** for Number of Random Numbers, select **Uniform** for Distribution, and type[18] **1** and **769** in the two boxes designated as Between. Under Output Options, click on Output Range and type G7 in the corresponding box.

If you were to click OK, you would see three columns of 20 random numbers uniformly distributed between 1 and 769; however, your numbers would be the same as everyone else's. As we want each student to create a different set of samples, there is one more step: you must supply a number in the box labeled Random Seed. To ensure that each of you has a unique (or nearly unique) random seed, please enter your birthday in the form ddmmy, where dd contains the one or two digits of the day of the month in which you were born, mm is a two-digit representation of the month (January = 01, February = 02, ..., December = 12), and **y** is the last digit of the year in which you were born. Here is how you would designate the random seed for various birthdays:

January 2, 1968	2018
July 17, 1966	17076
December 31, 1960	31120

- Enter this four- or five-digit number in the box labeled Random Seed, and click **OK**.

Now, to obtain the random sample:

- Place the cursor on cell **G28** (two rows below the bottom of the first column of random numbers).
- Type the formula **=INDEX(C$7:C$774,G7)** and press **Enter**. (Be careful to enter the $ signs exactly as shown.) If the random number in cell G7 was equal to or greater than 1, but less than 2, the number in G28 will be the first height listed in column C; if it was equal to or greater than 2, but less than 3, it will be the second, etc. Thus each of the 768 heights has an equal chance to be selected.
- Copy the formula in G28 down through row 47, and across to Column I, so that 20 rows and 3 columns are filled with sample observations. Those in Column G are sample heights, in H sample weights, and in I sample genders.

Now, for each of these three columns, compute the sample average, the sample standard deviation, the standard error, and a 68% confidence interval. Determine whether each of the three intervals covers the true average. Please bring printouts of all your computations to class. Among other things, your professor should check in class to see what fraction of 68% confidence intervals covered the true average.

---

[18] We really mean 769, not 768. The idea here is that every number produced, rounded down to the nearest integer, will be an integer between 1 and 768. For each random number drawn, each integer between 1 and 768 has a probability 1/768 of being selected.

Time Series

Introduction

A time series consists of observations on a single variable at discrete points in time, usually at equal intervals. Typical time series may involve:

- Elements of the national income and product accounts at yearly, quarterly, or monthly intervals (Gross domestic product, government spending, consumer price index, unemployment rate, etc.)
- Measures appropriate to individual business operations (monthly sales, machine efficiencies, defect rate, etc.)
- Financial market statistics (daily closing price of IBM stock)
- Meteorological data (daily high temperatures in Boston, monthly amount of precipitation in Chicago)
- Population data (U.S. population, or births, or deaths by year)

The order of the observations is an essential characteristic of each of these series.

We have already seen that some time series may exhibit trends and seasonals. For example, monthly retail sales in the United States are characterized by an upward trend and a pronounced seasonal, with a sharp peak in December and a trough in January and February. The upward trend may be explained, in part, by other time series: population growth, per-capita income growth, or inflation, for example. Fluctuations from the trend and seasonal effects may be explained, in part, as the effects of weather, consumer confidence, availability of credit, etc. Thus, it is quite natural to "explain" the behavior of a particular time series in terms of trend, seasonal, and the effects of other variables that are themselves time series. Unfortunately, those other variables, while potentially useful in explaining past values, may not be useful in predicting future values of the time series in which we are interested. For example, while bad weather may have reduced retail sales last month, unless we have a reliable forecast of the weather for next month, weather is not a useful variable for predicting future sales.[1]

Harvard Business School note 9-893-012. This note was prepared by Professor Arthur Schleifer, Jr.

[1] There are three important cases in which explanatory variables *should* be used in forecasting future values of a time series. First, if the explanatory variable is a "leading indicator" its value will be known in advance: permits for housing starts may be good predictors of refrigerator sales a year later. Second, some variables can be forecast with accuracy because their values can be controlled by the firm: changes in price, advertising expenditures, or packaging may contemporaneously affect sales of a product, but a company's decision to effect such changes will be known to a company forecaster in advance. Third, some explanatory variables may be more easily or accurately predicted than the time series in question: although their actual values will not be known in advance, their forecast values will. In effect, variables of this sort are like leading indicators. Although forecasts of elements of the national income and products accounts are subject to error, they may be much more accurate than forecasts of your company's sales, because professional forecasting organizations have devoted great effort to developing appropriate prediction models.

Because current values of explanatory variables are not always useful in predicting future values of a time series, forecasters are often restricted to looking at the information within the series itself, without relying on other explanatory variables. There is considerable information embedded in such series: we have seen that trends tend to continue and seasonal patterns tend to repeat. Thus it's a safe bet to predict that next January's retail sales will be substantially less than those of the preceding month, and it's probably reasonable to predict that they will exceed the preceding January's sales.

What if we have a time series with no discernible trend or seasonal?[2] Is there any information in the time series itself that will help us to predict its value in future periods? The answer is "Yes," but it takes some analysis to extract this information. We can think of the observations that constitute the time series as a sample from a process that generates data according to some (probabilistic) rule. We seek ways of using the sample data to identify the rule that governs the data-generating process. If we know what the rule is, we can then make (probabilistic) forecasts of future values of the time series.

The number of rules that could give rise to time series observed in the real world—even those with no trend or seasonal—is huge, and any attempt to cover the subject of time-series analysis in all of its richness requires a book, not a chapter. The purpose of this chapter is to introduce you to the concept of time-series analysis and to introduce two very simple rules for generating data that describe the way some important time series behave in the real world. At the end of this chapter, we'll give an indication of where and how this subject unfolds, but for now we shall introduce the two rules, then show how time-series data generated by one or the other of these rules can be identified and analyzed, and how appropriate forecasts can be made.

Concepts Used in Data generation Rules

Notation

We shall measure time at periods denoted by 1, 2, ..., t, ..., T. For example, period 1 might be midnight on January 1, period 2 might be midnight on January 2, etc. If we had a year's worth of data, T would be 365. Let's denote the value of the time series at time t by y_t: in the example above, the first observation (the value of the series on January 1) will be denoted by y_1, the last by y_{365}. We shall refer to the entire time series (one year's worth of daily data in the example above) as Y_t; the reason for the subscript t will become clear in a moment. Because the rule generating the time series is probabilistic—meaning that we won't be able to forecast the next value of the series with certainty—we need to introduce the notion of a random "disturbance" at time t, denoted by e_t. Each disturbance is, by definition, drawn from the same probability distribution, and its value doesn't depend on the value of any prior disturbances. (This definition is summarized by saying that the values of e_t are **independent** and **identically distributed**, sometimes abbreviated **iid**.)

[2] The assumptions of no explanatory variables, no trend, and no seasonal are less restrictive than they might appear to be. Even when we are analyzing time series that can be explained or predicted via trends, seasonals, or other explanatory variables, we are often interested in knowing what information, if any, is left in the series after these explanatory factors have been taken into consideration.

It is convenient to specify that the mean of e_t is 0. Thus, e_t could be +1 with probability 0.5 and –1 with probability 0.5; or it could be + 2 with probability 0.3, 0 with probability 0.1, and –1 with probability 0.6. Usually, we will specify simply that et has a normal distribution with mean 0 and some fixed standard deviation S—say S = 2.5.

Autocorrelation

A concept indispensable to the analysis of time series is **autocorrelation**. We have already looked at the idea of correlation between two variables, say x and y: we can estimate the correlation between x and y by plotting a scatter diagram; we can measure it by computing the correlation coefficient.

In time-series analysis, there is only one variable, but we can easily create new variables consisting of the old variable lagged one or more periods. If Y_t represents the original series, consisting of observations $y_1, y_2, ..., y_t, ..., y_T$, then Y_{t-1} will represent the same series lagged one period. If there are no observations prior to period 1, then the first value of Y_{t-1} will be missing, but the second will be y_1, the third y_2, and the last y_{T-1}. If the original series Y_t consisted of daily observations for a year, starting January 1 and ending December 31, then Y_{t-1} would have a missing first observation, its second observation would be the value of the series on January 1, and its last observation would be the value on December 30.

We thus have two variables, Y_t and Y_{t-1}, derived from a single time series. We can plot a scatter diagram of the two variables (only T–1 points can be plotted), and we can compute the coefficient of correlation between Y_t and Y_{t-1}. This coefficient is called the (first-order) **autocorrelation coefficient**.

There is no reason to restrict our analysis to one-period lags. A variable Y_{t-2} would start with two missing observations; the third observation would be the value of the series on January 1, and the last would be the value on December 29. A scatter diagram of Y_t versus Y_{t-2} would reveal whether there was any substantial second-order correlation, and the second-order autocorrelation coefficient would quantify it.

Two Data-Generating Rules

The Constant-Average Rule

One of the simplest rules by which a time series can be generated is the constant-average rule, which specifies that each value of the series is the sum of a constant, M, plus a disturbance; for the tth observation,

$$y_t = M + e_t \ .$$

Neither M nor the values of e_t are directly observable, but inferences about them can be made by analyzing the data.

Simulating the Series. What does a constant-average series look like? We can artificially simulate such a series by specifying a value for M, a probability distribution for e_t, and then drawing sample disturbances from this distribution. Suppose M = 10, T = 20, and e_t has a normal distribution with mean 0, standard deviation 2.5. Values of e_t should be such that each value is independent of prior values; in the long run, the average value should be 0, 68% of the values should be between –2.5 and + 2.5, 95% between –5 and + 5, 99.7% between –7.5 and + 7.5, the histogram should look appropriately bell-shaped, and the cumugram should be appropriately S-shaped.[3] Of course, in any sample, the actual values of e_t will fail to conform exactly to these criteria because of sampling error.

The first four columns of Table 3.1 shows values of the variable Y_t generated in this way. Figure 3.1 shows values of Y_t plotted as a time series.

Table 3.1

t	M	e_t	Y_t	Y_{t-1}	Y_{t-2}	Y_{t-3}	Y_{t-4}	Y_{t-5}
1	10	0.33	10.33					
2	10	2.25	12.25	10.33				
3	10	0.22	10.22	12.25	10.33			
4	10	0.13	10.13	10.22	12.25	10.33		
5	10	0.70	10.70	10.13	10.22	12.25	10.33	
6	10	–1.43	8.57	10.70	10.13	10.22	12.25	10.33
7	10	1.38	11.38	8.57	10.70	10.13	10.22	12.25
8	10	2.04	12.04	11.38	8.57	10.70	10.13	10.22
9	10	2.69	12.69	12.04	11.38	8.57	10.70	10.13
10	10	7.02	17.02	12.69	12.04	11.38	8.57	10.70
11	10	–2.03	7.97	17.02	12.69	12.04	11.38	8.57
12	10	–1.36	8.64	7.97	17.02	12.69	12.04	11.38
13	10	0.58	10.58	8.64	7.97	17.02	12.69	12.04
14	10	–0.55	9.45	10.58	8.64	7.97	17.02	12.69
15	10	–2.29	7.71	9.45	10.58	8.64	7.97	17.02
16	10	0.37	10.37	7.71	9.45	10.58	8.64	7.97
17	10	1.88	11.88	10.37	7.71	9.45	10.58	8.64
18	10	–0.58	9.42	11.88	10.37	7.71	9.45	10.58
19	10	–2.49	7.51	9.42	11.88	10.37	7.71	9.45
20	10	4.35	14.35	7.51	9.42	11.88	10.37	7.71
				Autocorrelation Coefficients				
Mean			10.66	–0.025	–0.213	–0.110	–0.139	–0.292
Std. Dev.			2.32					
Std. Error			0.52					

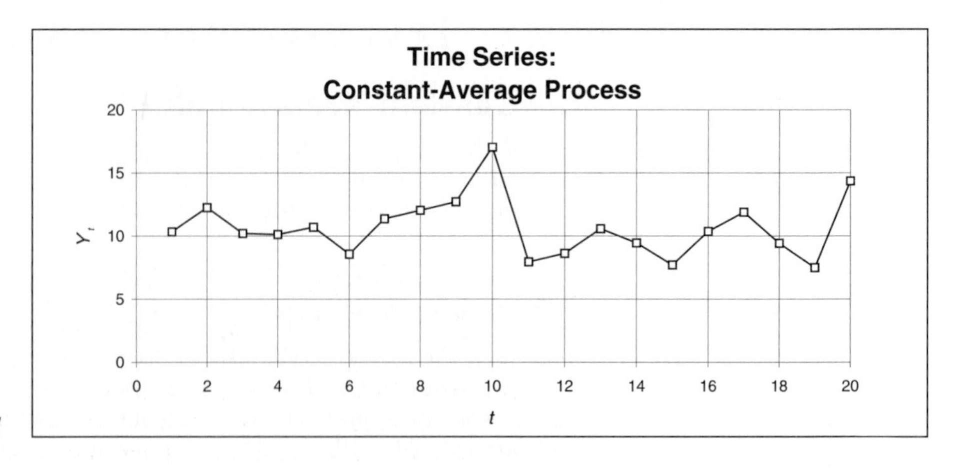

Figure 3.1

[3] There are many mechanical ways of drawing random disturbances that have these characteristics, but by far the easiest is to use the Excel 5.0 facility available by clicking Tools, Data Analysis, Random Number Generation, and specifying that the distribution is normal, that the mean is 0, the standard deviation is 2.5, and the number of random variables is 20 (the value of T).

Identifying the Rule. Now let's switch gears. Suppose we were presented with the 20 observations on Y_t given in Table 3.1. If we knew what those observations represented, we might be able to identify the rule governing the data-generating process. For now we will simply ask how, by examining the data, could we check a hypothesis that those observations came from a constant-average process?

If that hypothesis were true, each observation would differ from M by an amount that does not depend on the value of the previous observation. Therefore, a scatter diagram of each observation plotted against its predecessor should show no correlation (except for sampling error). In column 5 of Table 3.1 we show Y_{t-1}, the values of Y_t lagged by one period.[4] Figure 3.2 is a scatter diagram, showing values of Y_{t-1} on the horizontal axis and values of Y_t on the vertical. There is no discernible correlation. The coefficient of correlation between Y_t and Y_{t-1} (the first-order autocorrelation coefficient), based on the 19 observations for which both variables have values, is –0.025; it is printed at the bottom of the fifth column in Table 3.1. We have not introduced the specific technique for computing the confidence distribution of a correlation coefficient, but it can be shown that when the population correlation coefficient is 0 and the sample size is 19, a 68% confidence interval will cover sample correlation coefficients between –0.24 and +0.24, and a 95% interval will cover –0.45 to + 0.45; thus the observed sample correlation coefficient of –0.025 is by no means inconsistent with the hypothesis that in a sufficiently large sample, Y_t and Y_{t-1} would be uncorrelated.

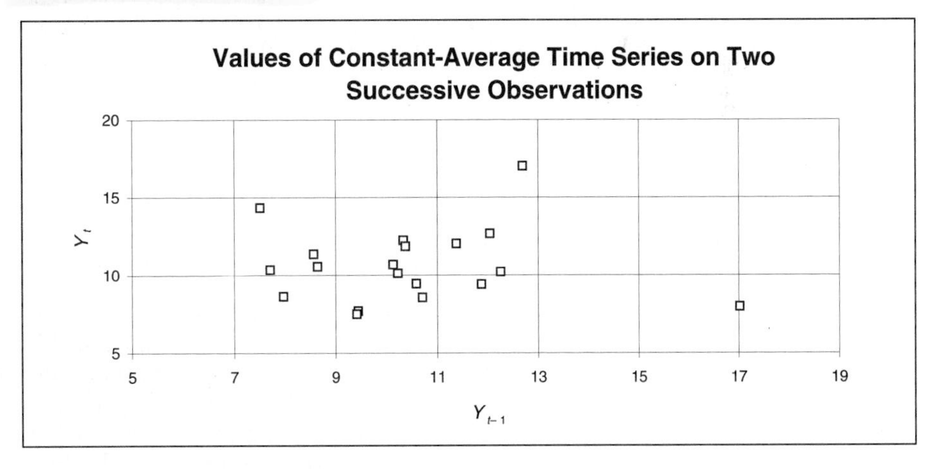

Figure 3.2

We could now go on to create new variables consisting of two-period lags Y_{t-2}, three-period lags Y_{t-3}, etc., and for each of these compute the appropriate-order sample autocorrelation coefficient: a second-order coefficient based on the correlation between Y_t and Y_{t-2}, a third-order coefficient, etc. These coefficients are computed up to five-period lags, and are shown at the bottoms of their appropriate columns in Table 3.1. As we would expect, they are sufficiently close to 0 to be consistent with a hypothesis that process autocorrelation coefficients of all orders are 0. This is the key to identifying a time series as one generated by a constant-average rule: for such a series, *sample autocorrelation coefficients of all orders will differ from 0 only because of sampling error.*

[4] These values are easily obtained in Excel 5.0 by copying the value of y_1 into an adjacent column one row down, and then dragging the Fill handle down.

Forecasting. Having identified the rule that governed the generation of the data, we can now try to make a forecast. If we knew that $M = 10$, and that the disturbances were normally distributed with mean 0, standard deviation $S = 2.5$, a "point" forecast for y_{21} would be 10, and a probabilistic forecast would be between 7.5 and 12.5 with probability 0.68, between 5.0 and 15.0 with probability 0.95, and between 2.5 and 17.5 with probability 0.997. Unfortunately, we don't know the value of M, the standard deviation of the disturbances, or even that the disturbances are normally distributed. But the data provide *estimates* of M and S: the sample mean $m = 10.66$ is not very far from the process mean $M = 10$; and the sample standard deviation $s = 2.32$ is not very far from the process standard deviation $S = 2.5$. If the disturbances were normally distributed, then the values of y_t in the sample of 20 observations should be approximately normally distributed; that this is so can be verified by plotting a cumugram of the actual values of y_t and of a normal distribution with mean of 10.66, standard deviation of 2.32. Figure 3.3 shows such a plot; the fit is quite good.[5]

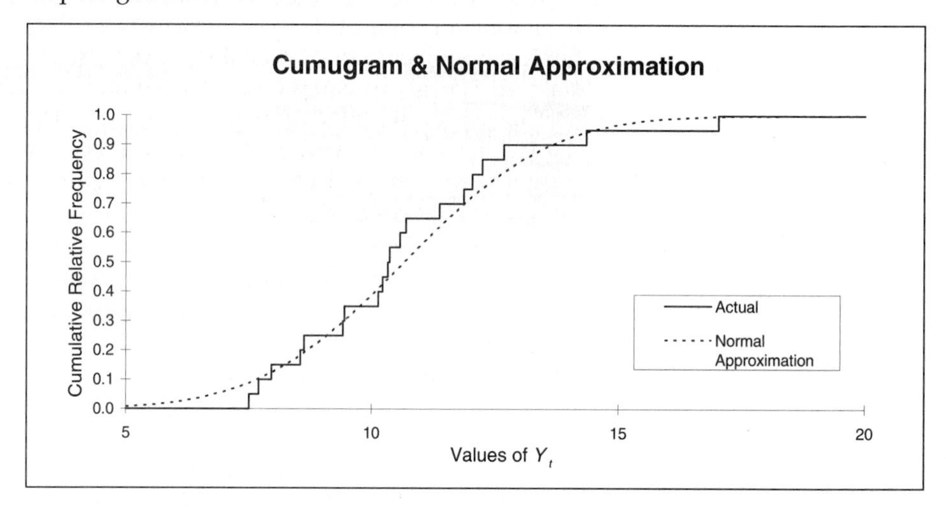

Figure 3.3

We can now either use the cumugram of actual values to make probabilistic forecasts or, more simply, use the normal approximation. Using the latter, we would get approximate probability intervals as follows: the value of y_{21} would be between 8.34 and 12.98 with probability 0.68; between 6.02 and 15.30 with probability 0.95; and between 3.70 and 17.62 with probability 0.997. These intervals are close to, but not exactly the same as, the ones we computed assuming we knew the values of M and S and the rule that governed the data-generating process for sure. In general, intervals computed in this way from sample data will be too narrow, on the average, because we have made four simplifying assumptions:

1. We are using the sample mean m instead of the process mean M.
2. We are using the sample standard deviation s instead of the process standard deviation S.

[5] Normal cumugram values corresponding to any given value of y can be computed by the Excel function NORMDIST(y,m,s,1).

3. We are assuming the disturbances are normally distributed, an assumption supported by, but not proved by, Figure 3.3.
4. We are assuming that the rule that governed the data-generating process was a constant-average rule, which is supported by, but not proved by, the autocorrelation analysis.

More sophisticated methods can make adjustments to take into account the first two simplifying assumptions, but only judgment can adjust for the third and fourth. For our purposes, just understanding that the intervals are somewhat too narrow is sufficient.

Suppose we want to forecast y_{22}, the value of y two periods ahead. If we knew the value of y_{21}, the value of y one period ahead, we could recompute m and s based on all 21 observations, and compute a point forecast and probabilistic forecasts based on these new statistics. But if we have only 20 observations in hand, and are looking two periods ahead, all we can do is use the values of m and s that are based on 20 observations: the forecasts for y_{22} and subsequent values of y are precisely the same as that for y_{21}.

There are thus two important characteristics of constant-average rules: both point forecasts and probabilistic forecasts of future values, no matter how far into the future, are all precisely the same; and all existing observations are equally important in determining forecasts of future values. The value of y_1 is as important in forecasting the value of y_{21} as is the value of y_{20}. As we shall see, these characteristics of constant-average rules do not apply to the next rule that we shall consider.

The Random-Walk Rule

A time-series observation generated by a random-walk rule is equal to its immediate predecessor plus a random disturbance:

$$y_t = y_{t-1} + e_t \quad .$$

If the first observation is y_0, then

$$y_1 = y_0 + e_1 \quad ,$$

and

$$y_2 = y_1 + e_2 \quad ,$$

etc. Unlike the constant-mean rule, the values of the disturbances e_1 through e_T are observable.

Simulating the Series. What does a random-walk series look like? We can artificially simulate such a series by specifying a starting value for y_0, a probability distribution for e_t, and then drawing sample disturbances from this distribution. Suppose $y_0 = 10$, $T = 20$, and e_t has a normal distribution with mean 0, standard deviation 2.5. The first three columns of Table 3.2 show values of the variable Y_t generated in this way. Figure 3.4 on the following page shows values of Y_t plotted as a time series.

Table 3.2

t	e_t	Y_t	e_{t-1}	e_{t-2}	e_{t-3}	e_{t-4}	Y_{t-1}
0		10.00					
1	0.33	10.33					10.00
2	2.25	12.58	0.33				10.33
3	0.22	12.81	2.25	0.33			12.58
4	0.13	12.94	0.22	2.25	0.33		12.81
5	0.70	13.64	0.13	0.22	2.25	0.33	12.94
6	–1.43	12.20	0.70	0.13	0.22	2.25	13.64
7	1.38	13.58	–1.43	0.70	0.13	0.22	12.20
8	2.04	15.63	1.38	–1.43	0.70	0.13	13.58
9	2.69	18.32	2.04	1.38	–1.43	0.70	15.63
10	7.02	25.34	2.69	2.04	1.38	–1.43	18.32
11	–2.03	23.31	7.02	2.69	2.04	1.38	25.34
12	–1.36	21.95	–2.03	7.02	2.69	2.04	23.31
13	0.58	22.53	–1.36	–2.03	7.02	2.69	21.95
14	–0.55	21.98	0.58	–1.36	–2.03	7.02	22.53
15	–2.29	19.69	–0.55	0.58	–1.36	–2.03	21.98
16	0.37	20.06	–2.29	–0.55	0.58	–1.36	19.69
17	1.88	21.93	0.37	–2.29	–0.55	0.58	20.06
18	–0.58	21.35	1.88	0.37	–2.29	–0.55	21.93
19	–2.49	18.86	–0.58	1.88	0.37	–2.29	21.35
20	4.35	23.21	–2.49	–0.58	1.88	0.37	18.86
			Autocorrelation Coefficients				
Mean	0.66		–0.025	–0.213	0.110	–0.139	0.881
Std. Dev.	2.32						
Std. Error	0.53						

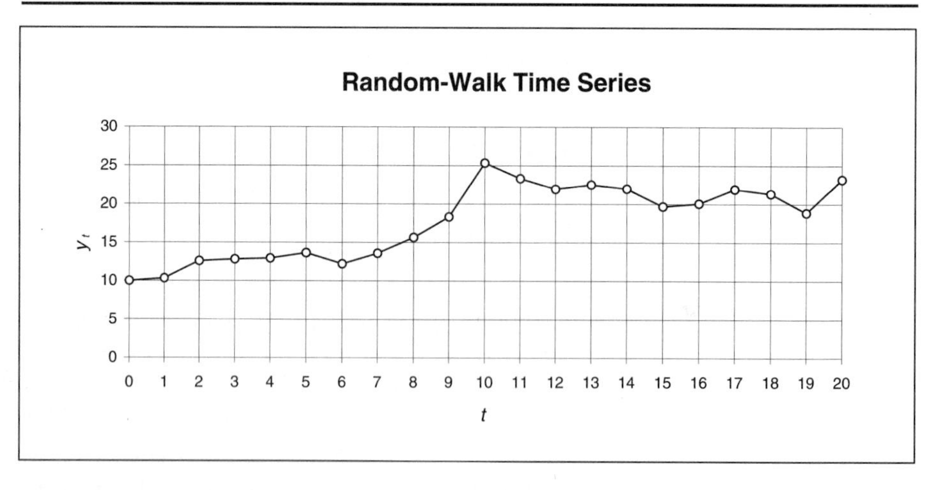

Figure 3.4

Identifying the Rule Given the 21 observations y_0 through y_{20} shown in Table 3.2, can we see whether a hypothesis that the observations were generated by a random-walk rule is consistent with the data? An immediate consequence of the rule

$$y_t = y_{t-1} + e_t$$

is that

$$y_t - y_{t-1} = e_t \ .$$

The difference between each observation and its immediate predecessor (often called the first difference) is just a random disturbance. Thus, first differences behave as if they were generated by a constant-mean rule with $M = 0$.

We already know how to analyze data generated by a constant-mean rule: we look at lags of one, two, and more periods, and compute sample first-order, second-order, and higher-order autocorrelation coefficients. Here the lags are not on the values of the series, but on their first differences. If the autocorrelations on these first differences can be shown to be equal to 0 except for sampling error, then the first differences are consistent with a hypothesis that they were generated by a constant-mean rule, and therefore the values of Y_t are consistent with a hypothesis that they were generated by a random-walk rule.

In columns 4–7 of Table 3.2 we show lagged first differences. (Column 2 shows values of the first difference itself.) At the bottom of columns 4–7 are the autocorrelation coefficients for the first differences. Just as was true in the analysis of the constant-average process, sample autocorrelation coefficients ought to be between –0.24 and + 0.24 with confidence 68% if the process had zero autocorrelation; all the coefficients are within that interval.

In Figure 3.5 we plot first differences lagged one period against "this period's" first differences. There is no discernible correlation in the scatter diagram. Notice, however, that if we plot the lagged values of Y_{t-1}, given in column 8 of Table 3.2, against current values of the series itself (Y_t), as in Figure 3.6, there is considerable correlation. (The correlation coefficient for one-period lags is 0.881.) Contrast this with the corresponding scatter diagram for the constant-average process.

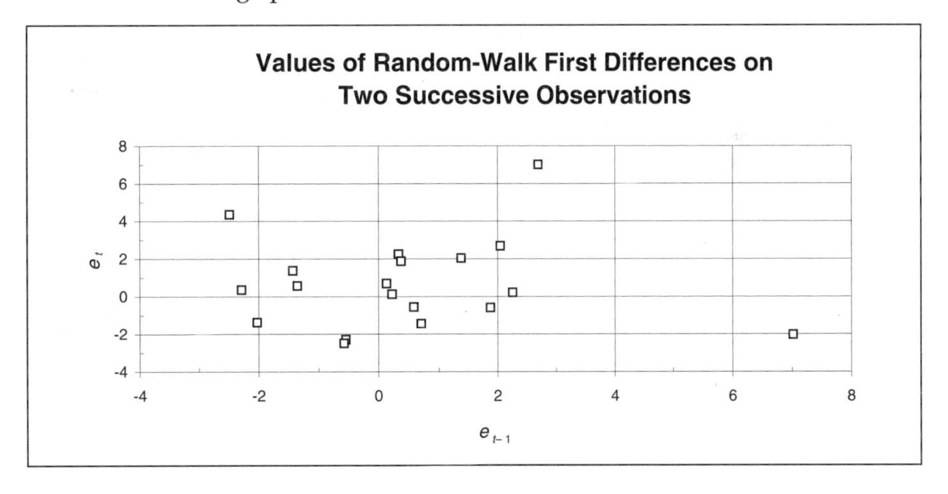

Figure 3.5

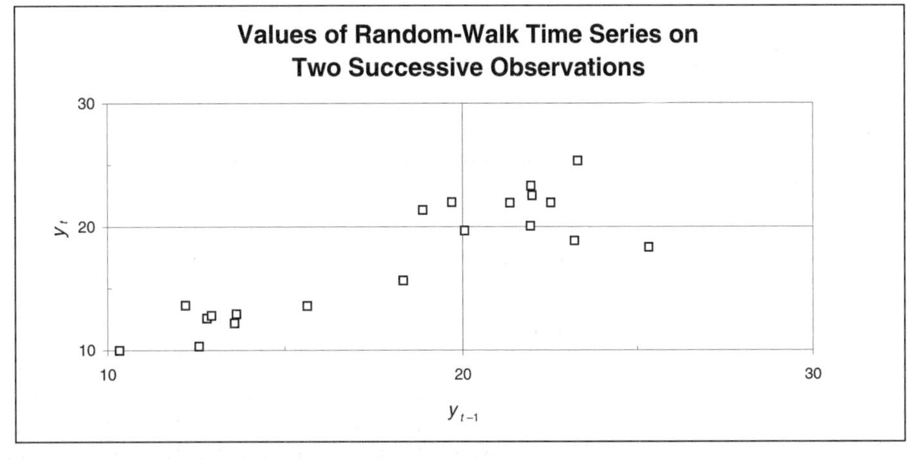

Figure 3.6

Forecasting If we know the values of y from y_0 through y_{20}, what is our best forecast for y_{21}? If we know that the data-generating process is a random walk, then

$$y_{21} = y_{20} + e_{21} \quad .$$

Because the mean of $e_{21} = 0$, the best point forecast for y_{21} is just the preceding value, y_{20}: the history of the process prior to the 20th observation is irrelevant to forecasts of the future. The same is true for point forecasts for y_{22} and subsequent y's: the best forecast is the last observed value.

Turning to probabilistic forecasts, y_{21} differs from y_{20} by the disturbance e_{21}. All the disturbances have been assumed to be independently and identically distributed. We can infer their distribution from a histogram or cumugram of their values (they are just the values of the first differences themselves) or, if we assume the disturbances are normally distributed with mean 0, we can estimate the standard deviation of the distribution by computing the sample standard deviation of the disturbances. For the data in Table 3.2, the standard deviation is 2.32. Thus, since $y_{20} = 23.21$, a probabilistic forecast for y_{21} is between 20.89 and 25.53 with probability 68%, between 18.57 and 27.85 with probability 95%, etc. As was true in the case of the constant-mean process, these intervals are somewhat too narrow, for all the same reasons.

A probabilistic forecast for y_{22} is a bit trickier. We know that

$$y_{22} = y_{21} + e_{22} \quad ,$$

and that

$$y_{21} = y_{20} + e_{21} \, ,$$

from which it follows that

$$y_{22} = y_{20} + e_{21} + e_{22} \quad .$$

What is the distribution of the sum of two iid disturbances? It can be shown that if each disturbance has a normal distribution, the sum also has a normal distribution. Furthermore, the mean of the sum is the sum of the means, or 0, and the standard deviation of the sum is $s\sqrt{2}$, where s is an estimate of the standard deviation of the distribution of any one disturbance.[6] Thus a forecast for y_{22} will lie in the interval $y_{20} \pm s\sqrt{2}$ with probability 68%, etc. For the data in Table 3.2, y_{22} will be between 19.96 and 26.46 with probability 68%, between 10.20 and 29.72 with probability 95%, etc.

These results can be generalized. If the last observed value in a random walk is y_T, then a probabilistic forecast of y_{T+n} will have mean y_T and standard deviation $s\sqrt{n}$: the further into the future you forecast, the greater your uncertainty.

[6] This follows from the fact that the standard deviation of the *mean* of a sample of size 2 is $s/\sqrt{2}$, and the *sum* of a sample of size 2 is just two times the mean.

Summary

These results are summarized in Table 3.3 below:

Table 3.3

	Constant Mean	**Random-Walk**
Process	$y_t = M + e_t$	$y_t = y_{t-1} + e_t$
Identification	All autocorrelation coefficients are 0, except for sampling error.	All autocorrelation coefficients of first differences are 0, except for sampling error.
Probabilistic Forecast *n* Periods Ahead*		
Mean	m (average of observed y's)	y_t (the last observed value of y)
Standard Deviation	s (sample standard deviation of the observed y's)	$s\sqrt{n}$ (where s is the sample standard deviation of the first differences)

* If the distribution of the disturbances is normal, then the forecast distribution is also normal.

What Comes Next?

We have discussed just two of the most important data-generating rules, shown how time series generated by these rules can be identified from data, and how point and probabilistic forecasts can be made. There are many other rules that generate time series that occur in practice. We are not going to investigate them in this chapter. Instead, we shall next turn to the analysis of time series—often series with trends and seasonals—where explanatory variables can be used to forecast future values. But before leaving the subject of forecasting without explanatory variables, let's take a brief look at how this subject would unfold if it were pursued further.

Having investigated constant-average and random-walk rules, the next kind of rule that would be worth exploring would be one where the (unobserved) average is not constant, but changes over time in random-walk fashion, and where the actual values of the series differ from this "moving" average by an iid disturbance. This rule for generating data is called a **moving-average rule**. It can be identified by examining the autocorrelations of first differences, and point forecasts can be made by constructing a weighted average of past observations, the weights being heaviest for the most recent observation, and declining "exponentially" for successively less recent observations. This forecasting mechanism is called **exponential smoothing,** a very robust technique that works quite well even for time series generated by rules that are not strictly of the moving-average variety.

Although we have restricted our attention to time series without trends or seasonals, we can take a "moving-average" time series and add to it a moving-average trend, and even a moving-average seasonal. Future values of such a series can be forecast by more complex exponential-smoothing techniques which smooth the seasonal, the trend, and the series itself. These techniques are often successfully applied to series generated by rules that are not strictly of the moving-average type.

A still more general and complicated set of data-generating rules assumes that each observation (or first difference) is a weighted sum of previous observations (or differences) plus a disturbance plus a weighted sum of previous disturbances.

Such processes can be identified by examining autocorrelations of the series, or of first differences, and formulas for forecasting future values are available. This very general time-series approach was pioneered by George E. P. Box and Gwilym Jenkins,[7] and is referred to as Box-Jenkins or ARIMA (Auto-Regressive Integrated Moving Average) methodology. The interested reader can find further discussion of these techniques in S. Makridakis, S. C. Wheelwright, and V. E. McGee, *Forecasting: Methods and Applications*, Second Edition. John Wiley and Sons, 1983.

Exercises on Time Series

1. Figure 3.7 shows annual average temperature in Boston from 1944 through 1990. The data are in worksheet BOSTEMP in workbook TIMSEREX.XLS. Given this set of data alone, forecast the annual average temperature for 1991 and 1992. How sure are you of your forecasts? What would your forecast and your forecast uncertainty be for 1993? Can you detect any evidence of global warming in the data?

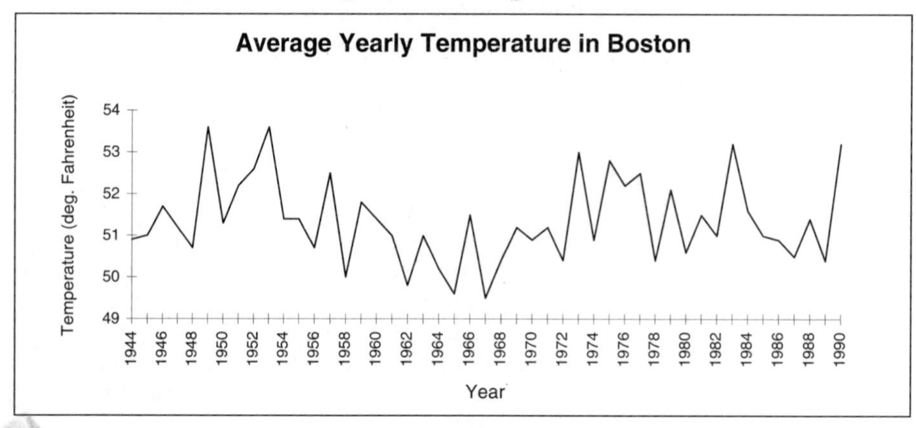

Figure 3.7

2. Figure 3.8 gives the end-of-month price of gold (dollars per troy ounce on the CMX) in monthly intervals from December 31, 1981, through November 30, 1992. The data are in worksheet GOLD in workbook TIMSEREX.XLS. Given this set of data alone, what would your forecast of gold price be for the end of December 1992? How sure are you of your forecast? What would your forecast and your forecast uncertainty be for December 1993?

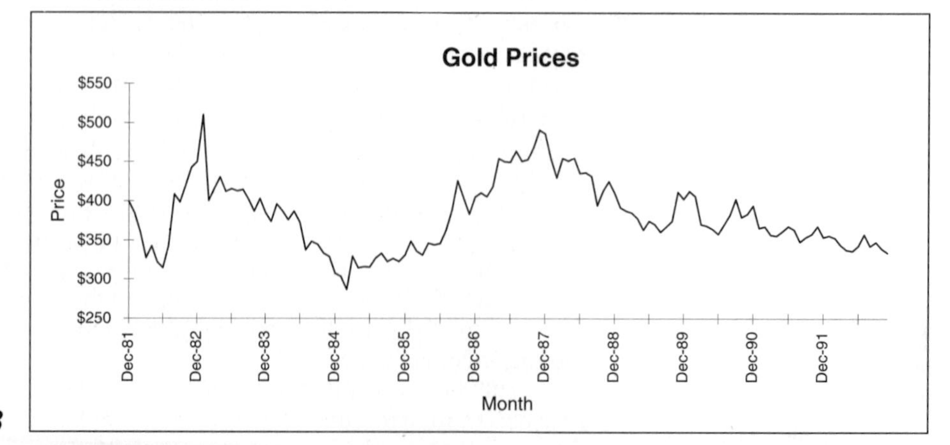

Figure 3.8

[7] G.E.P. Box, G. Jenkins, and G.C. Reinsel, *Time Series Analysis, Forecasting, and Control*, third edition. Englewood Cliffs, NJ: Prentice Hall, 1994.

THE BOSTON GAS COMPANY: WINTER, 1980–1981

On January 11, 1981, after a month of exceptionally cold weather, Massachusetts Governor Edward King made a public request for fuel conservation by the state's gas consumers and asked dual-fuel customers to switch to oil. January 12 and 13 were again extremely cold days, and on January 14, the governor ordered industrial and commercial users to turn their thermostats down to 55°. Despite King's request for conservation, Boston Gas Company continued to send out record quantities of fuel, as temperatures continued low. On Thursday, January 15, Martin Luther King Day (a school holiday in Massachusetts), Governor King announced that Boston Gas' supplies were dangerously depleted. He declared a state of emergency, and closed the area schools for the following day. This order permitted school buildings to cut back on their gas usage, but perhaps more importantly, it dramatized the seriousness of the gas situation, encouraging the conservation that Boston Gas and the governor had been urging. Along with King's declaration, the state's gas companies mounted a multi-media advertising campaign stressing conservation, and customers did begin to conserve fuel in a noticeable way.

Boston Gas' shortage of peak shaving fuel developed out of several concurrent and unexpected events. First, Boston experienced unprecedented cold during the fall and early winter. September 1, 1980, to January 31, 1981, was the coldest such period ever recorded in the Boston area. The 30 days between December 20 and January 18 were unusual both in their temperatures and in their timing. Late January-February cold spells were far more likely. Exhibit 1 shows the seasonal temperature pattern for Boston, based on daily temperature data from 1920 through 1981. To develop the curve, a centered 30-day moving average of mean[8] daily temperature over this (approximately) 60-year period was constructed. The frequency distribution of this 30-day moving average for any specified day of the year was based on the 60 data points for which the moving average was centered on the day in question, and from this distribution theoretical fractiles were computed.[9] The solid curve represents the median of this distribution, and the associated broken lines (moving outward from the median) indicate the 0.75 and 0.25 fractiles, the 0.90 and 0.10 fractiles, and the 0.99 and 0.01 fractiles of the moving-average temperature distribution. The median and 0.01 and 0.99 fractiles for December 1 through February 28 are repeated as dotted lines on Exhibit 2, where the heavy solid curve represents the winter of 1980-81, again using centered 30-day moving averages of temperature. The other plotted curves show comparable temperature patterns for four very cold winters: those of 1933–34, 1934–35, 1935–36, and 1947–48.

Harvard Business School case 9-182-196. This case was prepared by Alice B. Morgan, Research Associate, under the supervision of Professor Arthur Schleifer, Jr.

[8] Weather statisticians define "mean" temperature as the average of the day's highest and lowest temperatures.

[9] See the Appendix to this case for discussion of how the fractiles were developed.

Naturally the intense cold generated exceptional demand for heating fuel. In fact, not only were there more "degree days" than forecast in this period (degree days, defined later in this case, are a measure of fuel demand), but consumers were demanding heat at a greater rate than anticipated on the basis of past usage. The lower the temperature dropped outside, the warmer people seemed to want to be inside. Boston Gas had expected that sendout volume as a function of degree days would be lower than in the previous year, due to a perceptible trend toward conservation. Instead, when degree days were held constant, customers used more fuel in December and January, 1980–81, than they had a year earlier.

The final component in the company's supply shortfall was its supply sources themselves. Boston Gas bought different types of fuel from a variety of suppliers, under many different contract arrangements and at several different prices. (The complex supply picture is discussed more thoroughly later in this case.) The company's basic source of fuel, pipeline gas, could not be increased. Normally, pipeline gas was supplemented with a variety of other supplies, of which the most important was Liquefied Natural Gas (LNG). During warm weather, Boston Gas made its own LNG from pipeline gas and stored it in its own tanks. In addition, the company bought LNG from Distrigas of Massachusetts Corporation, which imported it by tanker from Algeria. A smaller backup source, SNG (Substitute Natural Gas), could be bought from one of Boston Gas' suppliers, Algonquin, or made by the company from propane at its Everett plant. Boston Gas could also use propane more directly, in a propane-air mixture. Under some circumstances, natural gas held in underground storage in New York and Pennsylvania could be sent via pipeline, but pipeline capacity imposed restraints on this alternative. Consequently, LNG was a major "peak-shaving" fuel, necessary to Boston Gas on very cold days. Exhibit 3 shows the role of different supply sources as degree days increase. Basic pipeline gas sufficed to meet 1980–81 levels of demand when the mean temperature was 45° or higher. Other supply sources were required as temperatures dropped. (Because of the complexity of the company's contractual arrangements, supplies were not necessarily used in the same order on every day.)

Looking back on the gas crisis, Boston Gas' President, John Bacon, remarked:

> People don't really understand how our supply works. We can't have a brownout, the way an electric utility can. As long as our system is operating, we must fill *all* of the demand. The gas is rushing through our pipes like a hurricane, and there's no way to reduce the pressure. If the pressure were to drop, air would enter the system—a very dangerous situation. The only way to meet, say, 90% of the demand is to cut off 10% of our customers. And people don't realize what that entails, either. We'd have to turn off whole communities by going out and closing the main valves. To turn the supply back on, we'd have to go house by house, relighting pilot lights—not a very quick procedure.

Because gas delivery was an all-or-nothing situation, with no "brown-out" or reduction in supply possible, a gas cutoff had very severe consequences. Boston Gas planned its supply so that there was little likelihood of a cutoff: the company consciously sought supplies from a variety of sources to avoid a crippling dependence on any one. Boston Gas served some commercial and industrial customers on an "interruptible" basis: gas service to interruptibles was used as a load–balancing device, and service could be shut off with minimal advance notice if supply became too tight. (Such customers could use oil under these circumstances.)

The interruptibles provided a demand cushion: the company planned on sufficient supplies to meet the needs of firm customers during a "design year" (as explained later in the case), but could sell excess to interruptibles. Boston Gas' main responsibility was to its firm residential, commercial and industrial customers. Its basic contingency planning anticipated cutoffs of interruptibles and dual-fuel firm customers (those who could switch to oil heat), voluntary conservation, and cutoffs of firm commercial and industrial customers (except for plant protection purposes), before any residential cutoffs would occur.

Events of 1980–81

Since 1968, Boston Gas had been using LNG to supplement pipeline gas, and since 1971 had been buying imported LNG from Distrigas. During the 1970s, both its pipeline suppliers, Tennessee Gas and Algonquin, experienced shortages of gas supply, and curtailed their pipeline transmission. Boston Gas compensated for the shortfall with LNG, as well as propane and SNG. When pipeline curtailments ended in 1979, the company used imported LNG as a source of base load supply during the summer, diverting its pipeline gas to fill storage facilities and to make LNG at the company's own facilities. During 1980, complications entered the Distrigas picture. The Algerian National Oil and Gas Company, Sonatrach, canceled its contract with El Paso Natural Gas Company, and removed a particular ship from service. The result was a shift in production and port facilities: Sonatrach had been filling Distrigas orders from its Skikda terminal, but now found it necessary to provide shipments from Arzew. LNG produced at Arzew had a higher heating content and was consequently vaporized into the Boston Gas system more slowly than LNG from Skikda, and the delivery schedule had to be revised accordingly. Negotiations to develop a delivery schedule satisfactory to Sonatrach, Distrigas, and its customers continued from April through August, and Boston Gas could not even estimate the size and frequency of shipments until early fall. During the summer, therefore, the company formulated its supply forecast without including any Distrigas LNG. But as the fall advanced, it appeared that LNG deliveries would be both regular and substantial, and Boston Gas factored into its supply forecast part of the anticipated Distrigas shipments. (The company still hesitated to count on all the contracted Distrigas LNG.) Jack McKenna, senior vice president, Operations, stated:

> In summary, we started the 1980–81 year with more LNG peak-shaving inventory than we thought would be necessary, and more than we had originally thought would be available, all because of the resumption of Distrigas deliveries. Given this, and the fact that we had arranged for other supplemental supplies (SNG and propane), we felt our peak-shaving needs were fully covered going into the winter heating season.

Exhibit 4 shows the supply sources planned for 1980–81 as of July, when Distrigas shipments were not included.[10] In September, Boston Gas included 6340 BBTU from Distrigas deliveries. In its October update, the company was anticipating 7000 BBTU of LNG from Distrigas through the end of June 1981.

---

[10] Gas was measured in cubic feet (cf) and its heating output was measured in British thermal units (BTU). One cf of natural gas provided approximately 1000 BTU. Boston Gas Company used M for thousands, MM for millions, and B for billions as its standard abbreviations for units.

The severe weather of 1980–81 meant that Boston Gas' reserves were shrinking more rapidly than expected, as extreme cold and unanticipated usage spurred demand. At first company officials felt that the increased usage levels might simply reflect a typical fall pattern: customers tended not to think about the costs associated with early winter fuel use. Sendout continued higher than expected, however, right into December. Usage did not drop off and it appeared that projected conservation would not be realized. In mid-December, supply problems developed at Exxon, the company's propane supplier. As a result, propane prices were scheduled to rise. To avoid the additional cost to its customers, Boston Gas arranged to substitute for contracted propane by purchasing some of Brooklyn Union Gas' Distrigas commitments during late December and January, and the company suspended deliveries under its Exxon contract subject to recall. In light of the termination of propane deliveries and the continuing high sendout, the company shut off its interruptibles on December 19. Cold weather, decreased conservation and high sendout continued, and new forecasts of demand showed that Distrigas supplies were now a necessary part of Boston Gas' fuel sources. Distrigas took delivery of an Algerian LNG shipment on December 26, with another expected on January 14.

On December 29, Boston Gas produced another forecast. Its assumptions were, first, that the increase in firm demand would continue higher than anticipated. Second, the company planned for "design" weather, that is, weather which was considerably colder than normal, likely to occur only once every 17 years. (The definition and development of design weather is discussed more fully below.) The weather had been running colder than design for most of December. The third assumption was that LNG shipments would continue to arrive, although the company was still not counting on the total quantities projected by Distrigas. At this point, the supply and sendout situation were being monitored daily, and at the end of December Boston Gas decided to ask Exxon to resume propane deliveries. Brooklyn Union had been unable to provide the contracted LNG from its December 26 Distrigas shipment because of the cold weather, and was now scheduled to use part of its January 14 Distrigas delivery to fulfill this obligation. In the end, Boston Gas was able to arrange a complicated pipeline exchange that permitted the Brooklyn Union gas to be delivered by mid-January, and Exxon resumed its propane deliveries on January 8.

Although supplies were tight, there would have been no problem in meeting demand if nature had not intervened, for a second time, at this point. In late December, a storm struck Arzew, the Algerian harbor from which Distrigas imported LNG. The storm damaged several facilities; more important, it sank a ship full of gasoline in the harbor, blocking all shipping traffic. It was not until January 5 that Boston Gas learned that Distrigas' ship was unable to enter Arzew harbor to load. The January 14 shipment of LNG was therefore rescheduled to arrive January 28. On January 7 Distrigas revised this arrival date to February 11. Now the company's forecasts indicated that if the weather continued at design levels (it was still running colder than design) there would be insufficient fuel supply to last beyond the end of January. Dual-fuel customers switched to oil heat on January 9, and Charles Buckley, vice president–Gas Supply, intensified the negotiations for additional supplies that he had started at the end of December.

Buckley was successful in purchasing LNG from Southern Energy Corporation, but ran into difficulties when he tried to arrange its transport from Savannah to Boston. A federal law, the Jones Act, made it illegal to transport LNG between American ports in a foreign-flag vessel, and the only appropriate American ships were two tankers presently located in Greece.

Given the approaching supply crisis, Buckley did not think the company could wait for an American ship to leave Greece, travel to Savannah, and then to Boston. Moreover, investigation revealed that it would be necessary to cut the ship's smokestack down by 5 to 6 feet to enable it to pass under the Mystic River Bridge to reach the Distrigas terminal. Buckley arranged for purchase of LNG on the spot market in Indonesia, and started an American ship to pick it up, at the same time entering negotiations with the U.S. Government for a waiver of the Jones Act. He knew that an Algerian ship was in Boston harbor for repairs, and was thus available to bring the LNG from Savannah if the legal obstacle could be surmounted. President John Bacon explained:

> We had the American ship waiting outside the Suez Canal while we talked with the government. It costs $500,000 to go through the Canal, and we didn't want to do it if we could get the Jones Act waiver.

The Jones Act waiver came through on January 12. The ship departed and ten days later was back with the LNG. Meanwhile, of course, the supply picture continued grim, as the abnormal cold persisted and sendout did not lessen. Jack McKenna, senior vice president, Operations, noted, "January 11 and 12 each had 57 degree days (+8° F), resulting in a combined sendout for those two days of 1059 BBTU, or in excess of the capacity of one LNG tanker. Our projections at this point showed we could not make it until the ship arrived." The company notified state officials of the impending shortfall, requesting that the governor urge customers to turn down their thermostats. The weather did not abate, and the request for conservation was ineffective, so the company asked Governor King to declare the emergency and close the schools.

When the extent of the crisis became clear, there was much public recrimination, and a strong concern over how a problem of such magnitude could have developed, and why it was not publicized sooner. In reviewing the events leading up to the emergency, Boston Gas Company officers felt that three sets of circumstances, each highly unlikely, had occurred simultaneously, creating a crisis which could not reasonably be anticipated. First, the weather was far colder than usual for Boston at any time of year, and the cold spell came earlier than such spells generally did. Second, a freak event disabled one of the company's supply sources at a time when it was urgently needed: the storm interfered with Distrigas shipments just when they had become critical elements in Boston Gas' supply configuration. Third, customers used more fuel per degree day than they had been using, reversing a strongly defined trend toward conservation. Jack McKenna commented:

> We've thought a lot about the high fuel usage. The cold came pretty early, so we lost some conservation there—when the weather is moderate at the start of the heating season, people will delay turning on their heat. Then people seemed to feel colder than usual—I think thermostats were being set higher as the cold persisted. And there's one more element. We estimate from a survey that 24% of our non-heating customers were using their ovens and ranges to provide heat during the cold spell. That's like adding 50,000 new heating customers!
>
> And it's constant, no cycling—you open the oven door and the heat keeps pouring out. Actually, it's not an unreasonable idea. The stove heat is 100% efficient, and costs less than half the price of oil. Of course, it isn't really safe. We did have one incident of a man and young child being asphyxiated. But some people got very sophisticated about it, using fans to spread the heat around. With a fan, you can heat a three-bedroom apartment pretty well with your oven. And that added three to four hundred thousand BTUs at the peaks—in fact, it was a real whammy!

President John Bacon added,

> We tend to be quite conservative about our supply arrangements—we know what it means to run out. But if we are too conservative, the consumer will pay more for his fuel than is necessary. Then the DPU will be after us on that account.

Shortly after the crisis ended, the Department of Public Utilities (DPU) began an investigation to determine how it had arisen and whether there was culpability on the part of any of the state's gas distributors.

Company Background

The Boston Gas Company was a wholly owned subsidiary of Eastern Gas and Fuel Associates, based in Boston, Massachusetts. Boston Gas was engaged in the distribution and sale of natural gas to residential, commercial, and industrial customers in Boston and 73 other eastern Massachusetts communities. Its 1,056 square miles of territory included nearly all of Boston, and was primarily a residential, trading, commercial, and light-manufacturing area. Exhibit 5 is a map of Boston Gas' territory. The system consisted of about 5,700 miles of mains, 400,000 services (pipes from mains to customers' premises), and 486,000 active customer meters; and the company employed approximately 1,900 people.

Boston Gas served several classes of customers. Interruptibles, discussed more fully below, could burn either oil or gas, and were supplied with gas at the company's discretion. Boston Gas guaranteed service to its firm customers, who might belong to the residential, or the commercial and industrial, category. Among residential customers, some used gas exclusively for nonheating functions (e.g., cooking, heating water, drying clothes), while heating customers used gas to heat their homes and for some nonheating functions as well. Heating customers' gas usage therefore varied considerably with temperature, while that of nonheating customers showed only minor variation with temperature. Although residential heating customers differed in their individual usage patterns, Boston Gas viewed average heating use per central-heating customer as a reliable statistic. Residential heating customers used, on the average, 130 MMCF (thousand cubic feet) of fuel per year.

Commercial and industrial users naturally varied greatly in size and in the way they employed their gas. Schools, hospitals, manufacturing facilities and commercial establishments could not be treated in terms of an 'average' customer, and were instead considered on the basis of gas volume billed. Because they used gas for so many different purposes, new customers in this class had to be analyzed individually to determine the relationship between their gas usage and changes in temperature.

Exhibit 6 shows several statistics on Boston Gas Company's customer base, including the number of residential heating and nonheating customers from January 1978 to July 1981, and sales of fuel, in MMCF, to both classes of residential customer, and to commercial and industrial users, for the same period. The sales figures represent actual billings, and it is important to note that residential nonheating customers received bills 6 times a year, based on meter readings at 2-month intervals, while residential heating customers, whose meters were also read at 2-month intervals, received 12 bills a year, with every other bill an estimated bill. Commercial and industrial customers were billed monthly, after monthly readings. Billing information was further complicated by the fact that there were 21 billing cycles, one terminating on each working day of the month. If meters could not be read on schedule for either residential or commercial customers, estimated bills were prepared.

During the mid-1970s, pipeline gas supplies had been constrained, and Boston Gas had accepted only very limited load additions. After 1977, additional supplies permitted increased load additions, and the high costs and sometimes uncertain supply of oil led to considerable demand for gas heat. In early 1980, Boston Gas eliminated its direct residential sales program, convinced that the company could use direct mailings to current nonheating customers as well as media advertisements to generate as many new heating customers as the system could accommodate.

In its service area, Boston Gas was a monopoly in a market dominated by fuel oil. It was regulated under Massachusetts law by the DPU with regard to its rates, territorial limits, financing, contracts, and accounting practices.

The 1980–81 Sendout Forecast

1. Weather

To understand how Boston Gas developed its forecasts, it is necessary to start with the company's weather analysis. Using 51 years (1923–1973) of weather data compiled by the National Oceanographic and Atmospheric Administration (NOAA), Boston Gas had developed a "normal" year, the average year for the period, by summing the degree days for each year and taking the mean. (Degree days were the number of degrees by which a day's mean temperature fell short of 65°. Days whose mean temperature exceeded 65° produced 0 degree days.) Over the 51 years, the average number of degree days per year was 5,758, with a standard deviation of 352. Exhibit 7 charts degree days per year (July 1 to June 30) from 1920–21 to 1980–81. For planning purposes, Boston Gas used the "design year," a year with much colder weather than normal. The design year had 6,300 degree days, a weather pattern likely to occur no more frequently than once in seventeen years. The design year included 25 days of temperatures 20° or less, totaling altogether 1,274 degree days, including a day when the temperature would reach –8°. Based upon its weather data, Boston Gas actually prepared design months, with specific daily temperatures, to reflect the usual monthly patterns of colder and warmer weather. Exhibits 8 and 9 show, respectively, the daily temperatures associated with the normal and the design year.

(According to Boston Gas, weather patterns show no correlation from one year to the next, nor does any short-term correlation appear until the level of sequential days. No cycles or patterns in temperature over the last 60 years have been identified.)

2. Supplies

Boston Gas' supply picture was extremely complex. There were several sources of supply, each making gas available at a different price, and each with its own capacity limits. At peak usage periods, demand outstripped the maximum that the pipeline could provide the system: as a result, the company had to maintain storage facilities, as discussed below. In striving to balance its supply and demand, the company's planners had to weigh the costs of extreme alternatives. On the one hand, they took into account the statistically unlikely possibility of weather cold enough to enforce a disruption of sendout. On the other hand, they bore in mind the cost of the additional storage facilities and stored gas required to reduce the probability of such a catastrophe. Through appropriate arrangements with its interruptible customers, Boston Gas endeavored to provide ample supplies for the coldest likely winters (as embodied in the design year), while still consuming all its contracted and stored fuel before the warmer weather began.

a. Pipeline Gas — The price of pipeline gas was regulated, and Boston Gas' two suppliers, Algonquin and Tennessee gas-transmission companies, were its least expensive fuel sources. Contracts with the two pipelines stipulated specific yearly maximum quantities available to Boston Gas. In addition, each contract provided for a Maximum Daily Quantity (MDQ), and the number of days in the year on which Boston Gas might take its MDQ. Boston Gas also had a Winter Supply contract (WS) with Algonquin, making pipeline gas available from mid-November through mid-April (at a price higher than the yearly contract called for): this 151-day contract stipulated an MDQ that could be taken any day; the total contract volume was 60 times the MDQ.

b. Substitute Natural Gas (SNG), Propane–Air — In addition to pipeline gas, Boston Gas bought SNG from Algonquin, and made its own SNG at its Everett facility, using liquid propane as feedstock. Propane could also be fed directly into the system, mixed with air, but this way of using propane depended to some degree on the current level of system demand. In general, SNG was the highest priced of Boston Gas' regular supplies.

c. Liquefied Natural Gas (LNG) — The other main peak-shaving fuel, and the most important, was LNG, either liquefied by Boston Gas itself, or purchased elsewhere. Distrigas was Boston Gas' sole LNG supplier, although LNG could be purchased on the open market from both domestic and international sources. LNG occupied only 1/600th the volume of its original gaseous state, and was readily interchangeable with pipeline gas. It could be vaporized into the Boston Gas system at very high rates, unlike propane-air. In 1980–81, Boston Gas relied on LNG for some of its supply on those days when degree days exceeded 34, as indicated in Exhibit 3. LNG was less costly than SNG or propane-air, but more expensive than pipeline gas.

d. Storage Facilities — Boston Gas stored some of its pipeline gas in four underground facilities rented from companies in Pennsylvania and New York. When the company did not require all its available pipeline gas for immediate sendout, for example during the summer, it diverted the gas to fill these facilities. In general, the storage facilities were full at the start of the heating season, and close to empty by the end of it. Pipeline gas not currently needed was also occasionally used to produce LNG at one of Boston Gas' two liquefaction plants, after which it would be stored in one of the company's four LNG storage tanks.

The contract with Distrigas provided for approximately monthly shipments. A ship required about 10 days to travel from Algeria to Boston, and a day or so for unloading, after which it started its return voyage. Distrigas' Everett terminal afforded temporary storage once the fuel was delivered. Half of each shipment had to be taken within ten days of arrival in port, while the rest had to be removed before the next ship arrived. Distrigas agreed to vaporize a specified amount directly into Boston Gas' system, and the rest was trucked to Boston Gas storage tanks. The company had to plan for sufficient space to receive the LNG, and had to vaporize as much as possible during periods when its own tanks were fairly full. Except for basic pipeline gas and an optional portion of its Exxon propane contract, Boston Gas was required to pay for its supplies, even if it failed to "take" them as scheduled.

The company was thus faced with the need to use efficiently its various types of supplies, while still meeting its contract obligations. Algonquin SNG and Winter Supply pipeline gas, available only during the heating season, were the first to be put into the system, as they had to be used every day to absorb the contract quantities. Boston Gas could then use its regular pipeline supply, SNG, LNG, and propane-air, as required. Since there was a contracted maximum available from the pipeline on the coldest days, it was necessary to accept more than was needed on warm days and store it (as LNG or in underground storage) for peak usage periods, or sell it to interruptible customers. Each day the company chose how much of its basic pipeline supply to use directly in the system, and how much to divert into storage or liquefy at one of its two liquefaction plants. When a Distrigas shipment arrived, the LNG had to be taken either by vaporization or into storage, at the contracted rate, thus temporarily affecting how much pipeline gas the company could accept. If Boston Gas was unable to send out all its planned supplies during the winter, it might find itself unable to honor some of its supply contracts during the spring and summer for lack of space in its pipeline and storage tanks. Yet the company was forced, by the extreme consequences of running out, to plan for greater firm sendout than it considered at all likely. To absorb the almost inevitable margin between its supplies and its firm sendout, the company sold fuel to interruptible customers who bought their gas with the understanding that they might have to switch to oil at any time. They paid slightly less than current oil prices, and they provided Boston Gas with a cushion for its potential surplus—a cushion that had recently brought in more revenue per cubic foot of gas than the company's firm customers.

Boston Gas thus planned its supplies to suffice for its firm customer base, including any new load, under design-year conditions. As the heating season progressed, the company checked its supply against the forecasted design demand, and used sales to interruptibles to equalize the two. In this way, Boston Gas strove to end the heating season with minimal inventories in underground storage, stored propane, and its LNG tanks (thus making space for summer refill), and to use its contracted volumes of Winter Service gas, its pipeline MDQ, its contracted SNG and propane, and its Distrigas quotas. During the spring and summer, Distrigas deliveries and pipeline gas were used to provide fuel for customers (including interruptibles) and to refill the LNG tanks and underground storage for the coming winter.

3. Conservation

In order to convert normal-weather or design-weather degree days into a sendout forecast, Boston Gas needed to know the rate at which its current customers were using fuel. To determine the current rate of fuel usage, Boston Gas began with data on the firm sendout for each month. The company's analysts established the number of degree days by which the month differed from a "normal" month (i.e., a month in the "normal" year as defined above). They then determined the month's "heating factor": the sendout per degree day, after elimination of the base load (used for heating water, cooking, and drying clothes, rather than for home heating). Using the heating factor, Boston Gas adjusted the sendout figure so that it represented what the sendout would have been, had the month been a normal month. By looking at normalized monthly sendouts, the company could compare usage year over year, even though the weather differed every year, and the customer base varied.

The company used this information to monitor trends in consumption. Since 1978, there had been a marked increase in conservation, and Boston Gas estimated that in 1979–80 residential customers had used about 7% less fuel than in the previous year (on a normalized basis). Additional data on conservation came from two sources. First, Boston Gas analyzed the billings history of a sample customer base, chosen for its comparative stability. The sample customer group of 115,000 had been identified in 1977, and by 1980 numbered roughly 90,000, half of whom heated their homes with gas. (Customers who moved or dropped out of the sample were not replaced.) Boston Gas read meters every two months, and sometimes was unable to complete a scheduled reading. As a result, the number of customers in each month's sample usually ran about 25,000. Analysts examined the sample customer usage over like time periods, taking into account variations in weather. The normalized sales rates were compared, and percent changes were noted. This ongoing survey revealed that conservation had been apparent in both 1978–79 and 1979–80.

In addition to its own research, Boston Gas had in 1975 arranged with Decision Research Corporation, a Lexington, Massachusetts, firm, to survey gas customers' current conservation activities and intentions. In early 1979, Decision Research updated its 1975 study, asking many of the same questions as had appeared in the earlier version. The results suggested a far greater recognition of the need for conservation, both from a financial and public-interest point of view. Moreover, insofar as intentions could be measured, the company's customers seemed to be aware of additional ways they could conserve, and to be willing to consider implementing some of them.

Walter Flaherty, the company's manager of Market Analysis and Rates, summed up the effects of conservation:

> Our firm sales had not increased since 1977 despite the addition of many new customers.

In July 1980, the Massachusetts Energy Siting Council had applauded the fact that Boston Gas was tracking customer conservation and incorporating it into anticipated sendout:

> The company is to be commended for the great strides it has taken in the past two years in its approach to measuring conservation impact on sendout requirements.

The Council went on to urge other local gas companies to give serious attention to the matter of growing conservation. If users were conserving more, the New England area could sustain more gas-heating customers, and could thus lessen its dependence on foreign oil, which the Siting Council considered distinctly in the public interest.

When Boston Gas prepared its sendout forecast in July 1980, the company estimated that gas prices were likely to rise about 16% during the coming year. Studies had shown a price elasticity for gas demand of 0.25 to 0.30. Local oil prices had risen precipitously (60% in 1979–80), and conservation measures and alternate heating sources (such as wood stoves and quartz heaters) were receiving considerable public attention. All these factors encouraged Boston Gas in its forecast of about 6% conservation for 1980–81, relative to 1979–80.

4. Sendout

Once Boston Gas knew the rate at which customers were likely to reduce their gas usage, the company determined how many new customers the system could accommodate, taking into consideration its supply configuration. In 1980–81, Boston Gas felt it could add some 16,000 new residential heating customers to its base of approximately 220,000. At an average load of 130 Mcf, this addition about equaled the planned firm commercial and industrial load increase of 2 Bcf. Calculations for the winter sendout were made using these anticipated load increases.

Boston Gas thus went into the winter with several forecasts. First, the company had a forecast for weather, in terms of the normal year and the extreme or design year, broken down into normal and design patterns or each month. In addition, there was a forecast for the rate of customer conservation of about 6% (normalized) over the preceding year.[11] Finally, the heating–customer base was forecasted to include the additions planned for the 1980–81 season.

Working from these subsidiary forecasts, Boston Gas prepared sendout forecasts for normal and design months. The forecast for each month depended not only on the anticipated weather patterns (as shown in Exhibits 8 and 9), but on the relationship between degree days and sendout characteristic of the month. The same number of degree days in November and in January did not generate the same sendout requirements, for the later winter months had a higher ratio of sendout to degree days. (That is, the heating factors rose in the later winter months.) There was no consensus as to why this should be the case. Among the explanations offered were that snow, more common in January and February, had a psychological effect on people, making them feel colder regardless of the actual temperature. Similarly, in late winter, people in New England were "tired of being cold," and might respond by setting their thermostats higher.

Whatever the reason, this situation necessitated that the forecast for each winter month take into account the varying rates of usage, as well as the planned size of the customer base and the anticipated conservation. The company's forecast for December and January 1980–81 was low, then, for two reasons. First, the weather was colder than design. Second, the expected conservation did not materialize—in fact, usage per degree day climbed, compared with that of previous years.

Special circumstances seemed to explain the anomalies of the 1980–81 winter. Customers' use of gas ovens to heat their apartments resulted in some substantial sendout increases. Even so, other cold periods in the past two years had not called forth the same volume of sendout, and there were numerous conjectures to explain what had happened this time. One school of thought held that the uninterrupted cold made people feel colder. Another noted that there was little temperature variation: not only were the daily mean temperatures low, but the extremes also were low. Or perhaps those conservation measures already taken represented a higher proportion of customers' potential conservation than the company had imagined. Finally, the reduced conservation might reflect a new phenomenon: the significant and growing difference between the prices of gas and oil. In any case, the conserving trend of the past two years seemed to have abated.

To develop Exhibit 10, daily firm sendout[12] data was divided by the number of residential heating customers. The resulting ratio is plotted as a function of daily average temperature for January 1, 1976, through July 31, 1981.

[11] Although Boston Gas did not maintain a data base to determine commercial and industrial customer conservation, the company's experience suggested that non-residential conservation paralleled that of residential customers.

[12] Daily firm sendout cannot be partitioned between residential and commercial customers.

Appendix: The Moving–Average Temperature Distribution

There were about 60 observations for each calendar day's centered 30-day moving average, so that, in principle, the distribution of this moving average for any calendar day could have been obtained from the historical frequencies. Such a procedure would have been tedious, and the values of the extreme (0.01 and 0.99) fractiles would have been heavily influenced by judgment, rather than data. Instead, a normal distribution was fit to each calendar day's frequency distribution by matching its mean and standard deviation, and from this normal approximation the fractiles were computed. In the charts below, the cumulative frequency of 30-day moving-average temperature for six different winter days is graphed with a solid line, against the dotted normal cumulative frequency distribution, to show how closely the two track.

Appendix Chart

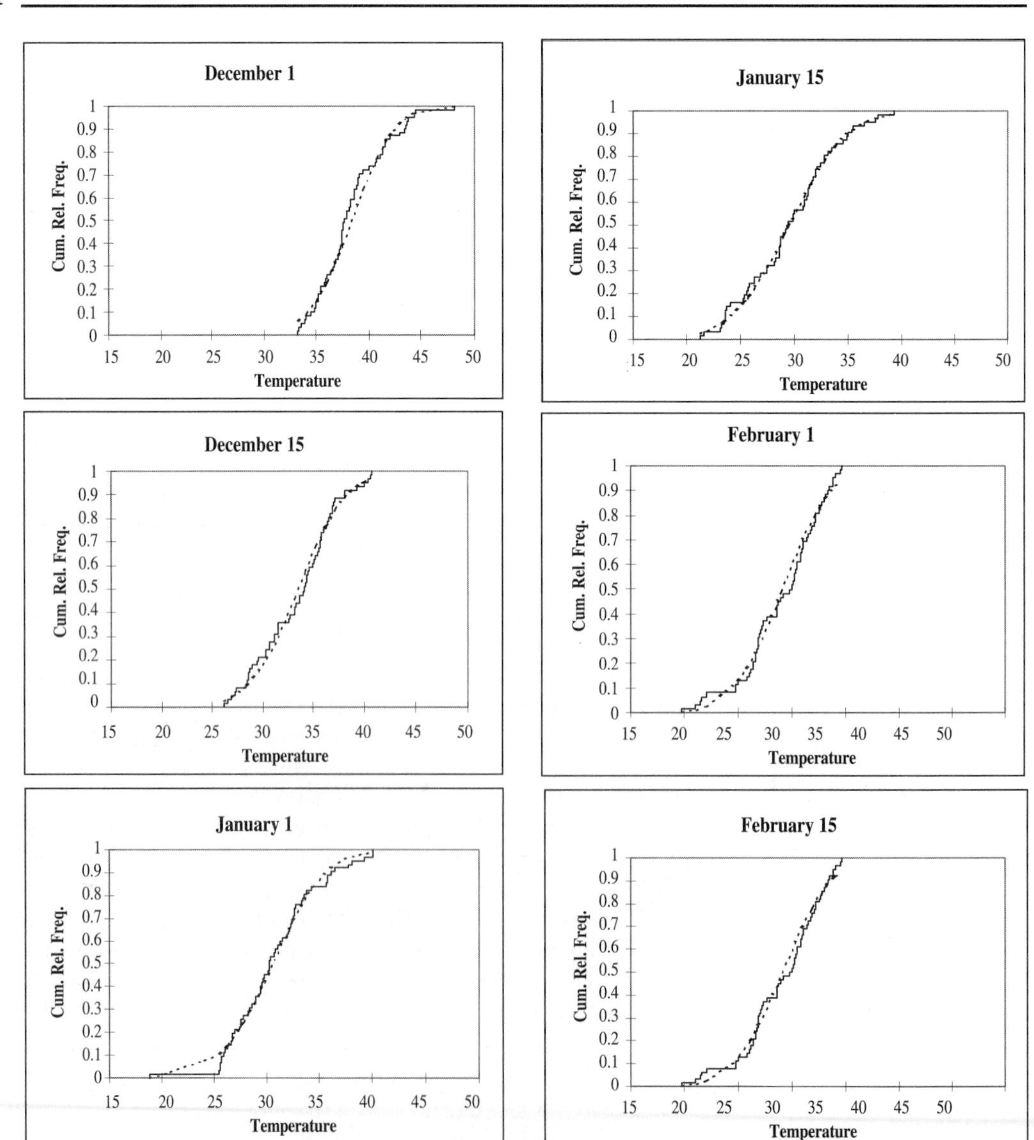

Exhibit 1

Annual Temperature Patterns: Fractiles of 30-day Centered Moving Average of Daily Temperatures

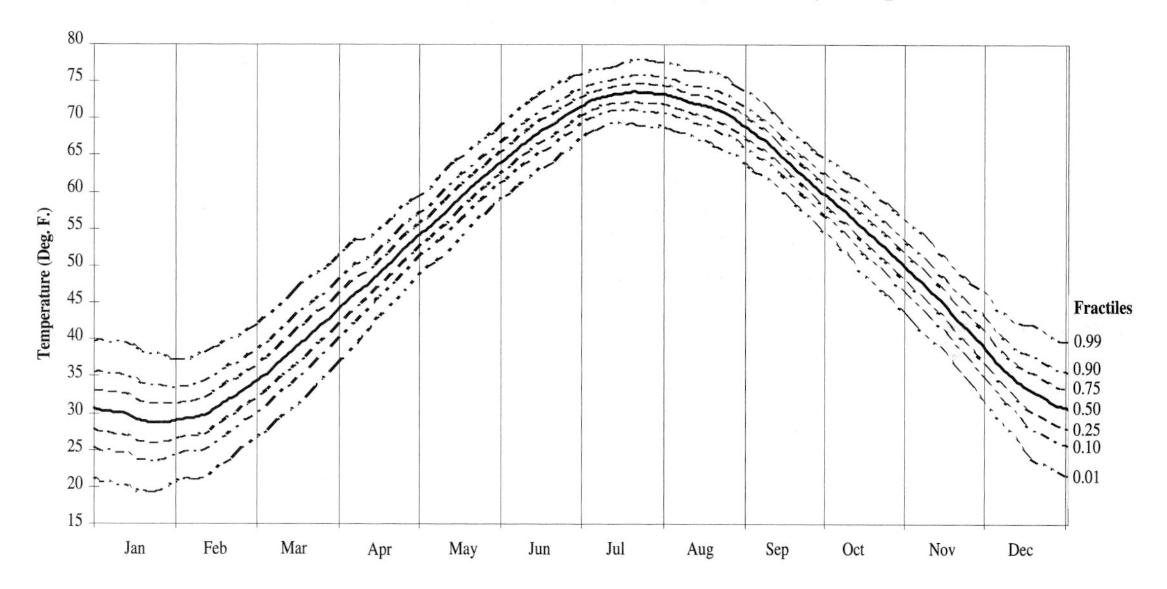

Exhibit 2

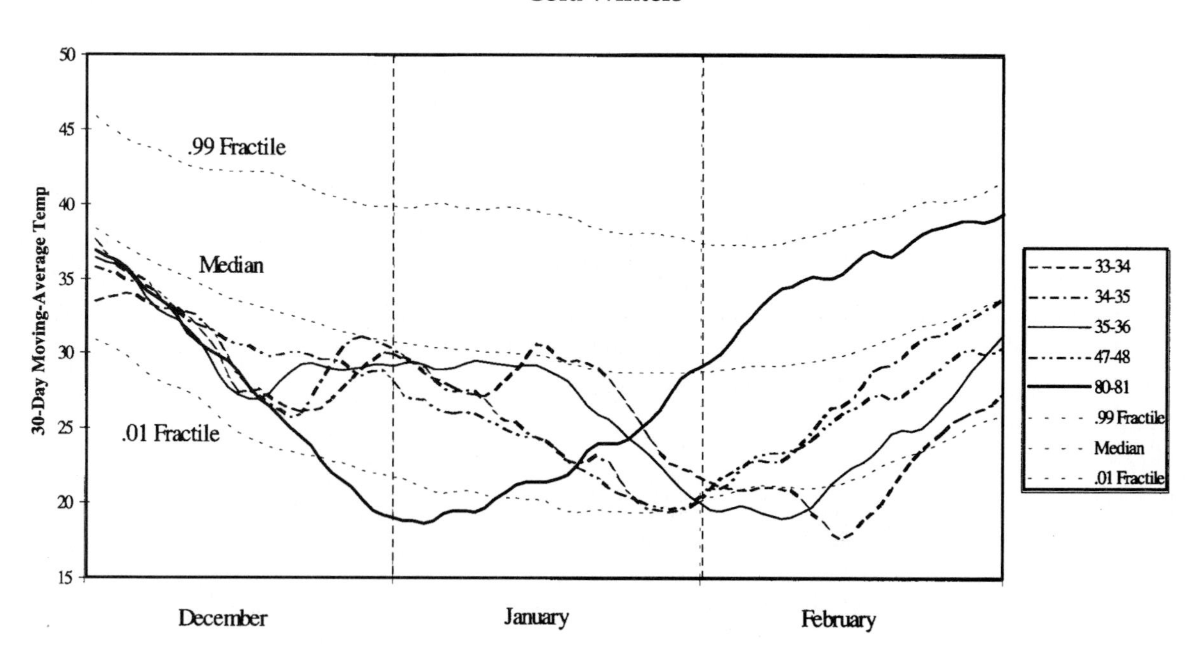

Exhibit 3

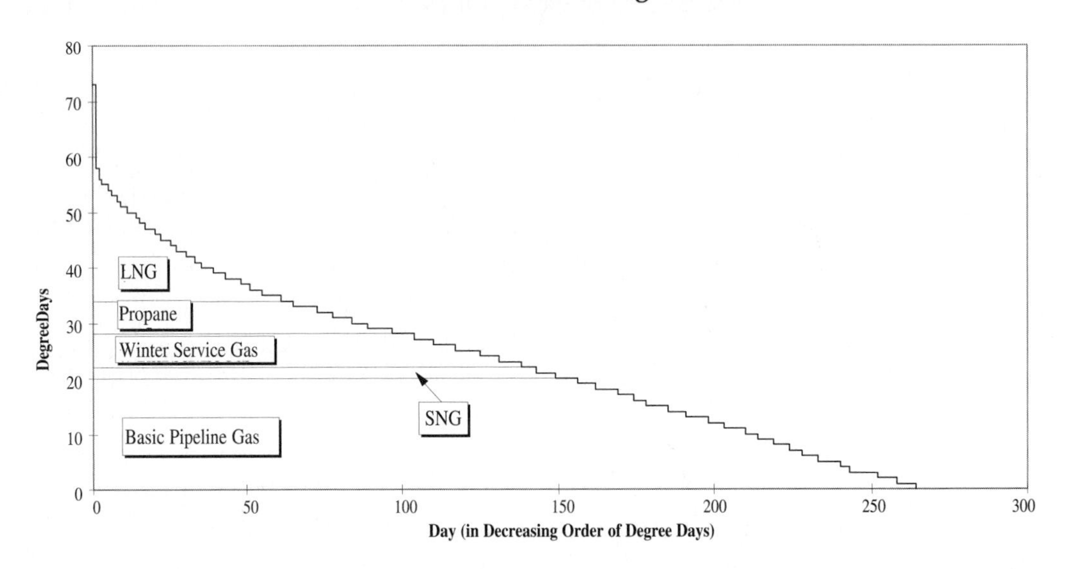

Exhibit 4

Planned Supply Sources as of July 1980

	MMCF
Algonquin Gas Transmission (natural gas)	39,377
Tennessee Gas Pipeline (natural gas)	21,956
Algonquin SNG (substitute natural gas from Algonquin)	1,844
Exxon propane (to be used for propane-air/SNG production by Boston Gas)	2,844
Boston Gas LNG	2,958
Distrigas LNG (already in storage)	568
Anticipated firm and interruptible sales	69,547
Balance (to storage)	**5,000**

Exhibit 5

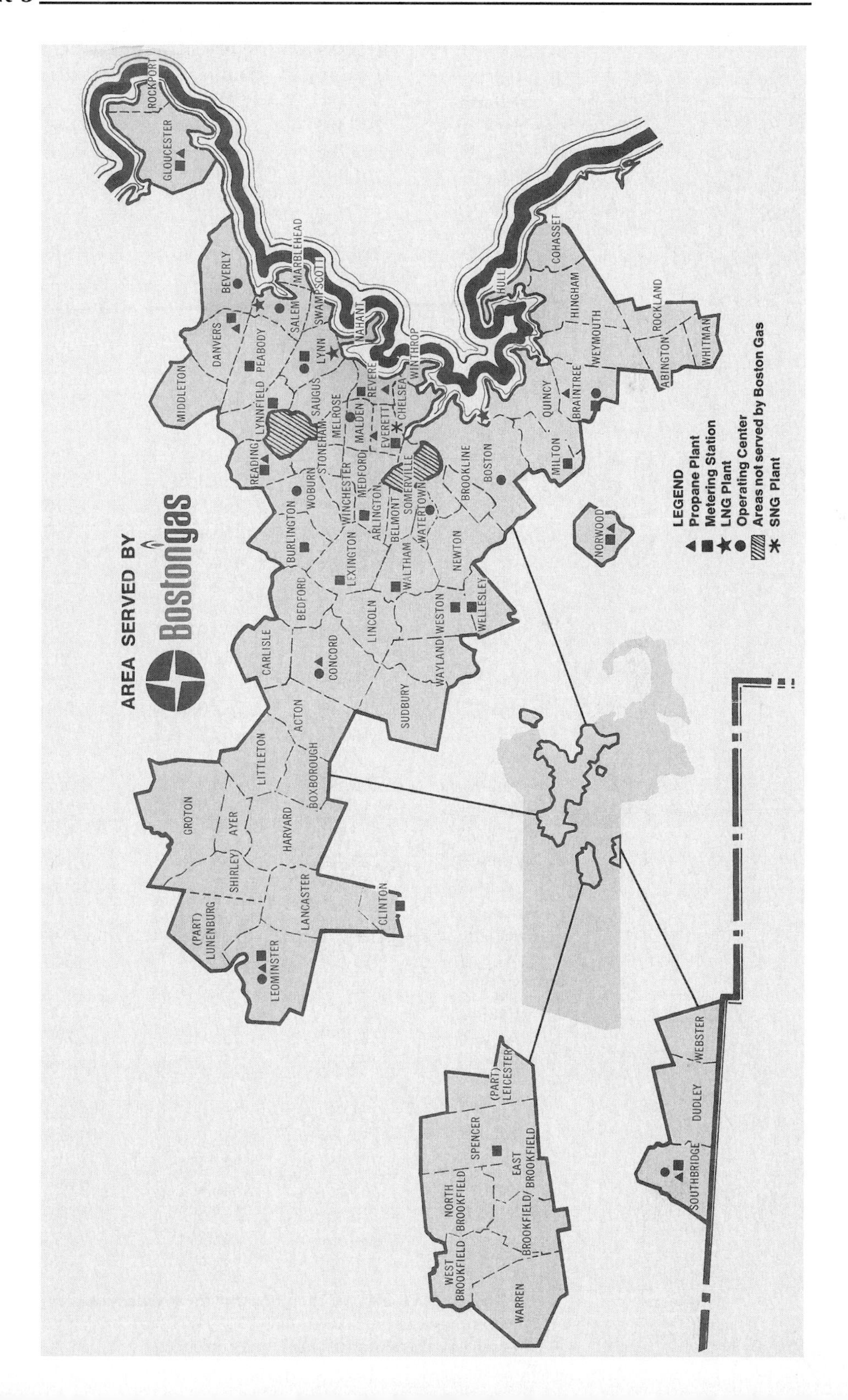
AREA SERVED BY
Bostongas
LEGEND
Propane Plant
Metering Station
LNG Plant
Operating Center
Areas not served by Boston Gas
SNG Plant
ROCKPORT
GLOUCESTER
MIDDLETON
DANVERS
BEVERLY
PEABODY
SALEM
MARBLEHEAD
SWAMPSCOTT
LYNN
NAHANT
LYNNFIELD
SAUGUS
READING
STONEHAM
MELROSE
MALDEN
REVERE
WINTHROP
EVERETT
CHELSEA
WOBURN
WINCHESTER
MEDFORD
SOMERVILLE
BURLINGTON
ARLINGTON
BELMONT
WATERTOWN
BROOKLINE
BOSTON
LEXINGTON
WALTHAM
NEWTON
BEDFORD
LINCOLN
WESTON
WELLESLEY
CARLISLE
CONCORD
WAYLAND
SUDBURY
ACTON
LITTLETON
BOXBOROUGH
GROTON
AYER
HARVARD
SHIRLEY
LANCASTER
CLINTON
(PART) LUNENBURG
LEOMINSTER
HULL
COHASSET
HINGHAM
WEYMOUTH
QUINCY
BRAINTREE
MILTON
ROCKLAND
ABINGTON
WHITMAN
NORWOOD
(PART) LEICESTER
SPENCER
EAST BROOKFIELD
NORTH BROOKFIELD
BROOKFIELD
WEST BROOKFIELD
WARREN
WEBSTER
DUDLEY
SOUTHBRIDGE

Exhibit 6

	Residential Customers		Sales in MMcf		
	Heating	Nonheating	Residential Heating	Residential Nonheating	Commercial & Industrial
Month					
Jan-78	201,433	270,816	4,802	544	3,094
Feb-78	201,765	271,159	5,012	575	3,063
Mar-78	201,987	271,002	4,681	563	3,175
Apr-78	201,911	270,886	3,158	529	2,407
May-78	201,202	270,459	2,203	483	1,716
Jun-78	200,552	269,636	980	438	1,086
Jul-78	200,173	268,333	726	389	788
Aug-78	199,797	267,728	644	356	756
Sep-78	200,060	266,979	786	350	857
Oct-78	200,973	267,601	1,492	400	1,167
Nov-78	202,075	267,977	2,171	441	1,639
Dec-78	203,205	268,481	3,553	476	2,602
Jan-79	203,995	268,741	4,529	535	3,106
Feb-79	204,747	269,048	5,320	571	3,487
Mar-79	205,324	268,896	4,326	562	3,007
Apr-79	205,654	268,433	2,990	509	2,159
May-79	205,221	267,414	1,665	456	1,440
Jun-79	205,074	266,171	959	422	1,008
Jul-79	205,115	265,112	681	378	767
Aug-79	205,691	263,970	675	349	749
Sep-79	206,863	262,345	705	351	808
Oct-79	208,622	262,051	1,358	386	1,121
Nov-79	211,006	262,038	2,274	441	1,701
Dec-79	214,252	261,226	3,175	479	2,200
Jan-80	216,517	260,754	4,562	536	2,955
Feb-80	218,184	260,215	5,236	555	3,538
Mar-80	219,999	259,100	4,607	560	3,107
Apr-80	220,480	258,738	3,105	495	2,251
May-80	220,621	257,616	1,957	466	1,635
Jun-80	220,899	256,113	977	415	989
Jul-80	221,270	254,905	720	372	835
Aug-80	221,500	253,864	640	323	763
Sep-80	222,429	252,691	673	315	836
Oct-80	224,379	252,253	1,338	354	1,148
Nov-80	227,295	252,005	2,779	424	2,041
Dec-80	229,632	251,763	4,266	487	2,939
Jan-81	231,921	250,958	6,553	576	4,143
Feb-81	234,996	248,760	5,443	544	3,396
Mar-81	236,481	247,524	4,230	501	2,974
Apr-81	236,954	246,901	2,922	452	2,245
May-81	236,819	246,592	1,975	427	1,553
Jun-81	235,996	245,394	855	371	999
Jul-81	235,007	243,979	653	317	763

Note: Boston Gas is required to maintain a content of 1000 BTU per cubic foot of gas. In fact, the content varies from 1005 to 1045 BTU per cubic foot. Sales data in this exhibit have been adjusted to reflect a content of 1000 BTU per cubic foot.

Exhibit 7

Degree Days per Heating Year (July1–June 30)

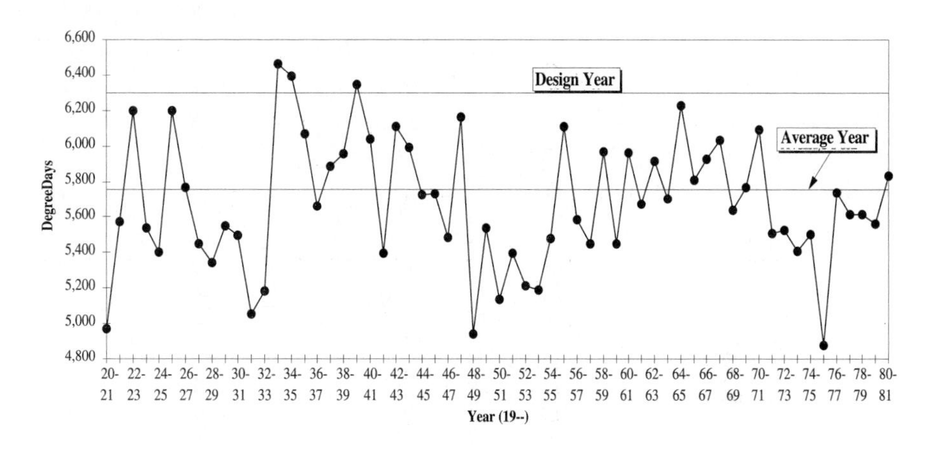

Exhibit 8

Normal Year Daily Weather Pattern in Degrees Fahrenheit
Based Upon 51 Years of Weather Experience, 1923–73 (5,758 Degree Days)

Day	*Jan*	*Feb*	*Mar*	*Apr*	*May*	*Jun*	*Jul*	*Aug*	*Sep*	*Oct*	*Nov*	*Dec*
1	28	28	39	37	54	61	77	76	75	62	52	29
2	39	32	36	38	44	58	75	73	80	72	54	34
3	46	40	38	45	46	53	72	71	68	60	64	40
4	43	20	32	50	50	60	69	77	71	59	50	39
5	30	15	26	43	58	56	71	80	65	55	59	47
6	23	10	19	36	61	68	66	84	70	54	51	52
7	28	21	24	40	63	66	64	86	69	57	49	42
8	31	31	25	44	50	69	62	78	73	61	55	40
9	24	41	30	47	54	70	68	74	74	65	56	49
10	19	38	41	46	48	67	72	72	72	53	54	48
11	32	29	44	41	57	61	70	69	67	66	46	38
12	30	24	34	43	58	64	74	68	64	63	44	34
13	38	18	28	48	51	68	79	66	68	68	36	32
14	32	30	29	56	62	73	84	69	66	57	40	35
15	40	33	34	52	56	70	81	73	62	54	42	29
16	35	42	40	44	55	74	78	75	66	59	43	23
17	26	38	48	62	52	76	74	72	64	64	48	17
18	18	46	50	58	53	68	76	68	60	51	53	22
19	22	36	58	49	67	63	73	65	65	49	46	30
20	24	34	40	45	56	58	70	64	58	47	41	36
21	33	22	35	40	60	62	68	62	55	48	37	45
22	34	26	36	42	68	65	66	61	61	52	34	41
23	36	35	37	49	64	71	67	67	59	46	38	37
24	44	26	39	42	70	78	70	63	62	50	31	31
25	42	27	41	50	60	74	78	70	56	56	33	27
26	36	25	37	54	56	72	82	76	53	48	42	16
27	20	36	45	60	54	80	77	82	60	43	45	12
28	16	30	47	51	66	78	78	79	63	50	39	21
29	0		52	65	71	73	76	70	54	52	34	28
30	14		45	53	69	81	72	74	50	44	26	27
31	36		42		76		75	66		40		32
					Total Degree Days							
	1,096	**987**	**844**	**520**	**238**	**54**	**4**	**10**	**94**	**321**	**608**	**982**

Exhibit 9

Design Year Daily Weather Pattern in Degrees Fahrenheit (6,300 Degree Days)

Day	*Jan*	*Feb*	*Mar*	*Apr*	*May*	*Jun*	*Jul*	*Aug*	*Sep*	*Oct*	*Nov*	*Dec*
1	25	24	37	34	53	63	77	76	75	62	50	26
2	37	29	34	36	42	58	75	73	80	72	53	32
3	44	37	36	45	46	53	72	71	68	60	46	33
4	41	15	29	49	50	60	69	77	71	58	49	37
5	27	12	25	41	57	56	71	80	65	54	44	45
6	19	10	15	45	59	68	66	84	70	52	50	42
7	24	17	20	37	63	66	64	86	69	57	47	40
8	28	30	21	42	49	69	62	78	73	61	54	37
9	20	39	27	45	53	70	68	74	74	65	55	47
10	15	36	39	44	48	67	72	72	72	57	53	46
11	30	26	42	38	56	61	70	69	67	66	44	35
12	27	13	31	41	57	62	74	68	64	63	42	31
13	35	14	25	47	50	64	79	66	68	68	33	29
14	29	27	26	55	62	73	84	69	66	56	38	32
15	38	30	32	51	54	70	81	73	62	54	40	26
16	32	40	35	42	48	74	78	75	66	59	41	19
17	22	35	46	62	51	76	74	72	65	65	46	12
18	14	40	49	59	60	68	76	68	60	50	52	18
19	18	36	47	48	67	63	73	65	56	47	44	30
20	20	31	38	43	56	58	70	64	57	45	39	33
21	30	18	32	38	60	62	68	62	54	46	34	43
22	31	22	36	41	68	63	66	61	59	51	32	39
23	40	32	34	47	64	71	67	67	58	44	35	34
24	39	22	37	40	70	78	70	64	62	50	28	28
25	40	23	39	47	60	74	78	70	55	52	45	23
26	33	21	34	52	55	72	82	76	51	48	42	11
27	16	33	43	59	54	80	77	82	60	41	43	7
28	10	27	45	50	66	78	78	79	63	52	36	17
29	−8		51	52	71	73	76	70	54	51	32	25
30	9		43	52	69	81	72	74	48	36	39	23
31	36		40		76		75	66		38		30
Total Degree Days	**1,194**	**1,081**	**927**	**568**	**253**	**57**	**4**	**9**	**112**	**346**	**664**	**1,085**

Exhibit 10

Daily Sendout per Residential Heating Customer as a Function of Temperature, 1/1/76–7/31/81

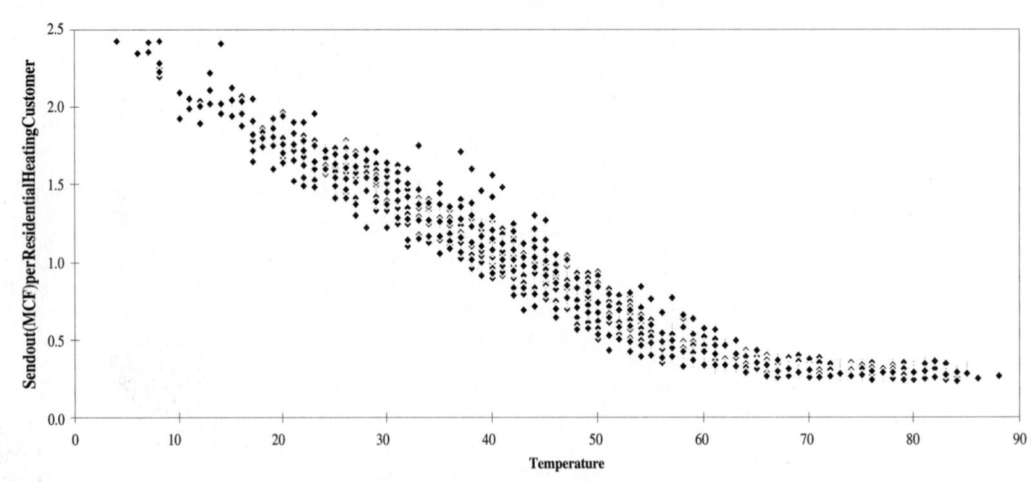

FORECASTING WITH REGRESSION ANALYSIS

Regression analysis[1] provides a powerful tool for developing forecasts of the future based on data from the past. It specifies the relationship between a variable of interest (the dependent variable) and other variables (the independent variables), and thus enables you to forecast a future value of the dependent variable when the values of the independent variables are known. In this chapter we discuss several ways of using observations of past values of a variable (and perhaps other "explanatory" variables as well) to make forecasts of its future values.

INDISTINGUISHABLE AND DISTINGUISHABLE DATA

▼ Forecasting Based on "Indistinguishable" Data

The following problem illustrates how data on past observations can be used to generate a forecast of some future observation. Imagine that you are interested in selling your house. To help you forecast its selling price, you obtain a sample (shown in Table 4.1) of the selling prices of ten houses sold in your city over the past year.

Table 4.1

Selling Prices of a Sample of Ten Houses

Selling Price
$109,360
$137,980
$131,230
$130,230
$125,410
$124,370
$139,030
$140,160
$144,220
$154,190

Let's assume that housing prices have remained stable over the last year, and that any house in your sample is as likely to be representative of the selling price of your house as any other—your house is **indistinguishable** from the houses in your sample. Then the selling prices of those houses would be relevant to your forecasting problem. Indeed, if you wanted to make a **probabilistic** forecast of your house's selling price, in the context of a decision problem, you would simply use the **frequency distribution** of the ten selling prices in your sample.

Quite often, however, we do not have a well-defined decision problem in mind, and want just a single number—a "point forecast" or "best guess." In terms of the original sample data, you might choose a central value such as the mean, median, or mode. The central value that we use in regression is the mean.[2]

Harvard Business School note 9-894-007. This note was prepared by Professor Arthur Schleifer, Jr.

[1] Other uses of regression analysis are discussed in Chapter 5.

[2] If we define a **residual** as the difference between a point forecast and an actual value, then (see the Appendix to Chapter 1) the mean minimizes the sum of squared residuals–it is a **least-squares** estimate. Regression estimates are also least-squares estimates, as we shall see.

The mean of your sample of ten houses is $133,618. How good an estimate is that? If all the prices in your sample were close to $133,618, you would feel quite confident that the selling price of your house would be in the vicinity of $133,618. You would be much less confident if the prices in your sample were widely dispersed. The sample standard deviation is a useful measure of dispersion, and serves as a benchmark relative to which other forecasting techniques can be compared. In this case, the sample standard deviation is $12,406.

Forecasting When the Data Are Distinguishable

Now suppose that your sample contains information not just on selling prices but also on the square footage of the ten houses, as in Table 4.2. The sample no longer represents houses that are indistinguishable: larger houses are **distinguishable** from smaller houses. If your house has 1,682 square feet of living area, it seems reasonable to confine your attention to "look-alike" houses. While no house in the sample is a perfect "look-alike," there are four houses that are more than 1,600 and less than 1,800 square feet, whose selling prices, you might conjecture, would be more like the selling price of your house than those that are much smaller or much larger. The average selling price of those four houses is $132,243, slightly less than the average across all ten houses, and the sample standard deviation is $8,513, less than the sample standard deviation for all ten houses. By restricting our attention to a subset of the data consisting of houses that are nearly indistinguishable look-alikes, we have slightly refined our point forecast, and have somewhat increased its accuracy.

Table 4.2

House Selling Price	House Size (sq. ft.)
$109,360	1,404
$137,980	1,477
$131,230	1,503
$130,230	1,552
$125,410	1,608
$124,370	1,633
$139,030	1,717
$140,160	1,775
$144,220	1,832
$154,190	1,934

In Table 4.3, we show the sample average and standard deviation for all ten houses, and then similar measures for three subsets of the data (or "cells") classified by size ranges. As you can see, the sample average goes up as size increases. For all three cells, the sample standard deviation is lower than it was when we looked at all ten houses together.

Sample Residuals and Their Standard Deviations

The sample standard deviation for the 1,600–1,799 cell—the appropriate cell for forecasting the selling price of your house—was $8,513, but was based on only four observations (and therefore three degrees of freedom). Does this properly measure our forecast uncertainty? Although the standard deviations for each cell are different, they are all based on very limited numbers of observations; could the differences have arisen by sampling error? Finally, it would be nice to have a measure of the overall efficacy of our partitioning the sample data into cells: have we really reduced the forecast uncertainty by this partitioning?

Table 4.3

	All Ten Houses	Size Range (sq. ft.)		
		1,400 – 1,599	**1,600 – 1,799**	**1,800 – 1,999**
Sample Average	$133,618	$127,200	$132,243	$149,205
Sample Standard Deviation	$12,406	$12,381	$8,513	$7,050
Number of Observations	10	4	4	2

To shed light on these questions, we start by defining a **sample residual** as the difference between an actual value of the dependent variable and its estimated or forecast value. The first house in our sample belongs to the 1,400 – 1,599 cell. Its selling price was $109,360. The average selling price of the four houses in that cell was $127,200, and that was our point forecast for each of those four houses. The residual is then $109,360 – $127,200 = –17,840. We can calculate residuals for the remaining sample observations in the same way: for each observation, we identify its cell, use the cell average as a point forecast, and compute as its residual its actual selling price less the forecast. In Table 4.4 we show residuals computed in this way for the ten houses in your sample. (Negative numbers are shown in parentheses.) Notice that the mean of the residuals is zero.[3]

Table 4.4

	SELLING PRICE	RESIDUALS	SIZE	
	$109,360	($17,840)	1,404	
	$137,980	$10,780	1,477	Size =
	$131,230	$4,030	1,503	1,400 – 1,599
	$130,230	$3,030	1,552	
AVERAGE	**$127,200**			
	$125,410	($6,833)	1,608	
	$124,370	($7,873)	1,633	Size =
	$139,030	$6,788	1,717	1,600 – 1,799
	$140,160	$7,918	1,775	
AVERAGE	**$132,243**			
	$144,220	($4,985)	1,832	Size =
	$154,190	$4,985	1,934	1,800 – 1,999
AVERAGE	**$149,205**			
	AVERAGE:	**$0**		
	SUM OF SQUARES:	**727,012,525**		
	DEGREES OF FREEDOM:	**7**		
	RSD:	**$10,191**		

If you believe that size is the only distinguishing factor—that no other independent variables can explain the variability in the residuals—and that residuals are not likely to be larger in magnitude, on average, in one cell than they are in another, then the residuals are indistinguishable. The residual of –$17,840 computed for the first observation, which belonged to the first cell, is just as relevant a measure of our forecast error as residuals computed for observations in the other two cells. Even though the three cell means are different, we use all ten residuals to derive an overall estimate of forecast error. If the residuals were all close to zero, you would feel quite confident about your forecast; if they were widely dispersed, you would feel less confident.

[3] The mean of the residuals in any cell must be zero, since each residual is the difference between a value in a cell and the cell mean. As a result, the mean of the residuals across all cells must be zero.

There are two tricks in computing the standard deviation of the sample residuals. One is a simplification. To compute any standard deviation, you take each value, subtract the mean, and square the difference. Since the mean of the residuals is zero, you can just square the residuals. You then add them, obtaining a "sum of squares," divide by the degrees of freedom, and take the square root. The second trick is computing the degrees of freedom. Each sample mean "uses up" a degree of freedom. In the partitioned housing data, there are three sample means. Hence there are only 10–3=7 degrees of freedom. At the bottom of Table 4.4, the various steps in computing the residual standard deviation (RSD) are shown in detail. The RSD is $10,191, a modest improvement over the sample standard deviation of $12,406.

Using the Data More Efficiently

If there were still other variables—age of house, lot size, neighborhood, etc.—that were related to selling price, we could in principle continue this process of partitioning our data into more narrowly defined cells. With our sample of ten houses, this is hardly feasible; with a very large sample, however, it would be practical to create "look-alike cells" that contained selling prices of houses that were alike with respect to several variables simultaneously. But, even with large samples, as the number of cells grows, the number of observations per cell and the degrees of freedom decline. On the one hand, a potential seller can identify a cell containing houses that look very much like her house; on the other hand, there will be very few such houses, and therefore her point estimate will be subject to large sampling error. And as the number of degrees of freedom declines, the denominator used in the calculation of the RSD gets smaller; although the numerator may also get smaller, the net effect may be for the RSD to increase—for your forecast uncertainty to get larger as you take into account more distinguishing factors.

In creating look-alike cells, we have used the data very inefficiently in two respects. First, we have ignored data on houses that are "almost like," but not "exactly like," yours. Wouldn't data about selling prices in the two size ranges adjacent to yours have some bearing on the selling price of your house? A second way in which we have used the data inefficiently is that we have partitioned it somewhat arbitrarily. Your house, with 1,682 square feet of living area, was treated as indistinguishable from houses of 1,600 or 1,799 square feet, but houses of 1,599 or 1,800 square feet were treated as irrelevant to your forecasting problem.

A way around both of these problems is to create a **model** that specifies the relationship between selling prices and the variables that help you forecast price—size, construction, lot size, age, etc. Selling price is the **dependent** variable; those variables that help you forecast price are the **explanatory** or **independent** variables.

A Regression Model

Let's first look at a model that relates selling price to size. While it is almost certainly true that some larger houses will sell for less than some smaller houses, it is reasonable to assume that as size goes up, selling price will go up **on average**. If each additional square foot increases average selling price by the same amount, the relationship between size and selling price is **linear.**

If each additional square foot increases selling price, but by less and less as size increases, the relationship is curvilinear, exhibiting **diminishing returns to scale**. If, on the contrary, each additional square foot increases selling price by a greater and greater amount, the relationship is again curvilinear, but this time exhibits **increasing returns to scale**. We can examine a scatter diagram to get clues about the relationship.[4]

Suppose we decide that the relationship is a linear one. We can express such a relationship as the following regression model:

$$y_{est} = b_0 + b_1 * x_1 \,,$$

where y_{est} is an estimate or point forecast of the actual selling price (y) of a house whose size (in square feet) is x_1, and b_0 and b_1 are constants (called **regression coefficients**) whose values are to be estimated from the data. Because we have assumed that the linear relationship between size and forecast selling price applies across all of the data, we can use all of the data to estimate the regression coefficients.

The regression coefficients are estimated by **least squares**. In principle, we could find the least-squares estimates by: (1) choosing arbitrary values of b_0 and b_1; (2) computing a forecast y_{est} of each house's selling price by multiplying its size by b_1 and adding b_0; (3) computing a residual $y - y_{est}$ for each house (the difference between its actual and its forecast selling price); and (4) squaring and summing all the residuals. We would then select new values of b_0 and b_1 and go through the same four steps, finally choosing as our least-squares estimates the values of b_0 and b_1 for which the sum of squared residuals is least. This process could be easily extended to problems which contain more than one independent variable.

Fortunately, we don't have to go through this tedious sequence of steps. Just as the mean is a least-squares estimate of a single variable, so there are formulas for obtaining least-squares regression coefficients, but these formulas require so much computation that the only practical way of solving regression problems is with a computer.

Regression programs may differ in how data are entered and how the output is expressed, but they all have certain elements in common.

INPUTS TO A REGRESSION ANALYSIS

You must supply the following information to the computer to perform a regression analysis:

1. *Identify the dependent variable.* The dependent variable is the variable that you want to forecast (selling price, in the example above).
2. *Specify the independent variable or variables.* The independent variables are those distinguishing factors that, in your judgment, help to explain the variation in the values of the dependent variable.
3. *Specify the relevant data.* Typically, your analysis will include all the data for which a specified relationship applies. Sometimes, however, it is hard to find a single relationship that links disparate groups of data. Suppose the house whose selling price we wanted to forecast were in Boston, and that our ten observations on house sales were for sales in Boston.

[4] How to detect and specify a curvilinear relationship, including increasing or diminishing returns to scale, is discussed later in this chapter.

If we could augment the data set to include houses in New York as well, we might find it difficult to specify a relationship between size and price that applied to both Boston and New York. We might choose to ignore the New York data, using only the Boston data to forecast Boston prices.

4. *Specify the nature of the relationship between the dependent variable and the independent variables.* This step requires some careful thought. As illustrated above, a forecast selling price based on the size of a house depends on the relationship you specify between size and price (such as linear, or with increasing or decreasing returns to scale). The issues involved in specifying such relationships are discussed in more detail later in this chapter.

5. *Provide values of the dependent and independent variables for the relevant observations.* You must provide these values to the computer in the form of a data file. In the forecasting problem, you would have to provide a file containing the selling price, size, and whatever other variables you deemed relevant for the ten houses in your sample.

Outputs From a Regression Analysis

The computer generates output from the input discussed above. In this chapter we discuss three types of output: regression coefficients, forecasts, and measures of goodness of fit.

Regression Coefficients

We saw in the last section that a regression model is a formula that relates a forecast or estimate of the dependent variable to the value(s) of the independent variable(s). The formula involves constants, called regression coefficients, whose values are estimated from the data. When there is only one independent variable and the relationship between the independent and the dependent variable is linear, there are two regression coefficients, b_0 and b_1.

Suppose we had the sample of ten houses shown in Table 4.2, with data on their selling prices and size. We perform a regression with selling price as the dependent variable and size as the independent variable (in the following discussion we refer to this regression as Model 1) and discover that $b_0 = 35{,}524$ and $b_1 = 59.69$. How do we interpret these regression coefficients? The value of b_1 tells us that if our regression model is specified correctly, each additional square foot of living space adds an average of about $60 to the value of a house. The "constant" term b_0 tells us that an average house with 0 square feet will sell for around $35,500. This interpretation of b_0 may seem utterly nonsensical, but let's not get distracted for the moment.

With just one independent variable, it is easy to visualize what is happening. Figure 4.1 is a scatter diagram of selling price vs. area. The straight line through the plotted points represents the "least-squares" fit to the data. It intersects the vertical axis at about 35,000 (the value of b_0), and has a slope of about 60 (the value of b_1): the line goes up from about $35,000 to about $95,000 over a range of 1,000 on the horizontal axis, or about 60 per unit increase in the independent variable.

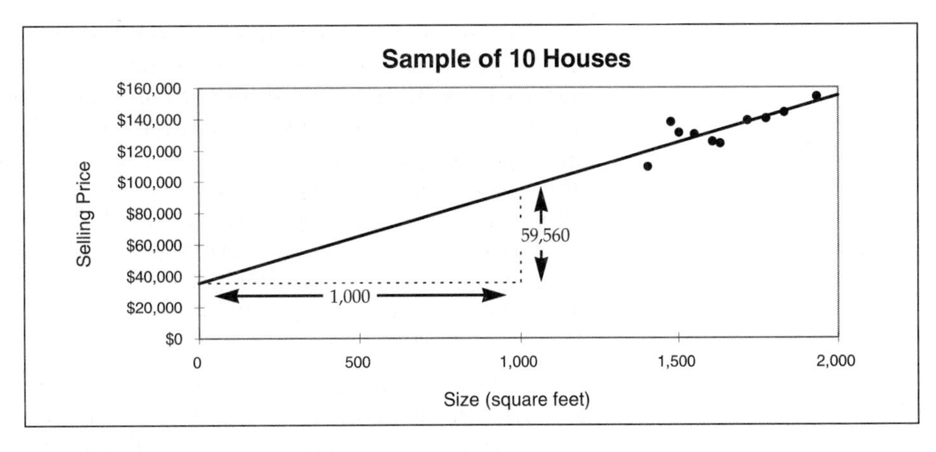

Figure 4.1

In any regression with just one independent variable (x_1), where the relationship between the dependent variable (y) and x_1 is assumed to be linear, least-squares estimates or forecasts y_{est} as a function of x_1 can be graphed as a straight line (the regression line), with the equation

$$y_{est} = b_0 + b_1 x_1 .$$

In this equation, b_0 is the "intercept," the height of the regression line where it intersects the vertical axis, and b_1 is the slope, the amount the regression line increases per unit increase in x_1.

Now let's introduce a second independent variable: the age of the house (in years), symbolized by x_2, values of which are shown in the third column of Table 4.5. Let us specify as our regression model (Model 2)

$$y_{est} = b_0 + b_1 x_1 + b_2 x_2$$

where the values of b_0 and b_1 will be different from the values they had in Model 1. When we perform the regression, the least-squares estimates of the b's are as follows: $b_0 = 4{,}045$, $b_1 = 86.84$, and $b_2 = -695.8$. (We summarize the results of Models 1 and 2 in Table 4.6.)

Table 4.5

SELLING PRICE	SIZE (SQ.FT)	AGE
109,360	1,404	20
137,980	1,477	2
131,230	1,503	5
130,230	1,552	4
125,410	1,608	23
124,370	1,633	34
139,030	1,717	25
140,160	1,775	23
144,220	1,832	28
154,190	1,934	25

Table 4.6

	MODEL 1	MODEL 2
CONSTANT TERM	35,524	4,045
COEFFICIENT FOR:		
SIZE	59.69	86.84
AGE		(695.8)

While it is much more difficult to visualize the relationship between y_{est} and the x's,[5] we can nevertheless interpret the b's. The value of 4,045 for b_0 is the forecast price for a brand-new house with 0 square feet of living area. Although this estimate of the constant term is absurd, it is less so than the constant term of around 35,000 in Model 1. We'll have more to say about this later.

[5] We could depict the relationship by means of a three-dimensional graph. Such a graph is hard to visualize and doesn't get us very far: if we perform a regression with three or more independent variables there is no way of graphing the relationship.

The value of 86.84 for b_1 means that if x_1 increases by one unit (one square foot) while x_2 remains constant, y_{est} will increase by 86.84: for houses of a given age, each additional square foot adds an average of $86.84 to their value. Notice that this is different from the regression coefficient of 59.69 associated with size in Model 1. We have learned one very important fact about regression: *the value of the regression coefficient associated with a given independent variable depends on what other independent variables are included in the model.*

Turning to the regression coefficient associated with age, we see that a one-unit increase in x_2, holding x_1 constant, *decreases* y_{est} by 695.8: two houses of equal size that differ in age by one year will differ in price by nearly $700 on average; the older house will tend to be cheaper.

When a regression model involves two or more independent variables:

- the constant term is an estimate of the value of the dependent variable when all of the independent variables are equal to zero.
- the regression coefficient associated with a given independent variable is an estimate of the amount by which the dependent variable will change when the independent variable in question changes by one unit, while all the other independent variables *in the model* are held constant.

Uncertainty in Regression Coefficients

The observations used in a regression analysis can almost always be thought of as a sample from some larger population or process; therefore the estimates derived from regression analysis are subject to sampling error. Just as the sample mean is an estimate of a population or process mean, so a regression based on sample data is an estimate of what the regression would be if we had in hand all the relevant data, not just a sample. In particular, each sample regression coefficient is a point estimate of a "true" regression coefficient. How much error there may be in this estimate is given by the regression coefficient's **standard error**. The formula for the standard error is fairly complicated, but virtually all computer programs compute and print its value.

In Chapter 2, "Sampling and Statistical Inference," we saw that the sample mean is an estimate of a population mean, and from this estimate and its standard error, you can derive a normal confidence distribution of the population mean, confidence intervals, and t values. By analogy, the estimate and standard error of a regression coefficient enable you to derive a (normal) confidence distribution for the "true" regression coefficient, confidence intervals, and t values.

In Model 1, the regression coefficient b_1 has an estimated value of 59.69 and a standard error of 15.10. With 68% confidence we can say that the "true" population coefficient is between 44.59 and 74.79, and 95% and 99.7% confidence intervals can be similarly constructed. The t value is $59.69/15.10 = 3.95$. Since the value of t exceeds 3, it is virtually certain that the "true" regression coefficient is positive.

In the same model, the constant term b_0 has an estimated value of 35,524 and a standard error of 24,933, implying a t value of 1.42. It is certainly possible that the "true" value of the constant term is close to zero, or negative. The "nonsensical" estimated value of 35,524 may have arisen only as a result of sampling error.[6]

---

[6] Another reason that the constant term may be misleading derives from our assumption that size and selling price are linearly related. For sizes in the range observed in our sample, an assumption of linearity may be quite justifiable, but if we were to extrapolate in either direction far outside that range we might find the relationship to be curvilinear. In particular, if there are diminishing returns to scale, the intercept of an appropriate curve might be quite close to zero.

Proxy Effects

We observed that b_1, the regression coefficient associated with size, had a value of 59.69 in Model 1 and a value of 86.84 in Model 2. Why are the values different? Recall that in Model 1 size was the only independent variable, whereas in Model 2 a second independent variable, age, was introduced. In Model 1, b_1 estimates the amount by which selling price increases when size increases by 1 square foot; in Model 2, it estimates the increase in selling price when size increases by 1 square foot *with age held constant*. Why is b_1 higher in this case? The argument is tricky, but let's take it step by step.

In Model 2 we saw that the relationship of age to selling price, when size is held constant, is negative ($b_2 = -695.8$): older houses tend to sell for less than newer houses of the same size. It also turns out that age and size are positively correlated; the correlation coefficient is 0.81. When age is left out of the regression, as in Model 1, size not only reflects its own relationship with selling price, but it also *proxies* for the relationship of age with selling price. Because the age-selling price relationship is negative, but age and size are positively correlated, the age effect for which size proxies is negative, and therefore b_1 turns out to be smaller than it was when the effect of age was held constant, as in Model 2.[7]

Proxy effects occur when two independent variables—say x_1 and x_2—(1) are correlated with each other; (2) are both related to the dependent variable (y), in the sense that both would have nonzero regression coefficients if both were included in the model; but (3) just one of them—say x_1—is included in a regression model, while the other is excluded. If all three of these conditions are satisfied, then x_1 will proxy for x_2, the proxy effect being greater the higher the correlation between x_1 and x_2 and the closer the relationship between x_2 and y. If both x_1 and x_2 are included in the model, their individual relationships with y will be correctly sorted out, but there may be other independent variables that we failed to include in the model for which the included variables proxy.

Observe that *proxy effects have absolutely nothing to do with sampling error*. As you add variables to a regression model, the effect of these added variables on a particular regression coefficient may be positive or negative,[8] and *may* greatly exceed anything that could possibly be attributed to sampling error. On occasion, you will find that a regression coefficient that was positive in a one-independent-variable model turns negative as you add other independent variables, or vice versa.

---

[7] You can relate the results of Models 1 and 2 exactly as follows. Perform a regression with x_2 as the dependent variable and x_1 as the independent variable. Let x_2 *(est)* be the estimated value of x_2 derived from this regression; we find that

$$x_2\,(est) = -45.24 + .03903x_1 \ .$$

Substituting this estimated value of x_2 for the actual value of x_2 in Model 2, we get

$$\begin{aligned} y_{es} &= 4{,}045 + 86.84x_1 - 695.8x_2 \\ &= 4{,}045 + 86.84x_1 - 695.8\,(-45.24 + .03903x_1) \\ &= 4{,}045 + 86.84x_1 + 31{,}478 - 27.16x_1 \\ &= 35{,}523 + 59.68x_1 \end{aligned}$$

which agrees with Model 1 except for a slight numerical rounding error.

[8] Whether the effect on a particular regression coefficient, say b_1, is positive or negative depends on (a) whether the added variable, say x_2, has a positive or negative regression coefficient in the model that includes it and x_1, and (b) whether the correlation between x_1 and x_2 is positive or negative.

Two Common Misconceptions

Notice that in Model 2 the dependent variable was measured in dollars, and the two independent variables were measured in square feet and years, respectively. *It is often incorrectly assumed that variables used in a regression must be measured in comparable units.*

We also saw in Model 2 that the two independent variables, size and age, were correlated. *A common misconception is that variables used in a regression must be uncorrelated.*

FORECASTS

Point Forecasts

In any regression model, a point forecast of the dependent variable can be made by multiplying known values of the independent variables by their respective estimated regression coefficients, adding the products, and finally adding the estimated constant term. In Model 2, a point forecast for a house that has 1,682 square feet and is 10 years old is:

$$y_{est} = 4{,}045 + 86.84*1{,}682 - 695.8*10 = 143{,}152.$$

Probabilistic Forecasts

When we perform a regression, we are attempting to take into account all distinguishing factors that can explain how a dependent variable varies in value. We only rarely can explain all such variation perfectly. What is left unexplained after a regression is performed is a collection of indistinguishable residuals—the differences between the actual values of the dependent variable y and the corresponding point forecasts y_{est}. These residuals are retrospective, in the sense that they come from data already in hand. We want to forecast a prospective value of the dependent variable—a value not yet known. Our point forecast will almost surely be in error. We want to quantify and attach probabilities to these prospective errors.

We can link the variability of our retrospective residuals with the uncertainty in our prospective forecast errors by the following logic. Because the residuals are indistinguishable, any one is as representative of the error we are likely to make on a given forecast as any other. For many purposes it is sufficient to assume that each retrospective past residual corresponds to a prospective forecast error whose probability is equal to the relative frequency of the residual in question.

When a probabilistic forecast is to be used as an uncertainty in a decision problem, this method is quite satisfactory. Table 4.7, for example, shows the data on ten houses given in Table 4.5, plus estimated values and residuals based on Model 2. If we wanted to make a probabilistic forecast for your 1,682-square-foot 10-year-old house, we would start with the point forecast of y_{est} = 143,152. Each residual added to the point forecast would give a possible forecast value of y, and since there are ten residuals, each such value would be assigned probability 1/10. Figure 4.2 represents the uncertainty that we would assign to the selling price of your house based on the Model 2 regression results.

Table 4.7

SELLING PRICE	SIZE (SQ. FT.)	AGE	y_{est} = 4,045 + 86.84*SIZE – 695.8*AGE	RESIDUALS
109,360	1,404	20	112,054	(2,694)
137,980	1,477	2	130,917	7,063
131,230	1,503	5	131,088	142
130,230	1,552	4	136,039	(5,809)
125,410	1,608	23	127,682	(2,272)
124,370	1,633	34	122,200	2,170
139,030	1,717	25	135,757	3,273
140,160	1,775	23	142,185	(2,025)
144,220	1,832	28	143,656	564
154,190	1,934	25	154,601	(411)

Sometimes you are not faced with a clearly defined decision, but a problem in which some uncertainty depends on your forecast. In this case you may want to reflect your forecast uncertainty in terms of confidence intervals. Standard practice is to *assume* that the sample residuals came from a population of normally distributed residuals, and thus to assume that the forecast errors are normally distributed, with mean zero and standard deviation equal to the estimated residual standard deviation, or RSD.[9] This in turn implies that the probability distribution for the forecast itself has mean y_{est} and standard deviation equal to the RSD. For example, the RSD for the sample of ten houses is 4,072. Thus a 68% confidence interval for our 1,682-square-foot 10-year-old house is between 143,152 – 4,072 = 139,080 and 143,152 + 4,072 = 147,224.

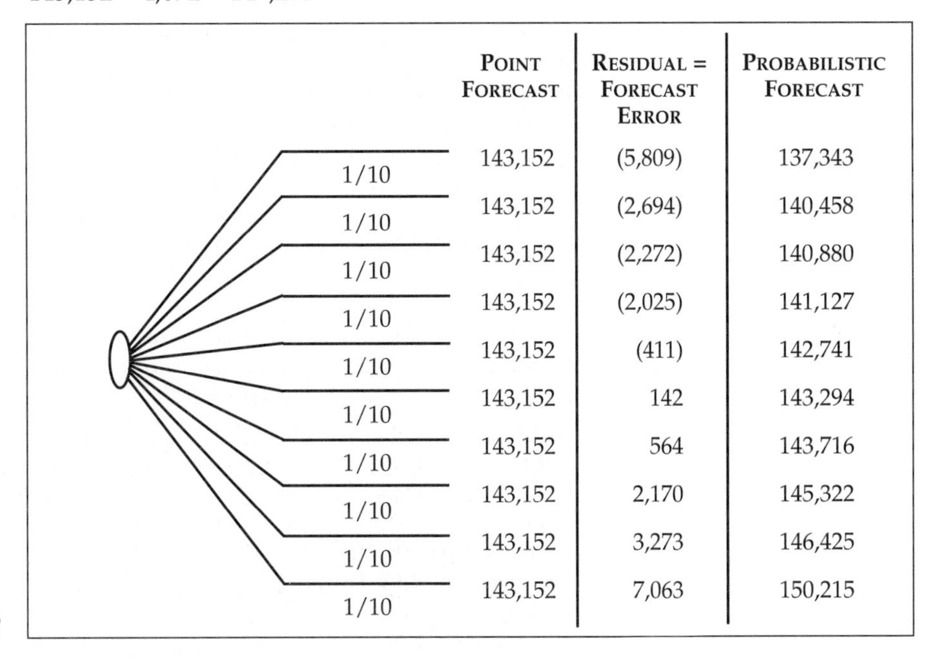

Figure 4.2

[9] In a subsequent section, we show how to compute the regression RSD.

A schematic of how we go from a regression to a probabilistic forecast is shown in Figure 4.3.

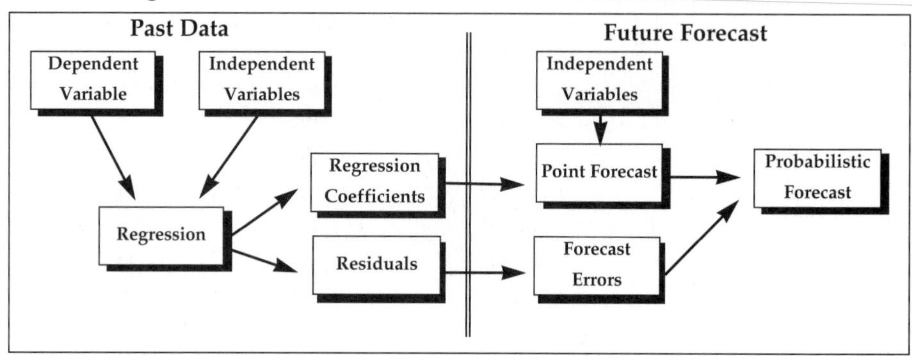

Figure 4.3

Additional Sources of Uncertainty

Both methods of deriving probabilistic forecasts from a regression assume that the only source of uncertainty is the variability of the residuals. We have swept under the rug two other sources of uncertainty that are always present, and in practice there are often still other uncertainties. All of these uncertainties combine to make the true forecast distribution more dispersed than the distributions computed above. The 68% confidence interval computed in the last section should really have a lower limit less than 139,080 and an upper limit greater than 147,224.

Although we can correct mathematically for some sources of uncertainty, at the cost of greatly complicating the formulas for forecast distributions, some of the uncertainty depends on pure judgment. Rather than make a partial correction, we will treat all additional sources of uncertainty judgmentally.

We begin with uncertainties that are a result of sampling error. We know that the regression coefficients are only sample estimates of the "true" population regression coefficients. Residuals are computed as if these sample estimates are the true regression coefficients. Similarly, the sample RSD is only an estimate of the "true" variability of the "true" residuals in the population, but we use it as if it is a value known exactly.

We may encounter sources of uncertainty other than sampling error. For example, the independent variables used in a forecast are assumed to have known values, but in some situations they may be estimates or forecasts whose values are uncertain at the time that the forecast is made. And often the most important source of additional uncertainty is whether the independent variables used are truly relevant, and whether the representation of the relationship between the dependent and the independent variables is correct: a probabilistic forecast is made with respect to a particular regression model, but there are often a number of other plausible models that may lead to different forecasts.

Thus, in addition to the uncertainty attributable to the variability of the residuals—the only uncertainty embodied in the forecast distributions derived in the last section—there are, or may be, at least five other sources of uncertainty arising from:

- uncertainty in the values of the regression coefficients, whose estimated values we are treating as if they are known with certainty
- uncertainty in the RSD, whose estimated value we are treating as if it is known with certainty

- possible uncertainty in the values of the independent variables used in a forecast
- possible uncertainty about which one of several competing regression models is correct
- the assumption that the residuals come from a population of normally distributed residuals

It is very difficult to quantify the degree to which a forecast distribution computed in either of the two ways described above understates our uncertainty, but the problem is less severe if:

- there are many observations
- there are not many independent variables
- the values of the independent variables used in the forecast are well within the range of the independent variables in the data from which the regression model was constructed[10]
- the values of the independent variables used in the forecast are known with certainty[11]
- the regression model has no plausible competitors
- the model fits the data well
- the sample residuals are approximately normally distributed[12]

MEASURES OF GOODNESS OF FIT

Regression programs provide various measures of goodness of fit, of which the two most important are the residual standard deviation (RSD) and the coefficient of determination, or R^2.

Residual Standard Deviation

The residual standard deviation is just what it sounds like—the estimated standard deviation of the residuals—the square root of the sum of squared residuals[13] divided by the degrees of freedom. When we partitioned the data into "look-alike" cells, the degrees of freedom were the number of observations (n) less the number of cells. Here the degrees of freedom are n minus the number of regression coefficients (including the constant term). Thus, in Model 2, there are n = 10 observations and three regression coefficients, so that there are seven degrees of freedom. The sum of squared residuals is 116,080,000, so that the RSD $= \sqrt{116{,}080{,}000 / 7} = 4{,}072$.

10 Extrapolating beyond the range of the data is dangerous for two reasons: (1) uncertainty in the regression coefficients becomes a more serious problem, and (2) a relationship between an independent and the dependent variable that appears to be linear over the range of data on which the regression was based may in fact be curvilinear; extrapolating a linear relationship in such a case can lead to serious errors.

11 Independent variables that are "leading indicators" are especially useful in this respect: if next period's stock price, for example, depends on this period's inflation rate, one-period-ahead forecasts can be made using known values of the independent variable. Unfortunately, such leading indicators are not easy to find.

12 A useful diagnostic tool is to plot a histogram of the sample residuals to see if the shape of the histogram is roughly normal.

13 Just as was the case when we partitioned the sample into look-alike cells, residuals from any regression fitted by least squares necessarily have mean 0. Hence, in computing the standard deviation, we can add up the squared residuals without first subtracting out the mean.

In comparing regressions having the same observations and the same dependent variable, the one with a lower RSD indicates a better fit. As you add independent variables to a regression model, the sum of squared residuals almost always decreases (at worst, it doesn't change), but the degrees of freedom also decrease. Because both the numerator and the denominator tend to decrease as you add variables, the RSD may go in either direction. If adding a variable causes the RSD to increase, you almost surely have too many independent variables in your model.

Coefficient of Determination (R^2)

While the RSD provides a good way of comparing different regression models in terms of goodness of fit, the measure is in units of the dependent variable, and it is hard to judge if an RSD of 4,072, say, indicates a good fit or a bad fit in an absolute sense. It is tempting to try to find an index that does not depend on the units of measurement of the variables in the regression. R^2 is such an index. As we shall see, however, it is a very fallible absolute measure of the goodness of fit.

R^2 measures the percent improvement in fit that a regression provides relative to a base case which assumes that the values of the dependent variable are indistinguishable. For the base case we compute the sum of squared residuals about the mean of the dependent variable; let's call this the "base-case sum of squares." If the sum of squared residuals from the regression—the "regression sum of squares"—is much less, the regression has explained much of the variability in the dependent variable, an indication of a good fit. If the regression sum of squares is almost as large as the base-case sum of squares, the regression has not explained much of the variability in the dependent variable: the fit is bad. R^2 measures the percent reduction in the base-case sum of squares achieved by the regression. Its formula is:

$$R^2 = (\text{base-case sum of squares} - \text{regression sum of squares}) / \text{base-case sum of squares} .$$

In the housing data, the base-case sum of squares for selling prices is 1,385,300,000. We have already seen that the regression sum of squares is 116,080,000. Thus $R^2 = 0.9162$.

An alternative way of defining R^2 is as the square of the correlation between the values of y_{est} and the true values (y) of the dependent variable. In the housing data, the correlation between the true selling price and the price estimated by Model 2 is 0.9572; its square is 0.9162.

Because the regression sum of squares will almost always decrease as you add independent variables, R^2 will always increase (or at worst remain unchanged) as you add variables. The computation of R^2 does not "penalize" you for using up degrees of freedom as you add variables.

"Adjusted" R^2

Some regression programs provide an "adjusted" R^2 instead of, or in addition to, the ordinary R^2. Adjusted R^2 does take into account the degrees of freedom used up in by the regression. Specifically, instead of using the raw base-case and regression sums of squares, it first divides each by its respective degrees of freedom. Thus, in the housing example:

$$\text{Adjusted } R^2 = (1{,}385{,}300{,}000/9 - 116{,}080{,}000/7)/(1{,}385{,}300{,}000/9) = 0.8923 .$$

(The first term—1,385,300,000/9—is the square of the sample standard deviation of the dependent variable; the second is the square of the RSD.)

Adjusted R^2 is always less than ordinary R^2. It always increases when the RSD decreases and vice versa. As you add independent variables, it may or may not increase; if it decreases, that suggests you are using too many independent variables in your model. If you have few degrees of freedom and a poor fit, adjusted R^2 may be negative.

Interpretation of R^2

Whether you use ordinary or adjusted R^2, you must remember that its interpretation is made relative to a base case that may itself be either a sensible way of forecasting the value of a future observation or a totally nonsensical way of forecasting. In the latter case, virtually *any* regression model is likely to produce a high value of R^2.

Here are a few benchmarks. You have data on the Standard and Poor's 500 stock index monthly closing price from January 1968 through March 1993. If you use time as an independent variable (January 1968 = 1, February 1968 = 2, etc.), $R^2 = 0.7586$; if you use last month's closing price as an independent variable, $R^2 = 0.9927$. You get high values of R^2 because the base case provides a nonsensical way of forecasting any particular month's S&P value: the "base-case" assumption that the values of the S&P 500 are indistinguishable over this twenty-five year period, when in fact they fluctuated between 63.54 and 451.67, had a pronounced upward trend, and a given month's price was generally much closer to the previous month's price than a price chosen at random, results in a base-case sum of squares which even the simplest regression models can reduce enormously.

Unfortunately, it does an investor very little good to forecast the *level* of the S&P 500. Investors make money by correctly forecasting *changes* in the level of a stock or an index. If you had a regression model for forecasting monthly changes in the S&P whose R^2 was only 0.05, you could, over the long run, do very well. Great value will accrue to an investor who can achieve even a small improvement over a base-case forecast that assumes that future changes will vary as past changes.

This should convince you that an R^2 of 0.99 does not necessarily indicate an extraordinary fit, nor does an R^2 of 0.05 mean that a regression is useless. It all depends on how difficult it is to do better than the base-case forecast.

TRANSFORMED VARIABLES

Transformations greatly increase your ability to specify relationships between a dependent variable and a number of independent variables. In this section we shall show how transformed variables can be created that:

- permit independent variables to be used that are not contemporaneous with the dependent variable in a time series.
- permit you to use ordinal and categorical variables as independent variables.
- permit you to express relationships between a dependent and independent variable that are curvilinear.

Lagged Variables in Time Series for Modeling Noncontemporaneous Effects

Suppose we believe that advertising affects sales. If we have a time series of a company's monthly advertising expenditures and unit sales, we could perform a regression with sales as the dependent and advertising as the independent variable, and see if there was any apparent relationship. Of course, we might want to include other variables, such as price, so that advertising does not inadvertently proxy for their effects.

On reflection, we might decide that although advertising this month might be related to sales this month, there might also be a carry-over effect from advertising expenditures made in previous months: last month's advertising might be related to this month's sales, and advertising expenditures two months ago might also have an influence, probably less, on this month's sales. We might believe that the effects of past advertising persist for some time.

Let y_t be our unit sales in month t, x_t be our advertising expenditure in the same month, and p_t be our average selling price in month t. Then our original data would consist of values of y_t, x_t, and p_t for as many months as we had data. To perform a regression in which sales depend on current and past values of advertising expenditure and current price, we would first compute **lagged transformations** x_{t-1}, x_{t-2}, etc., of x_t, and then run a regression with y_t as the dependent variable, and x_t, x_{t-1}, x_{t-2}, p_t, and additional lagged x's, if appropriate, as independent variables.

Values of lagged x's are computed in Table 4.8 for hypothetical data. Notice that each time you lag a variable an additional period, you create a missing value (denoted by #N/A in the spreadsheet). Since an observation used in regression must be complete, i.e., no value missing for any variable used in the regression, lagging a variable results in lost observations. A regression with x_t, x_{t-1}, and x_{t-2} as independent variables has four fewer degrees of freedom than one with just x_t as an independent variable: two are lost because there are two more variables; two more are lost because there are two fewer observations.

Table 4.8

Month (t)	Unit Sales (000) in Month t	Advertising Expenditures ($000) in Month t	Advertising Expenditures ($000) in Month $t-1$	Advertising Expenditures ($000) in Month $t-2$	Average Unit Price in Month t
Jan-92	1,137	1,144	#N/A	#N/A	$6.57
Feb-92	1,227	972	1,144	#N/A	$6.95
Mar-92	949	798	972	1,144	$6.54
Apr-92	842	861	798	972	$6.53
May-92	810	936	861	798	$6.64
Jun-92	707	770	936	861	$6.21
Jul-92	1,323	1,432	770	936	$6.78
Aug-92	1,471	1,330	1,432	770	$6.91
Sep-92	1,090	886	1,330	1,432	$7.04
Oct-92	890	996	886	1,330	$7.90
Nov-92	646	596	996	886	$6.09
Dec-92	757	774	596	996	$6.23
Jan-93	934	1,142	774	596	$6.62
Feb-93	1,071	932	1,142	774	$6.39
Mar-93	1,165	972	932	1,142	$6.04

We can lag not only values of an independent variable, but also values of the dependent variable. We might believe that sales levels tend to persist: this month's level is more likely to be high if last month's was high than if last month's was low. In that case, we might include y_{t-1} as an independent variable. If we believe that levels of sales in the more remote past tend to persist into the present, we could include values of y_{t-2}, y_{t-3}, etc. as independent variables as well. As before, each additional lag "costs" two degrees of freedom, one for the additional independent variable in the model, one for the observation lost in creating the lag.

Dummy Variables for Modeling Effects of Ordinal or Categorical Variables

Sometimes ordinal or categorical variables may be plausible explanatory variables in a regression. Selling prices of houses might depend on their condition; suppose, for each of the houses in your sample, we had an indication of its condition, on a scale from 1 to 5, 1 indicating "poor" and 5 indicating "excellent." Everything else being equal, we would expect that as condition gets better, selling price goes up, on average. But how do we deal with an ordinal variable like this in regression? If x represents the variable, and has possible values of 1 through 5, just including x in the model implies that, everything else being equal, the average difference in selling price between two houses rated 1 and 2 ("poor" and "fair") will be the same as for two houses rated 4 and 5 ("good" and "excellent"). But because the variable is ordinal, the difference between ratings of 1 and 2 may not measure the same difference in condition as the difference between 4 and 5. If we really believe the differences may be substantial, just using the rating scale x as an independent variable misspecifies the relationship between condition and selling price.

An even more serious problem occurs with categorical data. You might have a variable representing quality of construction, in one of three categories: frame, mixed frame and brick, and all brick. If we coded those categories 1, 2, and 3 respectively, could we just use the coded variable as an independent variable in our regression? Of course not! It might be the case that, all other things being equal, mixed frame and brick houses sell for more, on average, than either frame or all brick. Such a relationship could not possibly be revealed by including the coded variable in the regression.

Dummy variables provide a way of specifying ordinal and categorical relationships. Let's start with the simplest case, where there are just two categories (male vs. female, Republican vs. Democrat, yes vs. no, etc.). Suppose we code one of the categories 1 and the other 0; the actual assignment is arbitrary. In our housing example, suppose there were just two types of construction instead of three: frame houses (coded 0), and brick houses (coded 1). If we include this dummy variable as an independent variable in a regression, the corresponding regression coefficient tells us by how much brick houses differ from frame houses in selling price, on average, when the other independent variables included in the regression are held constant. A regression coefficient of 1,234, for example, implies that brick houses sell, on average, for $1,234 more than frame houses that are alike in all other respects measured by the other independent variables in the model. If the regression coefficient were (3,456), it would imply that brick houses sell for $3,456 less than frame houses.[14]

[14] Notice that this specification implies that whatever the average difference in price, it is the same for large houses as it is for small, the same for old houses as for new, etc. If you believe that the difference will be larger for large houses than for small houses, say, a different model specification is needed. However, we do not discuss such "interactive" relationships in this chapter.

Now let's return to the case where there are three categories of construction. We will create *two* dummy variables. One dummy will have value 1 if the house is mixed frame and brick, 0 if it is not (i.e., if it is *either* frame *or* all brick); the other will have value 1 if the house is all brick, 0 if it is not. In this specification, the third category—frame—will represent a "base case" as follows. Suppose the regression coefficients for the first and second dummies are 1,234 and (3,456) respectively. Remembering that we are talking about average relationships; with other variables included in the model held constant, these regression coefficients imply that, relative to the base case (frame houses), mixed frame and brick houses sell for $1,234 more, while all-brick houses sell for $3,456 less.

Suppose we had selected some other category—say all brick—as the base case. Then the regression coefficient for frame houses would be 3,456 and for mixed frame and brick 4,690 and we would reach the same conclusion as before: mixed frame and brick houses sell for $1,234 more, and frame houses for $3,456 more, than all brick. (The constant term would also change, decreasing by 3,456.) Although the choice of base case affects the values of the regression coefficients, it does not affect their interpretation.

In general, when a categorical or an ordinal variable has c categories, you can represent the effect of each category by defining c–1 dummy variables, use any one of the categories as a base case, and then use the c–1 dummy variables, along with whatever other independent variables are appropriate, in the regression.

Detecting, Specifying, and Interpreting Curvilinear Relationships: Exploratory Analysis

Suppose you suspect that in a model with more than one independent variable, the relationship between a particular independent variable (x) and the dependent variable (y) is curvilinear rather than linear. A scatter diagram of x against y might reveal such curvilinearity. But the relationship between x and y will be distorted by the proxy effects of the other independent variables on x.

A better way to detect curvilinearity under these circumstances would be to perform a regression using all the variables, compute the residuals, and plot the residuals (on the vertical axis) against x (on the horizontal). If this plot looks curvilinear, it suggests that the relationship between y and x, when all the other independent variables in the model are held constant, is curvilinear.

An even easier "quick and dirty" method of detecting curvilinearity is to include both x and a squared transformation of x (i.e., x^2), along with all the other independent variables, in the regression model. Thus, the model is specified as:

$$y_{est} = b_0 + a_1x + a_2x^2 + b_1x_1 + b_2x_2 + \ldots$$

where x_1, x_2, ... are the other independent variables and a_1 and a_2 are just the regression coefficients for x and for x^2.

For fixed values of the other independent variables, a graph of y_{est} as a function of x will be a **parabola**, a curve that either rises to a peak and then descends, or that descends to a trough and then rises. A value of a_2 clearly different from 0 provides evidence of a curvilinear net relationship between x and y; if, on the other hand, a_2 differs from 0 only by chance (sampling error), the data do not supply strong evidence of a curvilinear relationship.

Adding a squared transformation of an independent variable to the model thus provides an easy way of *detecting* curvilinearity, but *understanding* in what way the relationship between y and x is curvilinear is a trickier matter. To interpret the nature of the relationship correctly, two facts about parabolas are important to know:

1. If a_2 (the regression coefficient for x^2 in the previous equation) is *negative,* the parabola rises to a peak and then descends, while if a_2 is *positive,* the parabola first descends to a trough and then rises again.
2. The trough or peak occurs at the value of x for which $x = -a_1/(2a_2)$.

The behavior of the estimate depends on whether a_2 is positive or negative, and whether all, or nearly all, of the values of x are on one side or the other of the critical value $-a_1/(2a_2)$, or whether they substantially straddle that value. Table 4.9 shows six cases to consider:

Table 4.9

CASE #	VALUES OF X	VALUE OF a_2	BEHAVIOR OF y_{est}
1	$x < -a_1/(2a_2)$	$a_2 < 0$	Increases at a decreasing rate (Decreasing returns to scale)
2	$x > -a_1/(2a_2)$	$a_2 < 0$	Decreases at an increasing rate
3	x straddles $-a_1/(2a_2)$	$a_2 < 0$	Increases to max., then decreases
4	$x < -a_1/(2a_2)$	$a_2 > 0$	Decreases at a decreasing rate
5	$x > -a_1/(2a_2)$	$a_2 > 0$	Increases at an increasing rate (Increasing returns to scale)
6	x straddles $-a_1/(2a_2)$	$a_2 > 0$	Decreases to min., then increases

Figure 4.4 shows these cases. The arrows indicate the location of $-a_1/(2a_2)$ relative to the values of x in the data.

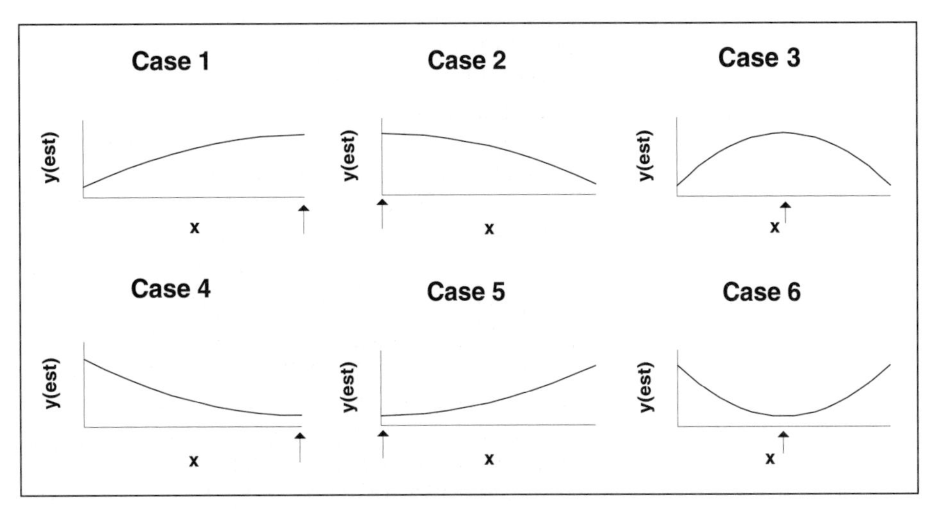

Figure 4.4

If, for example, $a_1 = 17.43$ and $a_2 = -2.367$, then $-a_1/(2a_2) = 3.682$ and if most of the values of x are above 3.682, Case 2 applies: y_{est} decreases at an increasing rate as x increases.

This method of analysis, it should be repeated, is *exploratory*: it will usually reveal curvilinear relationships and permit you to classify them appropriately. Three points are worth making:

1. When introducing a squared transformation (x^2), be sure to include the original value of x in the model as well.
2. The method discussed here will not reveal more complicated curvilinear behavior—for example, a relationship between y and x in which y first increases at an increasing rate and then at a decreasing rate as x increases.
3. Even if the general curvilinear relationship between y and x is detected and properly interpreted, the form of the model, using x and x^2 to capture the curvilinearity, may be inappropriate. In particular, curvilinearity is often a consequence of a multiplicative relationship between y and x, in which case modeling the relationship by using logarithmic transformations is appropriate. For details, see *Multiplicative Regression Models,* Chapter 6.

USING THE REGRESSION UTILITY

The regression utility supplied with this text lets you perform regressions on data in Excel files, view various outputs, and make forecasts of values of the dependent variable when corresponding values of the independent variables are known. Although Excel has an "add-in" regression feature, the utility is being made available because it is easy to use.[15] This section provides instructions for using the utility, and then guides you through worked examples that show both how the utility is used and how to specify and interpret the results of regression analyses.

Performing Regressions

1. Setting Up the Problem

Opening the Data File. Use the File Open commands in Excel to open the relevant data file.

Activating the Regression Utility. Follow the instructions distributed with the regression utility to activate it.

Setting Up the Data Range. Highlight the entire block of data you want to analyze. This block of data should include all observations of the variables you want to analyze.[16] In addition, you may include (only) one row of column labels. Column/variable labels appear in the regression output and are a convenient way of describing data. If your labels extend over several rows, you may want to insert a new row with abbreviated labels just above the first row of data.

To select the data range, click on the column label of the cell in the upper-left corner of the block and highlight the entire block of data. To highlight a large database quickly hold down the [Shift] and [Ctrl] keys (on DOS computers), or the [Shift] and [Command][17] keys (on Macs), then

---

[15] The Excel add-in requires that all independent variables be in contiguous columns, which necessitates considerable shuffling of columns before each regression is run. Also, it has no facility for excluding observations, and its output is hard to interpret. The regression utility avoids these limitations.

[16] If you have created new transformations of variables in your data file, these transformations should be included in your data range. If you think you may create new transformations, you can reserve an appropriate number of columns by specifying a data range that extends to the right beyond the existing data. (Alternatively, you can create new, transformed variables and then respecify the data range.)

[17] Throughout this text, whenever the DOS control key is mentioned, Mac users should interpret it as the command key.

press first the down arrow and then the right arrow. This sequence highlights the block of data, unless there are missing observations that cause the highlighting to stop prematurely.

Choose **Data** from the menu bar and **Regression** from the pull-down menu, then **Set Data Range** from the Regression menu. When prompted, "Does the top row of your data range contain column (or variable) labels?," click **Yes** if in fact you have included the labels in your data range, as suggested earlier. When the process is complete, your data should be surrounded by a box.

2. Setting Up the Dependent Variable Column

Click any cell in the column you want to be your dependent variable. Again choose **Data** from the menu bar and **Regression** from the pull-down menu, then **Set dependent variable column** from the Regression menu. You should then see the values in the dependent variable column in bold type.[18]

3. Setting Up the Independent Variable Column(s)

Click any cell in the first column you want to select as an independent variable. If you want to select more than one independent variable and they are in *adjacent* columns, you can highlight any row containing the range of columns. If you want to select *nonadjacent* columns, hold down the [Ctrl] key as you click a cell in each column you want to select as an independent variable. Once you have selected all the columns containing independent variables, choose **Data** from the menu bar and **Regression** from the pull-down menu, then **Set independent variable column(s)** from the Regression menu. You should see the values in the independent variable columns with a shaded background.

4. Performing the Regression

Before initiating the regression calculations, you may need to exclude observations.

Setting Up the Excluded Observations. The Regression utility will not calculate if you have observations with any values missing, so you will need to exclude observations containing such missing values. You may want to exclude observations for other reasons as well. If, after examining the data, you want to exclude observations from the database, hold down the [Ctrl] key and click a cell in each row you want to exclude. Choose **Data** from the menu bar and **Regression** from the pull-down menu, and choose **Set excluded observation(s)** from the Regression menu. Values of the variables in rows that were excluded will appear with strikeout lines through them.

Performing the regression. Choose **Data** from the menu bar and **Regression** from the pull-down menu, then **Perform regression** from the Regression menu. The statistics will be calculated for you. Once the regression has been performed, you will find yourself in the rows below your database looking at the output.

Outputs

A sample of the output that appears after you have chosen the **Perform regression** option is shown in Figure 4.5. This output was generated from the data set HTWT.XLS. Our model used weight as the dependent variable, and height and gender as the independent or explanatory variables.

[18] If you should happen to see "#####" signs, it means that your column width is too narrow. To widen the column, click on the letter of the column, select **Format** from the menu bar, **Column** from the pull-down menu, the **Autofit Selection** from the dialog box. Your column should now be wide enough to allow you to see the numbers.

	HT	Constant	M/F
Regr. Coef.	4.198	(133.7)	(17.86)
Std. Error	0.210	14.9	1.74
t value	20.0	(9.0)	(10.3)

Regression Number 1
Dependent Variable: WT

# of obs = ***768*** *Deg of F =* ***765***
R-squared = ***0.6092*** *Resid SD =* ***15.94***

Figure 4.5

The "constant" term will always appear in the dependent variable column.

5. Calculate Y_{est} and Residual Values

Once you have looked at the regression statistics, you may want to calculate Y_{est} (the regression estimates of the dependent variable) and residual values for the observations in your data range. This procedure can be time-consuming for large data sets. To initiate this process, choose **Data** from the menu bar and **Regression** from the pull-down menu, and **Calculate Y_{est} and residual values** from the Regression menu. The values of Y_{est} and the residuals appear in two columns to the right of the data range you selected.

6. View Charts

There are five different types of scatter diagrams to view. For charts involving Y_{est} and/or residual values, you must of course calculate those values first, using the procedure described above. The charts available are:

Y_{est} vs. Y_{act} (the actual value of the dependent variable)
Y_{est} vs. Residuals
Any X vs. Y_{act}
Any X vs. Residuals
Any X vs. any other X

To view the charts, choose **Data** from the menu bar and **Regression** from the pull-down menu, and **View charts** from the Regression menu. Within the dialog box, click on the chart you want to view. To move to a second chart, just repeat the preceding steps.

Forecasting

If you have a number of past observations on a dependent variable and a set of independent variables, and you want to forecast values of the dependent variable on future observations given specified values of the independent variables, do this:

1. Be sure that the values of the *independent* variables on future observations are appended to the past observations in the data file.
2. Leave the values of the *dependent* variable on these observations blank.
3. Set the data range to include both the past and future observations.
4. Set the dependent variable, then set the independent variables.
5. Exclude the future observations, using the **Set excluded observation(s)** option from the Regression menu.
6. Perform the regression.
7. Calculate Y_{est} and Residual Values.

The values of Y_{est} on the future observations are point forecasts for those observations. (Because the future observations were excluded, values of all the variables on these observations, including Y_{est}, will appear with a strikeout line.) Residuals on the missing observations have values that are completely arbitrary; they may appear as blanks or with some error code, which you can safely ignore.

Performing Another Regression

Reset Current Settings. Once you have finished reviewing the output from your first regression, you may want to run a second regression. Choose **Data/ Regression,*** then **Reset current settings** from the Regression menu. Often you will want to keep the data range, but select different dependent and independent variables. Depending on what you want to change, you can then select the ranges to clear:

- Dependent variable column
- Independent variable columns
- Excluded observations
- Clear all ranges

Once you have cleared the ranges, set your new variables and perform the regression again. Your regression output will again appear below your data set, but *above* your first regression. It will be numbered Regression Number 2. Again, if you want to calculate values of Y_{est} and residuals, or view charts requiring those columns, you will need to select those options again. Old values of Y_{est} and residuals will be overwritten.

Return to Prior Regressions. If you want to return to a prior regression, choose **Data/Regression**, then **Prior regressions** from the Regression menu. You will be prompted "Which one of your prior regressions would you like to retrieve?" Type in the number you want to retrieve. Keep in mind that you will need to choose **Calculate Y_{est} and residual values** again if you want to look at charts for these data.

View Statistics. This choice will allow you to move to the regression output portion of the spreadsheet. Just choose **Data/Regression** and then **View statistics** from the Regression menu.

Print Statistics. This choice allows you to print the regression statistics. Choose **Data/Regression** and then **Print statistics** from the Regression menu.

Doing Regression Analysis

We will now use the regression utility to analyze three data sets. You will learn how to perform a regression, understand regression output, diagnose violations of regression assumptions, transform variables, and make forecasts.

A good precursor to any regression analysis consists of doing the kinds of graphical analysis (scatter diagrams and time-series charts) discussed in Chapter 1. Many of these graphical analyses can be done within the regression utility.

▼ Example 1: Burlington Press

The Burlington Press publishes textbooks, primarily texts for junior high schools (seventh and eighth grade). As part of an analysis the company was carrying out in order to try to understand the market for their texts, the data in Figure 4.6 was collected on total purchases of texts for seventh grades by one city.

* This is shorthand for choose **Data** from the menu bar and **Regression** from the pull-down menu.

(The numbers give total purchases, not just Burlington's share.) Each seventh-grader received a set of texts from the school, used the books for the year, and then returned them. When new editions were brought out or when the schools conducted curriculum reviews, the texts for a particular subject might all be replaced at one time.

Figure 4.6

	A	B	C	D	E
1					
2					
3					
4				Purchased	
5			Year	Texts	Students
6			1967	2,111	2,000
7			1968	2,083	2,027
8			1969	2,264	2,050
9			1970	2,025	2,052
10			1971	2,303	2,061
11			1972	2,149	2,075
12			1973	2,177	2,079
13			1974	2,023	2,089
14			1975	2,178	2,091
15			1976	2,057	2,093
16			1977	2,371	2,131
17			1978	2,368	2,162
18			1979	2,439	2,194
19			1980	2,457	2,250
20			1981	2,764	2,292
21			1982	2,783	2,363
22			1983	2,596	2,412
23			1984	2,500	2,447
24			1985	2,598	2,470
25			1986	2,756	2,488
26			1987	2,457	2,502
27			1988	2,713	2,525
28			1989	2,748	2,567
29			1990	2,773	2,585
30					

How might you use this information to predict purchases of texts for the 1991 school year if the seventh-grade population (which was known quite closely six months before the start of school) was expected to be 2,600?

Preliminary Analysis

Data. Burlington_Press is a worksheet in workbook REGRUTIL.XLS containing values of Year, Number of Purchased Texts, and Number of Students. The data in the file run from 1967 through 1990. A model for predicting purchases of texts might be:

$$Texts = B_0 + B_1 * \text{Year} + B_2 * \text{Students} + \text{error} \qquad \text{(Model R1)} \ .$$

Since we want to forecast 1991 purchases, it is a good idea, before starting any analysis, to append a row to the data file consisting of what you know about 1991: the year and the number of students. Enter 1991 in the cell beneath "1990" (cell C31), and 2600 in cell E31. Leave cell D31, which represents the unknown Number of Purchased Texts for 1991, blank. The observation in row 31 is incomplete; we must remember to exclude it when we run the regression.

Dependent and Independent Variables. Clearly, we are interested in forecasting Number of Purchased Texts; this is, therefore, the *dependent variable*. We might initially believe that the Number of Purchased Texts will increase or decrease with time, with number of students, or both. As a preliminary step, motivated especially to activate the graphics capabilities of the regression utility, let's designate School Year and Number of Students as *independent variables.*

Running the Regression Utility. First, open REGRUTIL.XLS, and click on the Burlington__Press tab, and activate the regression utility. Now, click the label for the first data column **(C6)**. (Remember, we can use only one row of the column label.) To highlight the whole block of data, hold down the Shift and Control keys, then press [↓], then [→].

Click **Data/Regression** and then **Set data range**. Reply **"Yes"** to the next question; row 6 contains variable/column labels. Indicate the column containing the dependent variable by clicking any value in column D (Purchased Texts), then clicking **Data/Regression**, and on the **Set dependent variable column** from the Regression menu. The numbers in column D should now appear in boldface.

To indicate the columns containing the independent variables, click on any value in column C (School Year). Then while holding down [Ctrl] the move to any value in column E (Students) and click on that value. Both cells should be emphasized. Now click on **Data/Regression** and then on the **Set independent variable column(s)** option. The numbers in columns C and E should now appear with a shaded background.

The final step before performing the regression is to indicate the observation whose value is to be excluded. Move the cursor to row 31 (the observation for 1991), click any value in that row, click **Data/Regression**; then click the **Set excluded observation(s)** option. Row 31 now appears with strike-out bars through the numbers.

You have now told the regression utility everything it has to know; once again, click **Data/Regression** and then click **Perform regression**. After a moment you will see the regression output.

Charts. Before trying to understand the regression output, let's look at what graphics capabilities are now available. Click **Data/Regression** and then click **View charts**. A dialogue box appears, giving you options for the five different kinds of charts discussed above:

- Y_{est} vs. Y_{act}
- Y_{est} vs. Residuals
- Any X vs. Y_{act}
- Any X vs. Residuals
- Any X vs. any other X

As the values of Y_{est} and residuals have not yet been computed, only the third and fifth options are available at this stage. If you take the third option, you can, for example, look at a time-series chart of Year vs. Number of Purchased Texts, or a scatter diagram of Number of Students vs. Number of Purchased Texts. If you take the fifth option, you can look at a time-series chart of Year vs. Number of Students. After you examine these charts, you can print any "current chart" by clicking **File** and then clicking the **Print** option.

Computing Y_{est}, Residuals, and a Point Forecast. Click **Data/Regression** again, then click the **Calculate Y_{est}** and **residual values** option. After a short time two new columns (F and G) will be filled in with values of Y_{est} and residuals. Compare the values of Y_{est} with the values of Texts, year by year. They should be close. The difference between Texts purchased and Y_{est} is the residual. If the regression estimated values of the dependent variable perfectly, all values of Y_{est} would match the corresponding values of Y, and the residuals would all be zero.

Scroll down to row 31, the incomplete observation for 1991. You will see a value of 2,814 (displayed with a strikeout line) in column F. Given the regression model implied by the choice of dependent and independent variables (Model R1), this is the point forecast for the number of texts to be purchased in 1991.

Regression Output

Let's turn now to the regression output. You can scroll to it, or click **Data/ Regression** and then click the **View statistics** option. You will see lines of output labeled "Regr. Coef.," "Std. Error," and "t value." Below these are four other numbers, identified as "*# of obs*," "*Deg of F*," "*R-squared*," and "*Resid SD*." The output is shown in Table 4.10.

Table 4.10

Regression Number 1
Dependent Variable: TEXTS

	Year	Constant	Students
Regr. Coef.	8.128	(15,664)	0.8824
Std. Error	16.454	31,289	0.5803
t value	0.5	(0.5)	1.5

# of obs =	***24***	*Deg of F =*	***21***
R-squared =	***0.7638***	*Resid SD =*	***135.6***

Regression Coefficients. The numbers in columns C and E ("Year," and "Students") are associated with the independent variables. The numbers in column D are associated with the "constant term" in the regression equation; these constant-term values will always appear in the dependent-variable column, which simply is a convenient location in which to display them.

To interpret the first three lines of output, we must start with the realization that the regression that we performed assumed that the 24 observations in the data file were generated by a model of the form:

$$Texts = B_0 + B_1 * \text{Year} + B_2 * \text{Students} + \text{error}, \qquad \text{(Model R1)}$$

where the B's are constant but unobservable regression coefficients. From our sample of 24 observations, estimated values of the B's (denoted by lower-case b's) are obtained: b_0 = (15,664), b_1 = 8.128, and b_2 = 0.8824. Given these estimated regression coefficients, you should verify that the values of $Texts_{est}$ can be computed by the formula:

$$Texts_{est} = -15{,}664 + 8.128 * \text{Year} + 0.8824 * \text{Students} \quad .$$

For example, the value[19] of $Texts_{est}$ for 1967 is:

$$Texts_{est} = -15{,}664 + 8.128*1967 + 0.8824*2{,}000 = 2{,}089$$

and the forecast for 1991 is:

$$Texts_{est} = -15{,}664 + 8.128*1991 + 0.8824*2{,}600 = 2{,}813 \quad .$$

Similarly, the values of the residuals can be computed from the formula:

$$\text{Residuals} = Texts - Texts_{est} \quad .$$

For example, the value of the residual for 1967 is:

$$\text{Residual} = 2{,}111 - 2{,}089 = 22 \quad .$$

You can add the 24 residuals in column G, using the =SUM function, and verify that their sum is 0.[20] If, in column H, you compute the values of the squared residuals and add them, you will find that they add to 386,276.3. If the estimated regression coefficients had any other values than the ones displayed, this sum would be higher; in that sense, the coefficients are estimated by **least squares.**

The estimated regression coefficients are derived from a sample and are therefore subject to sampling error. The next two lines of output—the Standard Error and the *t* value—provide information about sampling error similar in nature to the information that was provided in Chapter 2 for evaluating sampling error associated with estimates of a population mean. Thus the estimated value of the regression coefficient for Year, 8.128, is our best estimate of the increase in the Number of Texts Purchased from one year to the next, given that the Number of Students remains constant. That number is very uncertain, however. Using the standard error to construct confidence limits, we can say, for example, that, with 68% confidence, the true value of the regression coefficient lies between –8.326 and 24.582. To some degree, this uncertainty is reflected in the low *t* value of 0.5. Therefore, we cannot be at all confident that the sign of the true regression coefficient is positive: from the data alone, it is far from certain that the number of texts purchased will increase over time unless the number of students increases.

Observations and Degrees of Freedom. Turning to the last four outputs, there were 24 observations used to estimate the two regression coefficients and the constant term; since three estimates were generated from the data, this leaves only 24 – 3 = 21 degrees of freedom.

Residual Standard Deviation. The residual standard deviation (often abbreviated as RSD) of 135.6 is an estimate of the standard deviation of the errors. It is computed as the standard deviation of the residuals in column G, "corrected for degrees of freedom." We have already observed that the sum of squared residuals is 386,276.3, and since the mean of the residuals is zero, this sum is also the sum of squared deviations from the mean. If we divide this sum by the number of degrees of freedom, 21, and take the square root, we get 135.6, the value given in the regression output.

---

[19] These numbers all happen to differ by 1 from the numbers printed for Y_{est} and the residuals by the regression utility. This is because of roundoff error. If you compute the numbers in Excel, pointing to the cells containing the values of the estimated regression coefficients, you will get numbers that agree with the values of Y_{est} and the residuals displayed by the utility.

[20] You may actually see a number like 8.6402E-12, which is "scientific notation" for 0.0000000000086402, which, accounting for roundoff error, is essentially zero.

***R*-squared (R^2).** Finally, *R*-squared (or R^2) is computed by taking the square of the ratio of two standard deviations. The numerator of the ratio is the standard deviation of Y_{est}; the denominator is the standard deviation of Y.

Use the =STDEVP function to compute the standard deviation of Y_{est} in cells F7 through F30; you should get 228.16. Now do the same for the values of Y in cells D7 through D30; the standard deviation is 261.06. Finally, compute $(228.16/261.06)^2 = 0.7638$, the value of R^2 reported in the regression output (Table 4.10). If Y_{est} perfectly predicts Y on all observations, then Y_{est} and Y will have identical standard deviations and $R^2 = 1$. If the regression has absolutely no predictive power, the values of Y_{est} will be the same on every observation: they will have values equal to the mean of Y, and their standard deviation will be 0. Thus in this latter case, $R^2 = 0$.[21]

Alternatively, R^2 can be computed as the square of the coefficient of correlation between Y and Y_{est}. If Y_{est} is a very good predictor of Y, then the two will be highly correlated and R^2 will be near 1. If the independent variables have no predictive power in relation to the dependent variable, then Y_{est} will be a poor predictor of Y, and their correlation will be near 0, as will R^2. The first chart available under the **View charts** option plots a scatter diagram of Y_{est} vs. Y. It provides a graphical view of the goodness of fit of the regression, which is summarized numerically by R^2.

In comparing competing regression models having the same dependent variable, a model with higher R^2 and lower RSD certainly indicates a better fit, and may indicate a better model, although it is certainly possible, by "fishing" through the data, to find independent variables that have no apparent relationship to the dependent variable but by chance are correlated with it in the data. Adding an independent variable, no matter how unrelated to the dependent variable, will never cause R^2 to decrease, but because the RSD is "corrected for degrees of freedom" and one more variable uses up one more degree of freedom, the RSD *may* increase. When the RSD increases when you add a new variable, you usually have evidence of "overfitting."

Except for the problem of overfitting, R^2 is a reasonable measure for comparing competing regression models, but just what constitutes a "good" R^2 depends on judgment. You should remember that the base case against which goodness of fit is measured is just using the mean of the dependent variable as an estimate of its value on all observations. In data with an upward trend, such as many economic time series have, this base-case measure is absurdly naive; any simple model which did no more than recognize the trend would do substantially better. Because it is so easy, when there is a pronounced trend, to do better than the base-case forecast, you should not be surprised to have a high value of R^2, nor should you conclude from the fact that R^2 is high that your model is necessarily a good one. If, on the other hand, you had a model that could predict daily change in stock prices that, without overfitting, had an R^2 of just 0.05, you could become very rich very quickly.

Transformations

We turn now to a discussion of transformations, one of the ways in which you can add flexibility to how you specify a model, and one of the useful tools for converting a model which violates some of the regression assumptions to one that conforms more closely to those assumptions.

---

[21] Statisticians call the square of the standard deviation the variance. Thus R^2 is the ratio of two variances, with that of Y_{est} in the numerator and Y in the denominator. For this reason, R^2 is sometimes defined as the fraction of the variance of Y that is "explained" by the regression.

Returning to the Burlington Press, you might notice, either by scanning column G or by producing a chart showing a time series of the residuals, that the values of the residuals from 1977 through 1983 were all positive, and that this coincided with a period when the number of students was increasing relatively fast. Reflecting on the process by which texts are purchased, you might conclude that the number of sets of books that the school system will own at the end of any school year will be equal to the number of students that year;[22] that one or more books in a set may be replaced at the end of the year because individual books were damaged or lost, or because a new textbook was adopted for the entire class; and that complete sets must be purchased for whatever increase in the number of students occurs between the end of one year and the beginning of the next. We have no information on replacement of individual texts or adoption of new texts, but we now have reason to believe that two of the drivers of text purchases are:

- students last year
- additional students this year

These variables are not in the original data file, but they can be derived from variables that *are* in the original file. In regression parlance, such derived variables are called **transformations.** In a time series, a transformation that creates a variable whose values are those of some other variable in a prior period is called a **lag** transformation. One that creates a new variable that represents the change in the value of some other variable over some period of time is called a **difference** transformation. Thus the variable Students Last Year is derived by **lagging** Students This Year by one year; the variable Additional Students is derived as the difference between Students This Year and Students Last Year.

To create these variables in Excel, first start with a clean slate by clicking on **Data/Regression**, then on the option **Reset current settings**, and then on **Clear all ranges** when the dialog box appears. We will create the lagged variable Students Last Year in column F, and the difference variable New Students in column G. Enter column labels of "Last Year" and "New Students" in cells F6 and G6, respectively. Next, click on cell **F8** and set the value in that cell equal to the number of students in the preceding year; i.e., enter the formula = E7. The cell should now contain the value 2,000, the number of students in 1967. Now click on cell **G8** and set the value in that cell equal to the change in the number of students between 1967 and 1968, i.e., enter the formula = E8 – F8. The cell should contain the value 27.

Now copy cells F8 and G8 down as far as the data go (from row 9 to row 31).

Regression Using Transformed Variables. We are now ready to run a new regression. The only trick is that in addition to excluding row 31 (the forecast), you must now exclude row 7 (the data for 1967, which is incomplete because we have no values for the newly created variables.) Before invoking the **Set excluded observation(s)** option, click on any value in row 7, scroll down to row 31 and, while holding down [Ctrl], click on any value in that row. To run the regression, you need to:

1. **Set Data range** (C6:G31)
2. **Set Dependent variable column** (column D)
3. **Set Independent variable column**(s) (columns F and G)
4. **Set Excluded observation(s)** (rows 7 and 31)
5. **Perform regression**

---

22 There might be more sets than students in a year where enrollment declined, but declining enrollment has not occurred in the data in hand.

The model we are estimating can be formally stated as:

$$Texts = B_0 + B_1{*}\text{Students Last Year} + B_2{*}\text{New Students} + \text{error} \quad \text{(Model R2)}$$

The regression output is shown in Table 4.11. What does it imply? First, comparing it with the output of Regression 1, shown in Table 4.10,

- the value of R^2 is higher (0.8479 vs. 0.7638),
- the RSD is lower (108.4 vs. 135.6), and
- the t values associated with the independent variables are much higher.

Table 4.11

Regression Number 2
Dependent Variable: TEXTS

	Constant	Last Year	New Students
Regr. Coef.	51.80	0.9905	5.910
Std. Error	274.75	0.1259	1.311
t value	0.2	7.9	4.5

# of obs =	***23***	*Deg of F =*	***20***
R-squared =	***0.8479***	*Resid SD =*	***108.4***

On the other hand, because we have incomplete data for 1967, there is one less observation, and hence one less degree of freedom. Also, the t value for the constant term suggests that the true value of the constant could easily be zero or negative, even though its sample value is positive. Nevertheless, the regression appears to fit the data much better than Regression 1, and it tells a simple story: on average, about one text is added to each set left by students last year, and new students acquire roughly six new texts; these two factors account for much of the variability in the number of texts purchased.

Can We Do Better? Before making a forecast, we might ask whether we should have included Year as one of the independent variables in Regression 2. We can easily add a variable by clicking on the columns for Year, Students Last Year, and New Students, invoking the **Set independent variable column(s)** option, and performing the regression. The model we are now estimating is:

$$Texts = B_0 + B_1{*}\text{Year} + B_2{*}\text{Students Last Year} + B_3{*}\text{New Students} + \text{error.} \quad \text{(Model R3)}$$

The results of Regression 3 are shown in Table 4.12. Compared with Table 4.11 (Regression 2):

- R^2 has increased very slightly (from 0.8479 to 0.8496);
- RSD has increased (from 108.4 to 110.6);
- because one more independent variable was included, the degrees of freedom decreased (from 20 to 19);
- the t values associated with the independent variables are much lower.

We can conclude that including Year as an independent variable has resulted in overfitting.

Table 4.12

Regression Number 3
Dependent Variable: TEXTS

	Year	Constant	Last Year	New Students
Regr. Coef.	6.760	(12,818)	0.7664	5.654
Std. Error	14.513	27,633	0.4980	1.446
t value	0.5	(0.5)	1.5	3.9

# of obs =	***23***	*Deg of F =*	***19***
R-squared =	***0.8496***	*Resid SD =*	***110.6***

The Forecast. Return to Regression 2 by invoking the **Prior regressions** option, and compute Y_{est} and the residuals, getting a point forecast of 2,701 for 1991; the confidence distribution for the forecast thus has mean 2,701 and standard deviation equal to the RSD, or 108.4. Assuming the residuals are normally distributed with mean given by Y_{est} and standard deviation equal to the RSD, we can state with 95% confidence that texts purchased in 1991 will be approximately between 2,701 – 2*108.4 and 2,701 + 2*108.4, (between 2,484 and 2,918). In practice, however, the interval should be made somewhat wider to account for uncertainties that we have not included in these calculations.[23]

▼ Dummy Variables

Dummy variables have just two possible values, 0 and 1 (see Chapter 1). They can be used to code any two-valued variable: men vs. women, Republicans vs. Democrats, buyers vs. nonbuyers, etc. Dummy variables are often used as independent variables in a regression analysis.

Example 2: HBS Students' Height, Weight and Gender

HTWT.XLS contains data on the heights (in inches) and weights (in pounds) of 597 men and 171 women (as reported by them) in a recent MBA class at Harvard Business School. In addition to the variables Height and Weight, the variable Gender, having value 1 if the person is a woman, 0 if a man, is included in the file.

Suppose we want to predict weight from height using the model:

$$WT = B_0 + B_1 * \text{HT} + \text{error} \quad . \qquad \text{(Model R4)}$$

Table 4.13 shows the results of a regression with weight as the dependent and height as the independent variable. Based on the regression output, we would forecast the weight of a person 70 inches tall as:

$$WT_{est} = -228.9 + 5.510 * 70 = 156.8 \quad .$$

[23] Sources of additional uncertainty come from: (1) using estimated (as opposed to true) B's, (2) inferring the residual standard deviation from a sample of residuals instead of from the true process that generated the residuals, (3) possible uncertainty about the values of the x's used in a forecast (are we sure that the number of students next year will be 2,600?), and (4) the possibility that the model is not correctly specified.

Table 4.13

Regression Number 1
Dependent Variable: WT

	HT	Constant
Regr. Coef.	5.510	(228.9)
Std. Error	0.178	12.4
t value	30.9	(18.5)

# of obs =	***768***	*Deg of F =*	***766***
R-squared =	***0.5554***	*Resid SD =*	***16.99***

Table 4.14 shows similar regression output when both height and gender (M/F) are included as independent variables using the model:

$$WT = B_0 + B_1{*}HT + B_2{*}\text{M/F} + \text{error} \quad . \qquad \text{(Model R5)}$$

Table 4.14

Regression Number 2
Dependent Variable: WT

	HT	Constant	M/F
Regr. Coef.	4.198	(133.7)	(17.86)
Std. Error	0.210	14.9	1.74
t value	20.0	(9.0)	(10.3)

# of obs =	***768***	*Deg of F =*	***765***
R-squared =	***0.6092***	*Resid SD =*	***15.94***

According to the output in Table 4.14, based on model R5, we would forecast the weight of a person 70 inches tall as:

$$WT_{est} = -133.7 + 4.198{*}70 - 17.86{*}0 = 160.2$$

pounds if the person were male, and as:

$$WT_{est} = -133.7 + 4.198{*}70 - 17.86{*}1 = 142.3$$

pounds if the person were female. Thus, women are forecast to be 17.9 pounds lighter than men of the same height, –17.9 being the regression coefficient associated with the dummy variable indicating gender. A comparison of Tables 4.13 and 4.14 clearly indicates that the regression that includes the dummy variable fits the data better and provides a more plausible model.[24]

We noted in Chapter 1 that one of the observations in the data file might be an outlier. If you compute the values of Y_{est} and the residuals associated with Regression 2, and then plot $\boldsymbol{WT_{est}}$ against the residuals using the **View Charts** option, (see Figure 4.7) you will notice one very negative residual, and if you now scroll through the data file, you will find a residual of –109 on line 577, associated with an observation for which a male student 72 inches tall claimed to weigh 60 pounds. This was obviously misreported or misrecorded, and should clearly be eliminated, using the **Set excluded observation(s)** option. After doing so, and rerunning the regression, we get the output shown in Table 4.15 (based on the Regression 5 model) that is a decided improvement in fit.

---

[24] Of course, both models have negative constant terms, implying that a sufficiently short person will have negative weight! This is merely an instance showing the dangers of extrapolating too far beyond the range of the data. If we had data that included infants, young children, and adults, we would almost surely find that the relationship between height and weight was curvilinear, and that the variability in weights was considerably higher at the tall end of the spectrum. Transformations of the variables would be required.

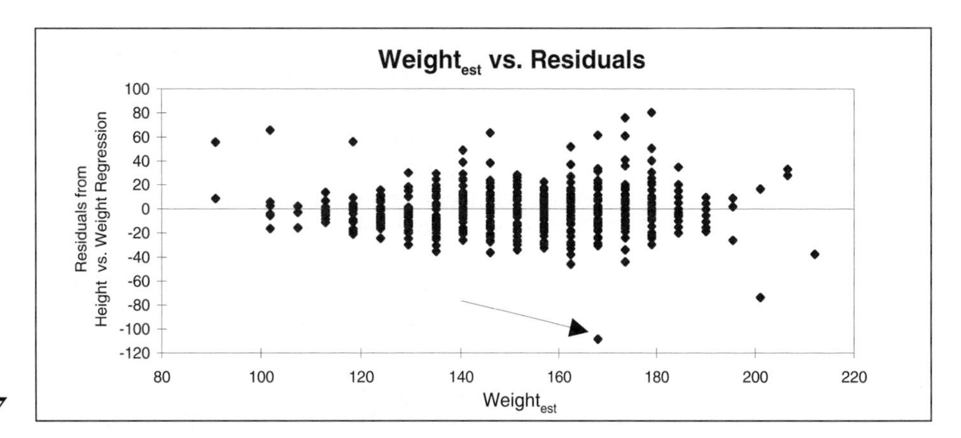

Figure 4.7

Table 4.15

Regression Number 3
Dependent Variable: WT

	HT	Constant	M/F
Regr. Coef.	4.224	(135.3)	(17.92)
Std. Error	0.204	14.4	1.69
t value	20.7	(9.4)	(10.6)

# of obs =	***767***	*Deg of F =*	***764***
R-squared =	***0.6263***	*Resid SD =*	***15.46***

▼ Categorical Independent Variables

Often we have independent variables that are **categorical** (see Chapter 1). Individuals' political-party preference might be recorded as Republican, Democratic, or Independent, with codes of 1, 2, and 3 assigned arbitrarily to the three categories; companies may be classified by SIC codes; items sold by a retailer may be characterized by color, coded in some systematic but arbitrary manner. Using these variables directly in a regression clearly makes no sense.

Standard practice is to convert each possible value of a categorical variable into a dummy variable. Thus, in the political-party preference case, we would create a dummy variable that had the value 1 if the person was a Republican, and 0 if not; another dummy variable having the value 1 if a Democrat, 0 if not; and a third dummy variable similarly representing an Independent. In a regression, party preference would be taken into account by using two of these three dummy variables, omitting the third. Any two will do. The reason for using one less than the number of possible values of the categorical variable, and interpretations of the output, will be made clearer by introducing an example.

Example 3: Brighton Catering Company

The president of the Brighton Catering Company was attempting to analyze data on labor costs of meals that the firm had prepared and served. Brighton's standard dinner included an appetizer, a main course, and a dessert. Customers could choose any one of three appetizers (fruit cocktail, shrimp cocktail, or melon with prosciutto). The firm also offered other options.

Customers might add a salad course, and they might include after-dinner drinks. The president wanted to adjust the firm's prices and, as a first step, wanted to understand the labor costs for different combinations of options. (Food costs would be considered separately.) Accordingly, careful cost studies had been conducted to determine the per guest labor costs of 35 different meals (with 100 to 150 guests each).

How could regression analysis be used to help predict the labor costs of a meal with shrimp cocktails, salad, and liqueurs? The data are in worksheet Brighton_Catering in workbook REGRUTIL.XLS. "Salad" and "Liqueurs" are each coded 1 if the item in question was served, 0 otherwise. "Appetizer" is coded 1 for Shrimp, 2 for Fruit Salad, and 3 for Melon and Prosciutto. We need to create a set of new variables to indicate whether a specific appetizer was served; we need to transform our categorical variable into three dummy variables, one for each category. In column G code 1 if Fruit was served, 0 otherwise (that is, if either Shrimp or Melon was served), and code the entries in columns H and I similarly for Shrimp and Melon.[25]

At the bottom of our data we entered values of the independent variables for three meals whose costs are to be forecast. In all three cases, a salad will be served, but not liqueurs, but the three cases differ with respect to appetizer.

In Table 4.16 we show the results of three regressions. In all three, Cost/Person is the dependent variable, and the three observations representing forecasts to be made are excluded. In the first regression, Salad, Liqueurs, and the first two of the Appetizer dummy variables—Fruit and Shrimp— are specified as independent variables. In the second regression, everything is the same except the last two of the Appetizer dummy variables—Shrimp and Melon—are specified, while in the third regression all three Appetizer dummies are used. Formally, the models can be stated:[26]

Regression 1:

$$\text{COST} = B_0 + B_1\text{*SALAD} + B_2\text{*LIQUEURS} + B_3\text{*FRUIT} + B_4\text{*SHRIMP} + \text{error} \quad , \qquad \text{(Model R7)}$$

Regression 2:

$$\text{COST} = B_0 + B_1\text{*SALAD} + B_2\text{*LIQUEURS} + B_4\text{*SHRIMP} + B_5\text{*MELON} + \text{error} \quad , \qquad \text{(Model R8)}$$

Regression 3:

$$\text{COST} = B_0 + B_1\text{*SALAD} + B_2\text{*LIQUEURS} + B_3\text{*FRUIT} + B_4\text{*SHRIMP} + B_5\text{*MELON} + \text{error} \quad . \qquad \text{(Model R9)}$$

[25] The Excel function that performs this coding is the =IF function. The entries in row 6 of columns G, H, and I are =IF (D6 = 1,1,0), =IF(D6 = 2,1,0), and =IF (D6 = 3,1,0) respectively, and these entries can be copied for all observations (including the forecasts).

[26] Although most of the regression coefficients in the three models are represented by the same symbol, the actual values of B_0 will differ in models R7, R8, and R9, as will those of B_1, etc.

Table 4.16

Regression Number 3
Dependent Variable: COST/PERSON

	Constant	SALAD	LIQUEURS	FRUIT	SHRIMP	MELON	Y(est)
Regr. Coef.	471.6	18.15	11.10	(374.7)	(354.6)	(368.0)	115
Std. Error	0.0	2.01	2.00	0.0	0.0	0.0	135
t value	#DIV/0!	9.0	5.6	#DIV/0!	#DIV/0!	#DIV/0!	122

# of obs =	***35***	*Deg of F =*	***29***
R-squared =	***0.8728***	*Resid SD =*	***5.796***

Regression Number 2
Dependent Variable: COST/PERSON

	Constant	SALAD	LIQUEURS	SHRIMP	MELON	Y(est)
Regr. Coef.	96.94	18.15	11.10	20.08	6.665	115
Std. Error	2.08	1.98	1.97	2.38	2.336	135
t value	46.7	9.2	5.7	8.4	2.9	122

# of obs =	***35***	*Deg of F =*	***30***
R-squared =	***0.8728***	*Resid SD =*	***5.699***

Regression Number 1
Dependent Variable: COST/PERSON

	Constant	SALAD	LIQUEURS	FRUIT	SHRIMP	Y(est)
Regr. Coef.	103.6	18.15	11.10	(6.665)	13.42	115
Std. Error	2.1	1.98	1.97	2.336	2.51	135
t value	49.1	9.2	5.7	(2.9)	5.4	122

# of obs =	***35***	*Deg of F =*	***30***
R-squared =	***0.8728***	*Resid SD =*	***5.699***

After each of the regressions was performed, Y_{est} and Residuals were calculated, and the values of Y_{est} were copied to the right of their respective regression outputs. In all three regressions of Table 4.16, the forecasts are identical. Why?

The interpretation of the regression coefficients of a system of dummy variables representing all but one of the possible values of a categorical variable is that the omitted variable represents a "base case" relative to which the effects of the other variables are measured. Thus, in Regression 1 (based on model R7), the base case is Melon; the regression coefficient for Fruit indicates that the estimated cost per person will be 6.665 (cents) less if the appetizer was Fruit instead of Melon, and similarly the estimated cost will be 13.42 more for Shrimp. It follows that the estimated cost of Shrimp will be 13.42 – (6.66) = 20.08 more than for Fruit.

Turning to Regression 2 (based on model R8) in Table 4.16 where Fruit is the base case, we find that the estimated cost of Shrimp is 20.08 more than Fruit, and Melon is 6.665 more; these results are completely consistent with those of Regression 1. Notice that the constants in Regressions 1 and 2 are 103.6 and 96.94, respectively, a difference of 6.66. How do you explain this difference? Can you now explain why both regressions give you the same forecast?

In Regression 3 (based on model R9) all three dummy variables representing appetizers were used. As you can see, the regression coefficients for the dummy variables representing appetizers are all large negative numbers, the constant term is a large positive number, but Shrimp is still 20.1 more costly than Fruit, and Melon is still 6.7 cheaper. Relative to Regression 2, where Fruit represented the base case, the constant is 374.7 higher. Again, the forecasts are identical with those of the preceding regressions.

Notice that in Regression 3 the standard errors of the regression coefficients are reported as 0.0 and the t values indicate an attempt to divide by 0. This should signal that some sort of error has taken place, and indeed one has. Although the computer has come up with a set of regression coefficients that provide forecasts identical with those of Regressions 1 and 2, the regression coefficients are not unique. For instance, if we added 10 to the regression coefficients for each of the appetizer dummy variables and subtracted 10 from the constant term, we would get identical estimates and forecasts. (Try it!) Indeed, if we added *any* constant amount to each of those regression coefficients and subtracted that amount from the constant term, the forecasts and estimates would be the same. In this formulation we have absolutely no idea what the "correct" values of the estimated regression coefficients should be. They are completely unstable, and their standard errors *should* be reported as "infinite."[27]

Regression 3 shows the results of a *misspecified* regression model. The problem is that one of the independent variables in the regression can be expressed as a *linear function* of one or more (in this case two) other independent variables in the regression. For instance,

$$\text{Fruit} = 1 - \text{Shrimp} - \text{Melon} \quad :$$

if the appetizer was a shrimp cocktail, then Shrimp=1, Melon=0, and the value of Fruit is necessarily 0, for example.

This is an extreme form of a problem that we shall encounter in less severe form in other regression analyses. It is called **collinearity** or **multicollinearity.** In its extreme form, it occurs when one independent variable is a linear function of one or more other independent variables in the regression. In that case, the regression coefficients are unstable, and the remedy is to drop one of the collinear independent variables from the regression. For that reason, when we create a system of dummy variables representing the various levels of a categorical variable, we always use *one less* dummy variable than the number of categories.

▼ Ordinal Independent Variables

Suppose we have as an independent variable in a regression the level of agreement that various respondents have with some statement, classified into five levels: Disagree Strongly, Disagree, Neutral, Agree, and Agree Strongly, with the possible responses coded 1 through 5. Should we use the variable, coded in this way, directly in the regression, or should we transform it? If you use it directly, you are implicitly assuming that the effect on Y_{est} of going from "Strongly Disagree:" to "Disagree" is the same as going from "Disagree" to "Neutral," etc. Can we tell if such an assumption is justified?

---

27 Many regression programs will simply refuse to give any regression output in situations like this, or will provide output that obviously makes no sense. In our case, the regression utility reports the standard error as 0.0 instead of "infinity."

One way to find out is to create a system of five dummy variables, one for each possible response, and to use four of them as independent variables. Suppose "Disagree Strongly" were the base case, and the estimated regression coefficients for the other responses were 3.5, 6.4, 9.3, and 13.2. That suggests that Y_{est} increases by (roughly) 3.3 units each time you move up one notch on the agreement scale. In that case, it would be reasonable for you to use the original ordinal variable, with values coded as 1 through 5, as an independent variable, thereby saving three degrees of freedom and simplifying the description of how the variables in the model are related. If, on the other hand, the estimated regression coefficients did not move up in roughly equal increments, you would undoubtedly fit the data better by using the four dummy variables, provided you had enough degrees of freedom.

Exercises on Forecasting with Regression

1. Use data file HOUSES.XLS to forecast the selling price of a 1,654-square-foot 34-year-old house. How sure are you about your forecast?
2. Produce scatter diagrams of Selling Price vs. Area, Selling Price vs. Age, and Area vs. Age. Perform a regression with Selling Price as the dependent and Age as the independent variable. Why is the regression coefficient on Age so small, as compared with its magnitude when Area is also included as an independent variable?
3. Data file ANSCOMBE.XLS contains data on a number of variables. Perform the following regressions:

	Independent Variable	Dependent Variable
a)	x_1	y_1
b)	x_1	y_2
c)	x_1	y_3
d)	x_2	y_4

In each case, plot a scatter diagram of the independent against dependent variable. What, if anything, could you do to obtain better regression results in each of the four regressions?

4. Using data file HBSMBA.XLS, develop a regression model that will predict students' first-year grade-point averages. Be prepared to discuss whether you would use such a model to screen students applying for admission.

CHEMPLAN CORPORATION: PAINT-RITE DIVISION

Daniel Williams, Paint-Rite's Director of Marketing, reviewed the data his staff had accumulated for him over the past several days. He was scheduled to present his division's paint sales forecast to corporate management for the following week, and he wanted to be certain that everything was in order. Dan was following a somewhat different route in developing a forecast for 1965's paint demand than had his predecessor, Dave Lebrun. Because his forecast concept represented something of a departure from the division's previous methodology, Dan was especially concerned to make clear how he had reached his forecast, and to have ready answers to any objections that might arise.

The Company

Paint-Rite was a division of the Chemplan Corporation, a substantial manufacturer of commodity and specialty chemicals. Paint-Rite produced both interior and exterior house paints, and, with $828 million in 1964 sales, was the largest of Chemplan's three divisions. The division had its own sales staff, which sold exclusively to distributors: they in turn sold paint to major construction firms as well as to hardware chains and independent retailers. Paint-Rite products thus ended up both on newly constructed houses and on older homes being refurbished.

Dave Lebrun, who had been Paint-Rite's Director of Marketing for several years before his recent retirement, had based his annual paint-demand forecast on the sales staff's assessment of the current business environment. Each sales manager polled his or her salespeople periodically, to get their views on likely requirements in different geographic regions. The sales managers passed on their information to Dave in the course of informal meetings on general marketing issues.

While Dave Lebrun's forecasting record had been satisfactory, Daniel Williams wanted to try out some new approaches. He hoped to be more analytical in his methodology, and planned to look carefully at the data representing Paint-Rite's sales history, as well as at other potentially relevant data series. Chemplan had just subscribed to the services of United States Economics, Incorporated (USEI), a recently founded firm which ran a large macroeconomic model of the United States, and which provided data series and forecasts for significant economic variables. Dan thought he might be able to use USEI's services to help him in his forecasting.

Harvard Business School case 9-191-090. This case was prepared by Professor Paul A. Vatter.

As a first step in developing the 1965 demand forecast, Dan looked at the division's sales history for the past 17 years, all the sales data that was readily available to him. He next thought about the final users of Paint-Rite products, and how he might approximate the trends in new housing and in repainting and remodeling. He conferred with a representative of USEI, who felt that new housing could be represented best by information about construction starts in the United States. (Paint-Rite's foreign sales were negligible.) To reflect the paint market for older homes, the USEI consultant suggested using data on the total home-improvement loans granted over the 17 years in question.

USEI was able to provide the time series which, in addition to Paint-Rite's sales, appear in Table A below. The firm also provided 1965 forecasts of 50.0 for the construction index, and $8.5 billion for home loans. Although only a few months of 1965 had passed, USEI's consultant expressed considerable confidence in the projections. Dan had plotted Paint-Rite's sales and each data series as a function of time (see Exhibits 1–3). Armed with this information, he had produced a forecast with which he felt comfortable. Now he had to explain it to Chemplan's management.

Table A

Year	Paint Division's Sales (Millions of Dollars)	Home Improvement Loans Granted (Billions of Dollars)	Index of Building Construction Started in U.S.
1948	$280.0	$3.909	9.43
1949	281.5	5.119	10.36
1950	337.1	6.666	14.50
1951	404.1	5.338	15.75
1952	402.1	4.321	16.78
1953	452.0	6.117	17.44
1954	431.0	5.559	19.77
1955	582.0	7.920	23.76
1956	596.6	5.816	31.61
1957	620.7	6.113	32.17
1958	513.6	4.258	35.09
1959	606.9	5.591	36.42
1960	628.0	6.675	36.58
1961	602.7	5.543	37.14
1962	656.7	6.933	41.30
1963	778.5	7.638	45.62
1964	827.6	7.752	47.38

Exhibit 1

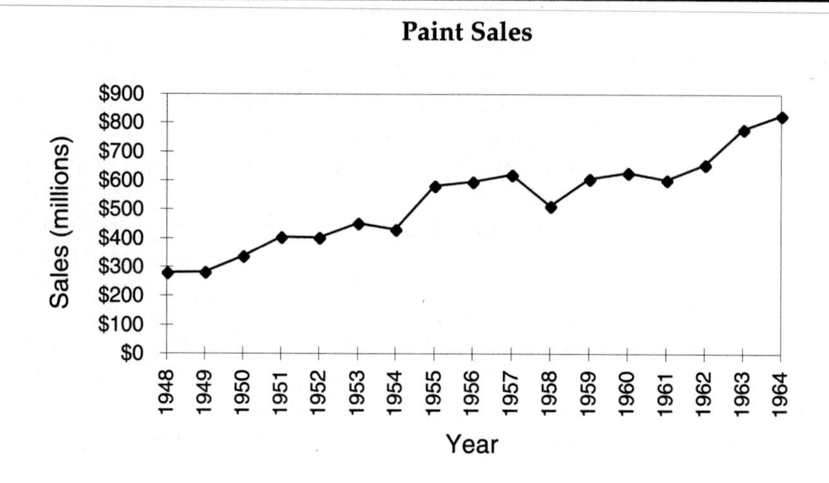

Exhibit 2

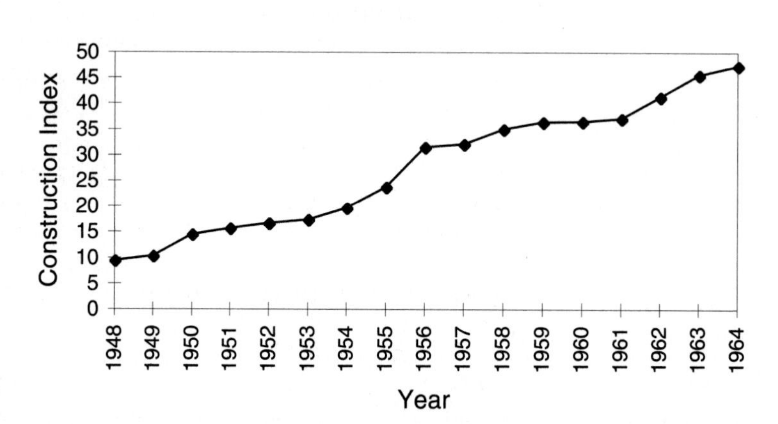

Exhibit 3

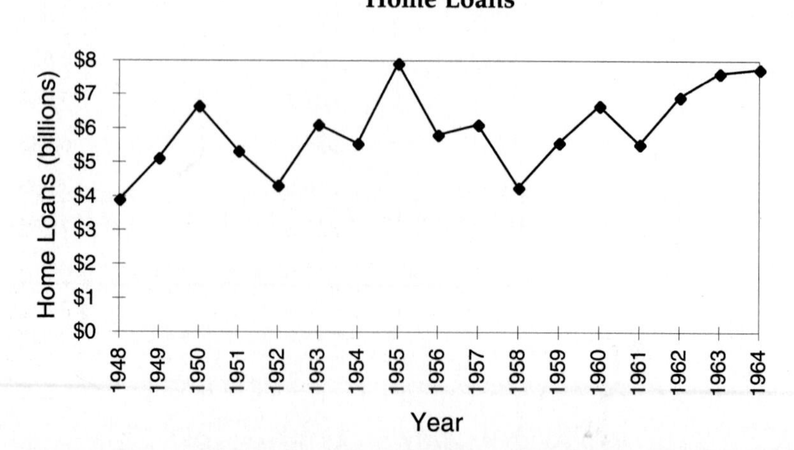

HARMON FOODS, INC.

John MacIntyre, general sales manager of the Breakfast Foods Division of Harmon Foods, Inc., was having difficulty in forecasting the sales of Treat. Treat was a ready-to-eat breakfast cereal with an important share of the market. It was the main product in those company plants that manufactured it. MacIntyre was responsible for sales forecasts from which production schedules were prepared. In past months, actual Treat sales had varied from 50% to 200% of his forecast. The greatest difficulty in preparing forecasts arose from the wide variability in historical sales. (See Exhibit 1. Sales were debited on the day of shipment; therefore, Exhibit 1 represents unit shipments as well as sales. Consumer Packs and Dealer Allowances, which are discussed later, are also shown in Exhibit 1.)

Manufacturing Problems

Accurate production forecasts were essential for the health of the entire business. The plant managers received these forecast schedules and certified their ability to meet them. A plant manager's acceptance of a schedule represented a promise to deliver: crews and machines were assigned, materials ordered, and storage space allocated to meet the schedule.

Schedule changes were expensive. On the one hand, the lead time on raw material orders was several weeks, so that ordering too little not only caused expensive shortages in lost production time but also disappointed customers. Reducing schedules, on the other hand, created a surplus of raw material. Lack of storage space required materials to be left on the trucks, railroad cars, or barges that brought them. Retaining these vehicles resulted in expensive demurrage charges.[28]

Overshadowing the storage problem was the problem of efficient use of the work force. Tight production schedules prevented unnecessary costs. Overtime was avoided because it was expensive and interfered with weekend maintenance. The labor force was highly skilled and difficult to increase in the short run. Layoffs, however, were avoided to preserve the crew's skills. This job security was an important part of the company's labor policy, and it created high employee morale. Thus, the production manager attempted to make production schedules efficient for a constant-size work force while using as little overtime as possible.

Harvard Business School case 9-171-248. This case was prepared by William Whiston.

[28] Demurrage charges are assessments made by a carrier against a consignee for delays in the unloading (or the initiation of unloading) of a transport vehicle. Usually one free hour is allowed after the normal unloading time for trucks. Railcars and barges have typical allowances of three days and one day, respectively, including unloading time. Charges for delays beyond these allowances range from $20 an hour for a truck and $32 a day for a railcar to $4,000 a day for a barge.

Advertising Expenditures

Inaccurate sales forecasts also reduced the effectiveness of Treat's advertising expenditures. Most of Treat's advertising dollars were spent on Saturday morning network shows for children. This time was purchased up to a year or more in advance and cost $80,000 per one-minute commercial. The brand managers in the Breakfast Foods Division, however, believed that these network programs delivered the best value for each advertising dollar spent. This opinion was based upon cost per million messages delivered, viewer-recall scores, and measures of audience composition.

Like many other companies, Harmon Foods budgeted advertising expenditures at a fixed amount per unit sold. Each year the monthly budgets for advertising were established, based on forecast sales. Brand managers tended to contract for time on network programs to the limit of their budget allowance. When shipments ran high, however, brand managers tended to increase advertising expenditures in proportion to actual sales. In such circumstances, they would seek contracts for time from other brand managers who were shipping below budget. Failing this, they would seek network time through the agencies, or if such time were unavailable, they would seek spot advertising as close to prime program time as possible. Thus, unplanned advertising expenditures could result in time that gave lower value per advertising dollar spent than did prime time.

Budgets and Controls

The controller of the Breakfast Foods Division also complained about forecasting errors. Each brand manager prepared a budget based on forecast shipments. This budget promised a contribution to division overhead and profits. Long-term dividend policy and corporate expansion plans were partly based on these forecasts. Regular quarterly increases in earnings over prior years had resulted in a high price-earnings ratio for the company. Because the owners were keenly interested in the market value of the common stock, profit planning played an important part in the management control system.

Discretionary overspending on advertising, noted earlier, amplified the problems of profit planning. These expenditures did not have budgetary approval, and until a new budget base (sales forecast) for the fiscal year was approved at all levels, such overspending was merely borrowing ahead on the current fiscal year. The controller's office charged only the budgeted advertising to sales in each quarter and carried the excess over, because it was unauthorized. This procedure resulted in spurious accounting profits in those quarters where sales exceeded forecast, with counterbalancing profit reductions in subsequent quarters.

The significant effect of deferred advertising expenditures on profits had been demonstrated in the past fiscal year. Treat and several other brands had overspent extensively in the early quarters; as a result, divisional earnings for the fourth quarter were more than $4 million below corporate expectations. The division manager, her sales manager, brand managers, and controller had felt very uncomfortable in the meetings that were held because of this shortage of reported profits. The extra profits recorded in earlier quarters had offset the shortages of other divisions, but in the final quarter, no division was able to offset the Breakfast Foods shortage.

The Brand Manager

Donald Carswell, the brand manager for Treat, prepared his brand's budget base—a set of monthly, quarterly, and annual forecasts that governed monthly advertising and promotional expenditures. These forecasts, along with forecasts from the division's other brand managers, were submitted to MacIntyre for approval. This approval was necessary because, in a given month, the salesforce could only support the promotions of a limited number of brands. Once approved, the brand managers' forecasts were the basis for MacIntyre's official forecasts.

The production schedule, in turn, was based upon MacIntyre's official forecast. This required mutual confidence and understanding. MacIntyre provided information on Harmon's and also competitors' activities and pricings at the stores. The brand managers furnished knowledge of market trends for their brands and their brands' competitors. The brand managers also kept records of all available market research reports on their brands and similar brands and were aware of any package design and product formulations under development.

As Treat's brand manager, Carswell knew that it was his responsibility to improve the reliability of sales forecasts for Treat. After talking to analysts in the Market Research, Systems Analysis, and Operations Research departments, he concluded that better forecasts were possible. Robert Haas of the Operations Research Department offered to work with him on the project. MacIntyre and the controller enthusiastically supported Carswell's undertaking. Although such projects were outside the normal scope of a brand manager's duties, Carswell recognized this as an opportunity to find a solution to his forecasting problem that would have company-wide application.

Factors Affecting Sales

Carswell and Haas delved into the factors that influenced sales. A 12-month moving average of the data in Exhibit 1 indicated a long-term rising trend in sales. This trend confirmed the A.C. Nielsen store audit, which reported a small but steady rise in market share for Treat and a steady rise for the commodity group to which Treat belonged.

Besides trend, Carswell felt that seasonal factors might be important. In November and December, sales slowed down as inventory levels among stores and jobbers were drawn down for year-end inventories. Summer sales were often low because of plant shutdowns and sales personnel vacations. There were fewer selling days in February. Salespeople often began the fiscal year with a burst of energy, jockeying for a strong quota position for the rest of the year. Carswell obtained data made available by the National Association of Cereal Manufacturers, showing seasonal effects on shipments of breakfast cereals in the United States. These indexes appear in Exhibit 2.

Nonmedia promotions, which represented about 25% of Treat's advertising budget, strongly influenced sales. The two main types of promotions were consumer packs and dealer allowances. Promotions targeted directly at the consumer were called consumer packs, so named because the consumer was reimbursed in some way for each package of Treat that was purchased. Promotions that sought to increase sales by encouraging the dealer to push the brand were called dealer allowances, so called because allowances were made to dealers to compensate them for expenditures incurred in promoting Treat.

Consumer packs and dealer allowances were each offered two or three times a year during different canvass periods. (A sales canvass period is the time required for a salesperson to make a complete round to all customers in the assigned area. Harmon Foods scheduled ten five-week canvass periods each year. The remaining two weeks, one at mid-summer and one at year-end, were for holidays and vacations.)

Consumer Packs

Consumer packs were usually a twenty-cent-per-package reduction in the price the consumer paid. The promotion could also be made as a coupon, an enclosed premium, or a mail-in offer. Based on the results of consumer-panel tests of all such promotions, however, Carswell was confident that these forms were roughly equivalent to the twenty-cent price reduction in its return to the brand. Consequently, he decided not to make a distinction among the different kinds of consumer packs. (Exhibit 1 shows the history of consumer pack shipments.)

Consumer packs, supporting advertising material, and special cartons were produced before the assigned canvass period for shipment throughout the five-week period. Any packs not shipped within this period would be allocated among the salespeople for shipment in periods in which no consumer promotion was officially scheduled. From a study of historical data that covered several consumer packs, Haas found that approximately 35% of a consumer-pack offering moved out during the first week, 25% during the second week, 15% during the third week, and approximately 10% during each of the fourth and fifth weeks of the canvass period. Approximately 5% was shipped after the promotional period. Because they saw no reason for this historical pattern to change, Haas and Carswell were confident that they could predict with reasonable accuracy future monthly consumer pack shipments.

Total shipments were favorably affected, of course, during the month in which the consumer packs were shipped. Because the consumer ate Treat at a more or less constant rate over time, Carswell was convinced that part of the increase in total shipments resulted from inventory build-ups by jobbers, stores, and consumers. Thus, he thought that the consumer packs might have a negative influence on total shipments in subsequent months as these excess inventories were depleted in the first, or possibly the second, month after the packs were shipped.

Dealer Allowances

Sales seemed even more sensitive to allowances offered to dealers for cooperative promotional efforts. Participating dealers received a $4 to $8 per case discount on their purchases during the allowance's canvass period.

The total expenditure for dealer allowances during a given promotional canvass period was budgeted in advance. As with consumer packs, any unspent allowances would be allocated to the salespeople for disbursement after the promotional period. The actual weekly expenditures resulting from these allowances followed approximately the same pattern as the one for the shipment of consumer packs. Consequently, Carswell believed that the monthly expenditures resulting from any given schedule of future dealer allowances could also be predicted with reasonable accuracy.

Dealers promoted Treat by using giant, spectacular end-of-aisle displays, newspaper ads, coupons, fliers, and so forth. Such efforts could affect sales dramatically. For example, an end-of-aisle display located near a cash register could do an average of five weeks' business in a single weekend. As with consumer packs, however, Carswell believed that much of the sales increase was attributable to inventory build-ups, and therefore he expected reactions to these build-ups as late as two months after the initial sales increase.

Actual expenditures made for dealer allowances from 1983 to 1987 appear in Exhibit 1.

Conclusion

Carswell and Haas felt that they had identified, to the best of their abilities, the important factors affecting sales. They knew that competitive advertising and price moves were important but unpredictable, and they wished to restrict their model to variables that could be measured or predicted in advance.

Haas agreed to formulate the model, construct the data matrix, and write an explanation of how the model's solution could be used to evaluate promotional strategies, as well as to forecast sales and shipments. Carswell and Haas would then plan a presentation to divisional managers.

Exhibit 1

MONTH	CASE SHIPMENTS*	CONSUMER PACKS (CASES)*	DEALER ALLOWANCE
Jan-83	#N/A	0	$396,776
Feb-83	#N/A	0	$152,296
Mar-83	#N/A	0	$157,640
Apr-83	#N/A	0	$246,064
May-83	#N/A	15,012	$335,716
Jun-83	#N/A	62,337	$326,312
Jul-83	#N/A	4,022	$263,284
Aug-83	#N/A	3,130	$488,676
Sep-83	#N/A	422	$33,928
Oct-83	#N/A	0	$224,028
Nov-83	#N/A	0	$304,004
Dec-83	#N/A	0	$352,872
Jan-84	425,075	75,253	$457,732
Feb-84	315,305	15,036	$254,396
Mar-84	367,286	134,440	$259,952
Apr-84	429,432	119,740	$267,368
May-84	347,874	135,590	$158,504
Jun-84	435,529	189,636	$430,012
Jul-84	299,403	9,308	$388,516
Aug-84	296,505	41,099	$225,616
Sep-84	426,701	9,391	$1,042,304
Oct-84	329,722	942	$974,092
Nov-84	281,783	1,818	$301,892
Dec-84	166,391	672	$76,148
Jan-85	629,404	548,704	$0
Feb-85	263,467	52,819	$315,196
Mar-85	398,320	2,793	$703,624
Apr-85	376,569	27,749	$198,464
May-85	444,404	21,887	$478,880
Jun-85	386,986	1,110	$457,172
Jul-85	414,314	436	$709,480
Aug-85	253,493	1,407	$45,380
Sep-85	484,365	376,650	$28,080
Oct-85	305,989	122,906	$111,520
Nov-85	315,407	15,138	$267,200
Dec-85	182,784	5,532	$354,304
Jan-86	655,748	544,807	$664,712
Feb-86	270,483	43,704	$536,824
Mar-86	365,058	5,740	$551,560
Apr-86	313,135	9,614	$150,080
May-86	528,210	1,507	$580,800
Jun-86	379,856	13,620	$435,080
Jul-86	472,058	101,179	$361,144
Aug-86	254,516	80,309	$97,844
Sep-86	551,354	335,768	$30,372
Oct-86	335,826	91,710	$150,324
Nov-86	320,408	9,856	$293,044
Dec-86	276,901	107,172	$162,788
Jan-87	455,136	299,781	$32,532
Feb-87	247,570	21,218	$23,468
Mar-87	622,204	157	$4,503,456
Apr-87	429,331	12,961	$500,904
May-87	453,156	333,529	$0
Jun-87	320,103	178,105	$0
Jul-87	451,779	315,564	$46,104
Aug-87	249,482	80,206	$92,252
Sep-87	744,583	5,940	$4,869,952
Oct-87	421,186	36,819	$376,556
Nov-87	397,367	234,562	$376,556
Dec-87	269,096	71,881	$552,536

* 1 case contains 24 packs

Exhibit 2

Seasonal Indexes for Breakfast Cereals Shipments

MONTH	INDEX
January	113
February	98
March	102
April	107
May	119
June	104
July	107
August	81
September	113
October	97
November	95
December	65

HIGHLAND PARK WOOD COMPANY

In early September 1987, George Simpson, sales manager of the Highland Park Wood Company, received an enquiry from Anne Butler, head buyer for Plainview Homes, a major Dallas-area homebuilder. Plainview was proposing to buy one million board feet of framing lumber at a price to be fixed now, for delivery six months later, in March. Butler explained that Plainview was planning to begin the construction, in early March, of 100 homes in a subdivision northeast of Dallas. Plainview would be willing to purchase all of its framing requirements from Highland Park if Simpson could quote an acceptable, firm price now. Because of recent soaring prices of building materials, Plainview was very cost conscious and desired to fix in advance as large a portion of its construction costs as possible.

Highland Park traditionally passed through to the customer its own cost of purchasing lumber thus eliminating the exposure both sides would face if the price were fixed at the time of the order. Butler, however, appeared to be asking Highland Park to assume this price risk by quoting in September for March delivery. She did indicate, though, that Plainview would be willing to pay more than the 5% margin which Highland Park usually commanded on its direct sales. At the same time, she hinted that Plainview expected a competitive price considering the size of the purchase.

Costing a Normal Deal

Highland Park was a wood retailer; it bought wood in bulk from a saw mill and carried it in inventory to satisfy customer needs. For a sufficiently large order Highland Park could arrange to have the wood shipped directly to the customer's job site.

The current mill price for the wood Plainview was requesting (Southern Pine #2, 2x4) was $279 per thousand board feet. As a "favored" customer of the mill, Highland Park would receive a 4% wholesaler commission but would have to pay $24 per thousand board feet for delivery from the mill. The delivery charge would be the same whether it were delivered to Highland Park's warehouse or directly to the customer's job site. Thus Highland Park's current delivered costs, before profit, was 0.96 x279 + 24=$291.84 per thousand board feet. For the usual retail order, Highland Park would add a markup of 20%, but for a direct shipment order the markup was only 5%. Had Plainview been asking for immediate delivery the company procedure would suggest a price of $291.84 x 1.05 or $306.43. Plainview could not accept current delivery since it did not have its own storage facilities which explained Plainview's proposal to Highland Park.

Harvard Business School case 9-190-013. This case was prepared by Professor David E. Bell.

Buy and Hold

One viable possibility was for Highland Park to buy the wood now and store it until the spring. Apart from the nuisance of taking delivery there was the holding cost to be considered and the additional trucking cost from Highland Park to Plainview's job site in March. The additional trucking cost Simpson estimated as $6 per thousand board feet, but the other costs were less clear. Highland Park usually figured storage costs at an annual rate equal to the prime interest rate plus 6%. This was broken down as Highland Park's cost of short-term capital (prime plus 2%), together with factors to cover taxes (1%), insurance (1%) and depreciation (2%). With the prime currently at 11% this implied a holding cost for six months of 8½% or about $26 per thousand board feet.

Simpson was reluctant to go with this alternative partly because an order of this size threatened to strain Highland Park's storage capabilities but also because he believed Plainview would balk at the price associated with this strategy.

Wait and Buy

The storage problem could be avoided if Highland Park simply waited until March to order the wood from the mill. However this would entail Highland Park bearing a substantial price risk if prices rose between September and March. A review of past prices confirmed Simpson's fears. (See Exhibit 1, columns 2 and 3.) Had Highland Park undertaken such a strategy in 1975, for example, they would have had to pay a price 35% higher in March (1976) than had prevailed in September (1975). On the other hand, in many years the price was substantially lower in the spring than it had been in the fall, which would have allowed Highland Park a substantial profit.

Hedging

Simpson had ruled out the possibility of trying to pass along the price risk in this deal to a third party such as a bank or a mill. In particular he had ruled out the possibility of hedging in the futures market. Indeed, if it had been possible to hedge away the risk by forward buying, Plainview could have done this directly themselves.

Forecasting

There was an active forward market in Hem-Fir, a wood grown almost entirely in the western part of North America and not completely substitutable for Southern Pine. Hem-Fir prices and Southern Pine prices were often quite different and did not always fall and rise together. However, Simpson watched the forward prices for Hem-Fir quite closely in the belief that they offered some insight into likely trends in the market conditions for wood in general. Perhaps significantly the forward price for March delivery of Hem-Fir was now considerably lower than the spot price (see Exhibit 1, columns 4 & 5).

Simpson had asked his assistant to see if there was a useful way to use available information to get a better feel for the likely price of Southern Pine in March. The assistant had compiled the data shown in Exhibit 1 but had not been able to interpret their usefulness.

The Decision

Simpson was perplexed by the historical pricing details his assistant had collected. He was intrigued by the Plainview proposition and regarded the advance purchase with fixed price guarantee as a marketing experiment and potential source of competitive advantage. At the same time he realized that his company's competitive position had not slipped sufficiently to warrant substantial risk taking.

Exhibit 1

Spot and Forward Prices for Southern Pine and Hem-Fir
($ per thousand board feet)

	Southern Pine Spot Prices		Hem-Fir		
			March	Spot Prices	
Year	Sept	March	Forward	Sept	March
1971	135	147	101	108	118
1972	153	175	131	147	183
1973	201	158	121	163	168
1974	112	119	128	126	125
1975	127	171	146	140	165
1976	187	183	173	180	195
1977	264	226	193	218	235
1978	225	237	196	246	238
1979	303	210	235	293	210
1980	197	214	191	194	195
1981	170	203	176	178	173
1982	191	280	159	163	222
1983	222	258	195	189	227
1984	202	212	146	177	178
1985	212	244	145	188	220
1986	215	242	172	232	238
1987	277		182	240	

Explanation:

Spot. The price of Southern Pine at the beginning of September 1971 was $135 per thousand board feet. The price of Southern Pine at the beginning of the *following* March was $147. Similarly, the price of Hem-Fir at the beginning of September 1971 was $108 per thousand board feet. The *following* March, the price had climbed to $118.

Forward. In early September 1971 the current price of Hem-Fir if delivered in March 1972, was $101 per thousand board feet. (For the purpose of this case, it suffices to think of these figures as market forecasts, as of September of the given year, for Hem-Fir prices the *following* March.)

CENEX

In April of 1983, the United States government announced an increase in the federal gasoline tax from 4¢ per gallon to 9¢ per gallon. The 5¢ increase was motivated by the need to fund the extensive overhaul that the nation's highway system then required.

This prompted the management of Cenex (short for Farmers Union Central Exchange Incorporated) to consider a thorough examination of their product mix. Cenex was unusual among farmers' cooperatives in having its own oil refinery that could convert crude oil into such products as diesel fuel, gasoline, and asphalt. Asphalt is a by-product of all crude-oil refining but the amount varies substantially with the source of the oil. Some Canadian oils were especially rich in asphalt (e.g., Cold Lake Blend). Although Cenex had not been buying such oil, a switch to asphalt-rich oil would result in increased production of asphalt and decreased production of gasoline.

The recent tax increase suggested that demand for asphalt would increase dramatically over the next few years as a result of the highway maintenance programs, and that demand for gasoline could fall as a result of the increased price. The price increase was exacerbated when many states raised their own tax on gasoline to fund increased local maintenance projects.

Background on Cenex

Cenex was a 60-year-old vertically integrated oil company based in St. Paul, Minnesota, owned by over a thousand local cooperatives in 15 states in the northwest United States, which in turn were owned by over a quarter of a million farmers. Cenex owned a refinery near Billings, Montana, with a crude-oil capacity of almost 42,000 barrels per day. The company purchased crude oil from producers in Wyoming, Montana, and Canada. Annual sales exceeded 750 million gallons of petroleum products, most of which went to owner cooperatives.

Production Planning

Exhibit 1 shows a schematic diagram of the refinery. At the heart of production planning lay a linear program that, supplied with the price of various crude oils and the prices and demand levels for end products, provided the most profitable mode of operation at the refinery, indicated the types of crude oil that should be purchased, and specified the end products that should be produced. A mainframe computer was dedicated to solving the LP, which was typically run several times per day to respond to refinery and marketplace developments, both of which were highly volatile. The model was also run on occasion for strategic planning purposes. Such a strategic issue arose in late 1985, when Cenex's 1983 decision to expand asphalt production was being reexamined.

Harvard Business School case 9-189-038. This case was prepared by Professor David E. Bell.

The Asphalt Issue

Although Cenex's production mix had always varied from month to month as a function of price and demand fluctuations, a major shift in operations was authorized in 1983 in order to increase asphalt production by making minor plant modifications and substantially changing crude-oil procurement strategies.

As part of the 1985 reexamination, Kevin Lindemer, Petroleum Marketing Analyst at Cenex, assembled some relevant data (Exhibit 2) and constructed several regression models to forecast asphalt price and consumption. Kevin wondered which of his models would be most useful for forecasting purposes and how they could help with the current decision. He was concerned that his data, based on national statistics, would not reflect local conditions in Cenex's trade area.

Forecasting Oil and Gas Prices

The LP model required forecasts of crude-oil prices and the retail price of gasoline. Cenex had regularly hired consultants to provide short- and long-term price forecasts for crude oil, diesel oil and gasoline. For 1986, they were predicting as a baseline scenario a price of $25 per barrel for crude oil and of 90¢ per gallon for gasoline (before federal and state taxes). For the purposes of sensitivity analysis, they also estimated a "best-case" scenario of $30 for oil and 95¢ for gasoline. By examining historical price patterns, Kevin added a worst-case scenario of $20 for oil and 85¢ for gasoline. Cenex judged that in the case of the baseline scenario, government highway spending would grow at 3% over 1985, in the best case at 4%-5%, while the worst case would imply no growth.

Exhibit 1

Sketch of Oil Refinery

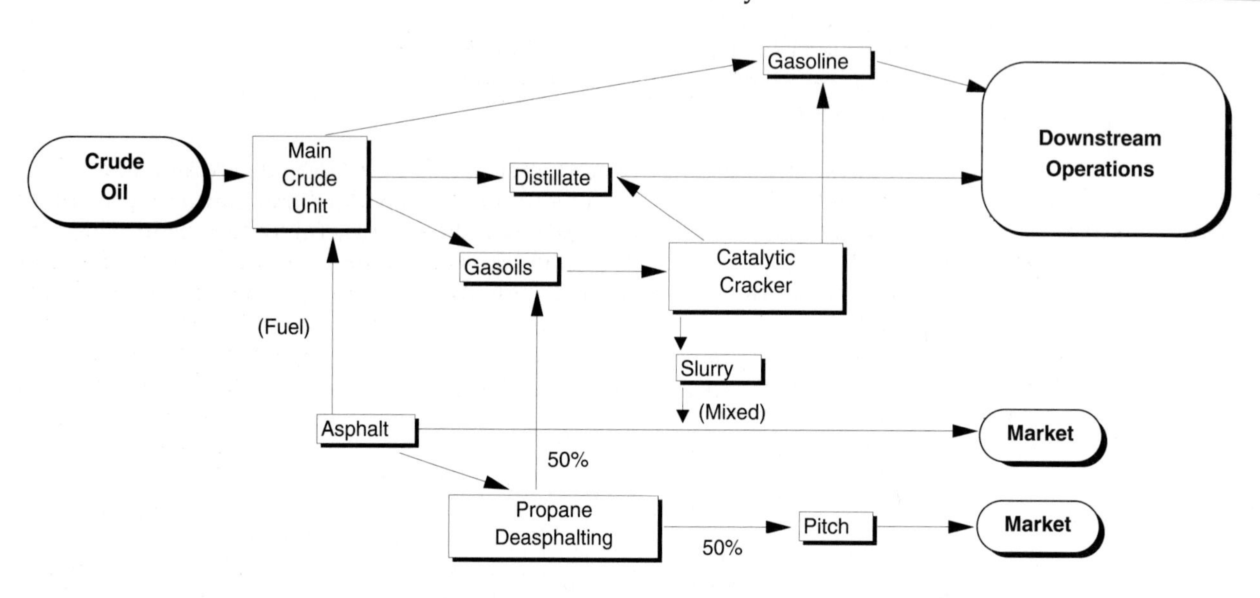

Note: 1) Asphalt may be sold as is, but is mixed with slurry to achieve consistency needed by a given customer. Or it may be used as a fuel within the plant. Or it may be further refined into gasoils and pitch. 2) Distillate includes diesel fuel, jet fuel, kerosene and other heating oils.

Exhibit 2

Asphalt Model Input Data
(All values in nominal dollars)

Year	*Producer Price Index*	*Asphalt Price (per ton)*	*Asphalt Consumption (Millions of bbl)*	*Sand/Gravel Price Index*	*Resid. Constr. (billions)*	*Highway Maint. & Repair (billions)*	*Retail Gasoline Price (cents/ gallon)*	*Refinery Acquisition Cost of Crude Oil (per bbl)*
1963	94.5	$22.21	124	97.3	$28.70	$2.9	20.11	$2.89
1964	94.7	$21.94	127	97.1	$30.53	$3.1	19.98	$2.88
1965	96.6	$22.91	135	97.5	$30.24	$3.3	20.70	$2.86
1966	99.8	$22.61	141	98.1	$28.61	$3.5	21.57	$2.88
1967	100.0	$22.67	138	100.0	$28.74	$3.8	22.55	$2.92
1968	102.5	$22.58	148	103.2	$34.17	$4.0	22.93	$3.17
1969	106.5	$22.27	152	106.7	$37.21	$4.2	23.85	$3.29
1970	110.4	$22.40	163	112.6	$35.86	$4.7	24.55	$3.40
1971	114.0	$28.56	167	121.9	$48.51	$5.1	25.20	$3.60
1972	119.1	$29.61	172	126.9	$60.69	$5.4	24.46	$3.58
1973	134.7	$29.98	191	131.2	$65.09	$5.9	26.88	$4.15
1974	160.7	$58.27	176	148.7	$56.60	$6.6	40.41	$9.07
1975	174.9	$69.33	152	172.3	$51.89	$7.3	45.44	$10.38
1976	183.0	$67.41	151	186.7	$68.59	$7.7	47.44	$10.89
1977	194.2	$70.84	159	199.0	$92.47	$8.6	50.70	$11.96
1978	209.3	$77.22	174	217.7	$110.43	$9.8	53.09	$12.46
1979	235.6	$93.15	174	244.0	$117.23	$10.6	74.33	$17.72
1980	268.8	$134.35	145	274.0	$101.15	$11.4	107.35	$28.07
1981	293.4	$163.74	125	296.3	$100.05	$12.2	116.32	$35.24
1982	299.3	$160.65	125	310.9	$85.39	$13.3	107.01	$31.87
1983	303.1	$158.00	136	313.3	$126.55	$14.2	95.36	$28.99
1984	310.3	$168.15	149	325.7	$155.15	$15.0	92.06	$28.63
1985	308.8	$179.60	155	336.2	$158.82	$16.0	90.14	$26.76

CFS Site Selection at Shell Canada Ltd.

Ian Jacobs, Manager of Convenience Food Stores at Shell Canada Ltd., reviewed the store-site selection model that he was about to present to management. The model predicted sales of existing stores quite well, and identified those sites that should never have been built upon. He was concerned, however, about the utility of the model for the selection of future sites because, by 1985, convenience food stores (CFS) at gasoline sites had become the single largest capital investment budget item in oil products marketing at Shell Canada. In theory, CFS investment offered higher rates of return than any investment alternative in oil products marketing. But a post audit of Shell's experience with its 40 sites in Western Canada showed that they did not in fact generate a uniformly high rate of return. Indeed, a substantial fraction did not clear the corporate hurdle rate. Even more troubling, revenues varied by a factor of seven across stores of the same size and layout.

Top management at Shell had decided that continued investment in CFS should be suspended until a methodology for choosing CFS sites had been developed. During the past two years, Strategic Planning Associates (SPA), an international management consulting firm, had been conducting a marketing strategy review for Shell Canada in oil products marketing. As part of this broader study, SPA was asked by top management to assist Jacobs in developing a model for CFS site selection.

Canadian Oil Products Marketing in the 1980s

The early 1980s had been a disastrous period for oil refining and marketing companies throughout the world. In Canada, the five major oil companies had refining and marketing divisions that had averaged only 2.5% return on assets employed. Several factors contributed to this profitability problem.

The primary cause was the world oil shocks of 1973-74 and 1979-80. As oil-product prices skyrocketed, demand for oil products fell dramatically. Moreover, as volume fell and oil products became increasingly commoditized, the basis for competition in the industry shifted from service to price. Thus, gross margins shrank at the same time that oil-products volume fell.

This pressure on profitability coincided with a dramatic rise in the real-estate value of gasoline station sites. Sites that ten years before had been profitable as stations would in many cases become more valuable now if the gas station were torn down and the land developed for alternate uses. The industry, which heretofore had evaluated a location for its potential performance as a gasoline-only retailing site, now began to consider the optimal configuration of a site, which might include a number of ancillary retailing services such as CFS, car wash, and automotive lubrication.

Harvard Business School case 9-189-026. This case was prepared by James Robo, a second-year student, under the direction of Professor Arthur Schleifer, Jr.

In this highly competitive environment, businesses such as CFSs became economically attractive investments for a number of reasons. First, a CFS could be erected on an existing gasoline-station site without the additional purchase of land, giving a distinct advantage to CFSs at gasoline sites over stand-alone CFSs. Second, there was a significant synergy between food and gasoline sales at these hybrid sites: the addition of a CFS increased gasoline volume by ten percent, on average. Finally, a single employee could operate both the gasoline and the food pay points, so that addition of a CFS required very few incremental employees.

The prevalence of CFSs on gasoline station sites in some regions of the United States attested to the attractiveness of the CFS-gasoline link. Because Canada had significantly lower penetration of CFSs, an opportunity arose for oil-products marketers such as Shell to pre-empt the entry of CFS chains into the Canadian marketplace.

Convenience Food Stores at Shell Canada

Despite this window of competitive opportunity, Shell's late entry into the marketplace seemed to pose a problem. Shell's forty sites in western Canada did not compare with the many hundreds of sites of several of the major Canadian CFS chains, four of which had more than 400 stores each. This lack of scale seemed to place Shell at a competitive disadvantage in purchasing, store operating experience, and consumer brand awareness.

Further analysis of the competitive environment by Jacobs and the SPA caseteam revealed that regional scale, rather than national scale, was the factor driving industry economics. Even though Shell was at a scale disadvantage at the regional level as well, this disadvantage would be overcome by rapid expansion of Shell's CFS network in the west.

Shell's Existing Site Selection Methodology

The success of the expansion strategy would depend critically on proper site selection. In the past, Shell had tended to select sites with high gasoline volume. Judgmental forecasts of sales showed almost no correlation with actual sales at maturity (3-5 years of operation). Furthermore, there was surprisingly little correlation between gasoline volume and food sales (see Exhibit 1), so that the strategy of choosing sites with high gasoline volume was ineffective. It was imperative to identify those factors that accounted for the seven-fold variation in sales volume across Shell CFS sites (all of which had the same square footage of retailing space), so that a site-selection methodology could be devised that would systematically choose sites with high sales potential.

Developing a Site-Selection Model

Jacobs and the SPA caseteam decided to take a two-pronged approach to studying the issue of CFS site selection. They would conduct research on Shell's existing CFS customers and then analyze the data generated by the post audit of Shell's existing sites. If a model that explained the historical sales variation could be developed, it could be applied to predict future sales of proposed sites.

The caseteam hypothesized that the variables that could explain this large sales variation would include: number and proximity of competitors; population, income level, and age distribution in the relevant trade area; traffic volume and gas volume at the site; age of the gasoline station; hours of operation; type of neighborhood; and proximity to schools and apartment buildings. Unfortunately, very little of this information was readily available. Moreover, the caseteam was unsure of what constituted the relevant trade area for a convenience food store, nor did they have a good understanding of how competition affected CFS sales.

To answer these questions, they undertook customer research and competitor surveys at thirty-four of Shell's forty sites. The first surprising finding was that only 14% of Shell's CFS food sales were to customers who purchased gasoline at the time of their CFS visit, which suggested why food sales and gasoline volume were uncorrelated. By asking customers to point out on a map where they lived or worked, the caseteam was able to determine the relevant CFS trade area. Nearly 2/3 of Shell's CFS customers lived or worked within a one-mile radius of the CFS, with about half living within a half-mile radius.

By placing schools, apartment buildings, competitors' locations, and major "physical barriers" on these same customer-research maps, the caseteam was able to judge how these factors affected sales. While anecdotal explanations could be found for almost every site's sales performance, no consistent patterns across the sites were discernible. The analysis depended on other factors not on the map.

Accordingly, data on trade area demographics, traffic, type of neighborhood, and competition were collected, compiled and entered into a computer for regression analysis (see Exhibit 2 for a description of the variables and the data). After a month of regression modeling, the SPA caseteam returned to Jacobs with a model that explained 69% of the sales variation using only five variables: population in the trade area weighted by distance from the site, type of neighborhood, and three dummy variables having to do with competition: complete absence of competition in the quarter-mile trade area, presence of competition directly across the street, and presence of competition within the quarter-mile trade area. (See Exhibit 3 for a full description of the five variables and the model results.)

Jacobs' Concerns

Jacobs was impressed by a calculation (see the end of Exhibit 3) that showed that if the model worked as well in predicting the performance of prospective sites as it had in explaining the performance of existing sites, Shell could increase its after-tax profit by $13 million. Nevertheless, several questions remained in his mind. Did the size and signs of the coefficients make intuitive sense? Did it even matter whether the signs made sense since this model was to be used purely for predictive purposes? Could a model developed from historical site performance be used to predict proposed new site performance? More immediately, should Jacobs apply this model to the three proposed CFS sites in Alberta and halt construction on any whose predicted sales volume fell below the level needed to clear the corporate hurdle rate? These were the questions that troubled Jacobs the most as he prepared to present the caseteam's findings to top management.

Exhibit 1

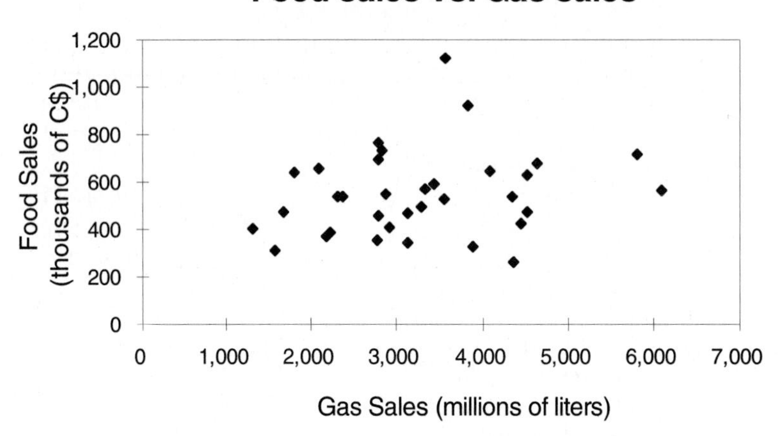

Percent of variance of food sales explained by gas sales = 4%

Source: Shell Canada

Exhibit 2

Description of Variables

Site Number:	Identifier coded with values from 1 to 34
City:	City code: CAL = Calgary, EDM = Edmonton, SKN = Saskatoon, REG = Regina, VAN = Vancouver, WIN = Winnipeg
Sales:	1985 Sales volume (thousands of Canadian dollars)
Population:	Number of people, classified into three clusters:
< ¼:	within a circle ¼ mile radius centered on the site
< ½:	within a ring greater than ¼ mile but less than ½ mile from the site
< 1:	within a ring greater than ½ mile but less than 1 mile from the site
CFS Competition:	Number of competitive convenience food stores, classified into four clusters:
AS:	across the street
< ¼:	within a circle ¼ mile radius centered on the site, but excluding stores across the street
< ½, < 1:	rings, as above
Supermarket Competition:	Like CFS competition, but for supermarkets
No Comp. < ¼:	Coded 1 if there is no CFS or supermarket competition within ¼ mile of the site, 0 otherwise
Gas Vol.:	Volume of gasoline sold at the site in 1985 (in thousands of liters)
<24 Hrs./Day:	Coded 1 if site is *not* operated 24 hours per day, 0 if it *is*
Suburb:	Coded 1 if the site is in a suburban neighborhood, 0 otherwise
Urban:	Urban site, classified into one of three categories:
Res.:	Coded 1 if the site is in an urban residential neighborhood, 0 otherwise
Inner:	Coded 1 if the site is located in a downtown (inner city) urban business district
Comm:	Coded 1 if the site is located on an urban commercial strip ("strip mall")

EXHIBIT 2 CONTINUED

Avg. Inc.:	Average per capita income (thousands of Canadian dollars per year), classified into two clusters:
< ¼:	within a circle of ¼ mile radius centered on the site
< ½:	within a ring greater than ¼ mile but less than ½ mile from the site
Males 20-34:	Percentage of the population living within ¼ mile of the site that is male and between ages 20 and 34
Pop > 65:	Percentage of the population living within ¼ mile of the site that is over age 65
Traffic:	Number of cars/day traveling past the site
Age:	Coded 1 if the site is less than 3 years old, 0 otherwise
Pop gradient:	The sum across the three population clusters surrounding the site of average sales per capita in each cluster, multiplied by the population in that cluster. Survey data showed that 26% of customers (and, presumably, 26% of sales) come from the ¼-mile circle. Thus, at Site #1, 26% of the $1,122,000 in sales was generated in the ¼-mile circle, or $291,720 in total, or $291,720/2,310=$126 per capita. Averaging these per-capita sales across all 34 sites, the ¼-mile circle averaged $70 per capita, the ring from ¼ to ½ mile averaged $22.8, and the ring from ½ to 1 mile averaged $6. The population gradient was computed by multiplying these average sales per capita by the population in the appropriate cluster, adding these products across the three clusters, and dividing by 1000.

The figure below depicts the various geographic segments described above, and shows the estimated average annual sales per capita in each segment.

Estimated Average Sales per Capita per Year in Rings Around Store Site

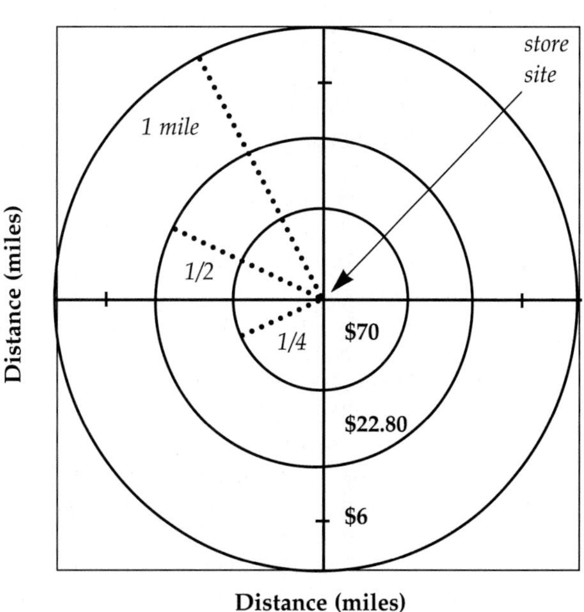

Exhibit 2 Continued

Data on Existing CFS Sites

Site			Population			CFS Competition				Supermarket Competition				No Comp.	Gas	<24 Hrs./		Urban			Avg. Inc.		Males	Pop		Store	Pop
#	City	Sales	<1/4	<1/2	<1	AS	<1/4	<1/2	<1	AS	<1/4	<1/2	<1	<1/4	Vol.	Day	Suburb	Res	Inner	Comm	<1/4	<1/2	20-34	>65	Traffic	Age	Grad
1	CAL	1,122	2,310	6,470	9,355	0	0	1	3	0	0	0	0	1	3,551	0	1	0	0	0	30.4	30.4	23.1%	1.6%	4,000	0	365.3
2	SKN	681	1,140	6,515	12,150	2	0	1	2	1	1	0	0	0	4,625	0	0	0	0	1	30.6	30.6	13.7%	4.4%	19,425	0	301.2
3	WIN	571	3,300	8,495	25,944	0	1	1	2	0	0	1	5	0	3,331	0	0	1	0	0	15.3	16.7	8.4%	13.4%	23,226	0	580.4
4	REG	357	2,240	5,725	16,255	2	0	2	1	0	0	3	0	0	2,765	0	0	0	0	1	35.4	35.4	10.8%	11.6%	12,040	0	384.9
5	SKN	458	1,640	3,400	12,850	1	0	0	1	1	1	1	1	0	2,770	0	0	0	0	1	28.4	28.4	13.5%	12.6%	14,950	0	269.4
6	WIN	592	3,735	10,013	29,281	1	0	0	1	1	0	0	1	0	3,429	0	0	0	1	0	20.3	17.6	20.2%	31.3%	19,750	0	665.4
7	VAN	733	7,235	16,000	14,515	2	1	2	0	1	1	2	0	0	2,827	0	0	0	1	0	17.2	17.2	22.8%	16.5%	22,800	0	958.3
8	EDM	644	3,079	3,718	22,952	0	3	0	2	0	0	0	0	0	1,798	0	0	0	1	0	18.6	17.0	27.6%	8.7%	25,900	0	438.0
9	VAN	567	1,400	9,075	27,350	0	0	0	6	0	0	0	8	1	6,094	0	0	0	0	1	26.4	26.4	14.2%	11.0%	41,100	0	469.0
10	WIN	528	2,772	5,280	7,571	0	0	1	3	1	0	0	0	0	3,548	0	1	0	0	0	26.7	25.6	16.2%	3.0%	5,274	0	359.9
11	WIN	471	2,576	3,776	20,609	0	1	0	5	0	0	1	1	0	3,130	0	0	0	0	1	24.5	30.1	9.8%	7.6%	16,097	0	390.1
12	EDM	411	1,749	4,440	15,326	1	0	0	5	0	0	0	5	0	2,900	0	0	1	0	0	29.6	31.3	18.1%	8.7%	30,200	0	315.6
13	WIN	478	1,596	5,580	21,528	0	1	1	3	0	1	1	3	0	4,507	0	0	1	0	0	21.7	23.6	11.4%	15.5%	17,621	0	368.1
14	EDM	498	1,904	4,120	9,356	1	1	1	0	1	0	0	3	0	3,289	0	0	1	0	0	24.5	28.0	15.5%	6.1%	24,600	0	283.4
15	EDM	662	3,040	6,045	19,560	0	0	2	3	0	1	1	0	0	2,080	0	0	0	1	0	17.7	26.0	21.6%	18.7%	25,800	0	468.0
16	EDM	390	1,178	2,308	16,935	0	0	2	1	0	1	1	0	0	2,215	0	0	0	0	1	21.4	23.4	15.6%	16.1%	30,500	0	236.7
17	EDM	543	2,745	5,400	8,908	2	0	0	2	1	1	0	1	0	2,301	0	0	0	0	1	22.3	23.5	15.8%	5.6%	10,650	0	368.7
18	CAL	551	2,225	6,675	12,515	0	0	1	0	1	0	0	4	0	2,865	0	0	0	0	1	29.3	29.3	11.0%	7.4%	29,700	0	383.0
19	EDM	699	1,881	5,394	21,825	0	2	0	1	0	0	2	1	0	2,771	0	0	0	0	1	20.7	22.4	18.0%	14.1%	28,800	0	385.6
20	EDM	372	1,105	3,316	9,777	1	0	0	2	0	0	1	1	0	2,175	1	0	0	1	0	26.8	33.5	16.3%	9.3%	11,300	0	211.6
21	EDM	540	1,965	4,455	17,372	0	0	1	5	0	1	0	3	0	2,352	1	1	0	0	0	38.6	38.9	14.9%	2.2%	4,200	0	343.4
22	VAN	346	3,135	5,720	13,635	0	0	1	1	1	0	0	3	0	3,119	1	0	0	0	1	30.6	30.6	9.2%	18.2%	17,400	0	431.7
23	VAN	719	1,920	4,395	22,215	0	0	1	2	0	0	1	1	1	5,795	0	0	1	0	0	26.5	26.5	11.3%	15.3%	26,400	0	367.9
24	WIN	768	4,240	8,840	46,365	0	1	0	5	0	1	1	2	0	2,781	0	0	0	1	0	12.5	14.3	17.7%	14.3%	40,965	0	776.5
25	EDM	313	2,955	3,760	10,895	1	0	1	2	1	0	1	0	0	1,556	1	0	0	1	0	22.5	19.8	27.5%	12.9%	26,900	0	357.9
26	CAL	474	1,270	2,730	4,260	0	0	0	0	0	0	0	0	1	1,665	1	1	0	0	0	33.6	33.6	25.9%	1.0%	4,000	0	176.7
27	CAL	541	2,008	3,642	15,125	0	0	0	3	0	0	1	1	1	4,346	0	0	0	0	1	37.9	37.9	13.7%	6.5%	47,700	0	314.3
28	VAN	650	1,915	3,375	7,865	0	0	0	2	0	0	1	1	1	4,078	0	0	0	0	1	35.7	35.7	12.8%	15.9%	19,500	0	258.2
29	WIN	630	2,012	4,707	16,711	0	0	0	3	0	1	0	1	0	4,516	1	0	1	0	0	30.9	32.1	9.6%	27.6%	16,472	0	348.4
30	CAL	329	2,025	4,535	5,128	0	0	0	2	1	0	0	0	0	3,870	0	1	0	0	0	40.4	40.4	20.0%	1.1%	4,000	0	275.9
31	WIN	428	1,400	3,855	11,185	0	0	0	2	1	1	0	2	0	4,447	0	0	0	0	1	25.1	25.1	12.1%	21.4%	27,087	0	253.0
32	REG	924	1,825	3,710	15,010	0	0	0	2	0	0	0	1	1	3,817	1	1	0	0	0	30.2	30.2	13.7%	1.7%	8,700	0	302.4
33	WIN	267	2,108	3,955	18,390	2	0	0	4	0	1	1	1	0	4,348	0	0	0	0	1	17.5	23.7	13.8%	20.6%	13,035	1	348.1
34	CAL	403	1,463	1,428	1,559	0	0	0	0	0	0	0	0	1	1,308	1	1	0	0	0	22.3	22.3	21.8%	4.5%	4,000	1	144.3

Exhibit 3

Variables in Model

Dependent Variable: Sales
Independent Variables:
Pop Gradient: See Exhibit 2
Across Street: Number of competitive CFS stores plus supermarkets across the street from the site
Comp < ¼: Number of competitive CFS stores plus supermarkets within ¼ mile of site (includes those across the street)
Pop < ¼ & no comp: Number of people within a circle of ¼ mile radius centered on the site having no competitive stores within the circle (= pop < ¼*no comp < ¼)
Sub ¼-1: Number of people within a ring greater than ¼ mile but less than 1 mile from a suburban site, weighted by average sales per capita [= suburb*(0.0228*pop < ½ + 0.006*pop < 1)]
Observations in Model: All except the two sites whose age was less than three years

Regression Number 1
Dependent Variable: SALES

	Constant	Pop Grad	Across St	Comp<1/4	Pop<1/4&no comp	Sub 1/4 - 1
Regr. Coef.	241.9	0.3713	(83.15)	100.2	0.1702	0.8660
Std. Error	59.3	0.1323	30.29	29.6	0.0318	0.3116
t value	4.1	2.8	(2.7)	3.4	5.4	2.8

# of obs = **32** *Deg of F =* **26**
R-squared = **0.6905** *Resid SD =* **106.7**

Model Results

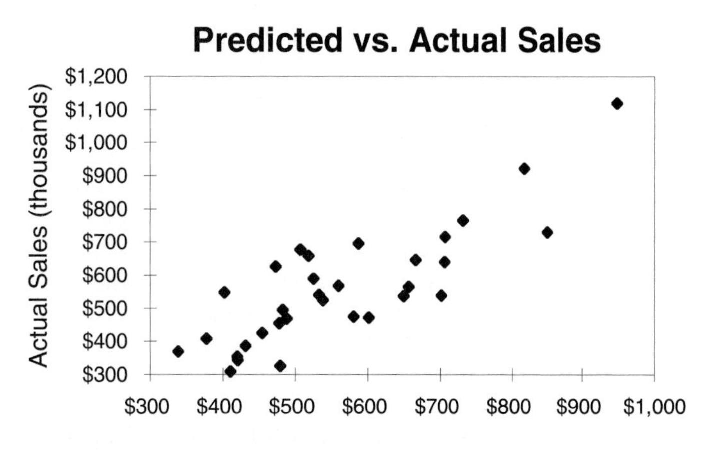

- If the model had been used before the CFS network was built, and only th sites with predicted sales over $600,000 had been selected, then the average store performance would have been $125,000 higher.
- If the model is applied to the next 50 stores, the potential benefit will be:
  50 Stores*$125K Higher Sales*30% Gross profit = $2M of Increased Profits per Year.
- That cash flow translates into a Net Present Value of $13M after-tax.

Firestone Tire & Rubber Company: The Industry Replacement Passenger Tire Forecast

In 1980, looking ahead five years, Firestone Tire & Rubber Co. forecast a replacement tire market of 105 million units. Meanwhile Goodyear foresaw a 1985 replacement market of 138 million tires. Firestone's gloomy predictions attracted considerable comment in the press. The difference between Firestone and Goodyear, *Fortune* observed, was equivalent to "the output of five $200 million radial plants working around the clock" (*Fortune*, Oct. 20, 1980, 115-6).

In recent years Goodyear had taken a generally optimistic view of the market, adding $1.7 billion in manufacturing capacity between 1974 and 1981. At Firestone, on the other hand, the new president and CEO, John Nevin, had closed seven bias-ply plants in the United States and Canada since assuming his position in late 1979. The forecast of the Tire Statistical Committee of the Rubber Manufacturers Association (RMA), about 133 million replacement tires in 1985, was close to Goodyear's. (Each year the Tire Statistical Committee produced forecasts for each of the next four years. Exhibit 1 compares RMA forecasts with actual replacement-tire shipments for 1973-80.)

In late 1980, the American tire industry was just beginning to recover from setbacks that had led most major producers to make substantial changes in long-term strategy. Competition from abroad had increased sharply in the previous decade, while growth in demand had slowed. The shift to radials, which offered better gas mileage and longer tire-life, spread from Europe to the United States. In the early 1970s American companies could not yet produce the new tires, and Michelin established a dominant position in the United States market. (The New York Times estimated Michelin's North American investment during the 1970s at more than $1 billion.) Concurrently, Japan's Bridgestone became a world-market presence and a potential threat in the United States. Exhibit 2 shows passenger-tire shipments for United States consumption from 1970 to 1980, for domestically manufactured and imported tires, in both original-equipment (OE) and replacement categories.[29]

The first radials offered by United States makers were of poorer quality than Michelin's, and the rate of premature failure was exceptionally high. Firestone was forced to recall its "500" series, and the company replaced 11 million tires between 1978 and 1981. Meanwhile, manufacturers continued to produce bias-ply and belted-bias tires, bringing on price wars, huge inventories, and ultimately excess manufacturing capacity. With gasoline and car costs rising, people drove less, further depressing demand.

As the decade ended, United States tire manufacturers acted to regain competitiveness. Of the five major tire companies, three turned to diversification: Uniroyal and Goodrich in chemicals and plastics; General Tire

Harvard Business School case 9-182-152. This case was prepared by Alice B. Morgan, Research Associate, under the supervision of Professor Arthur Schleifer, Jr.

[29] Discrepancies between replacement-tire shipment figures in this exhibit and in Exhibit 1 are due to differing estimates by Firestone and the RMA of non-RMA shipments.

in chemicals, defense, and entertainment through its unconsolidated subsidiary RKO. After a $120 million loss in 1979, Uniroyal sold its European tire operations and closed two of five North American plants. Goodrich pulled out of the OE market entirely, planning to produce only high-performance tires for the more lucrative replacement market. General, which relied heavily on the OE market for its tire-related business, lost $23 million in the first nine months of 1980.

Only Firestone and Goodyear, then, continued tire operations as their basic business. Goodyear went after the OE business that the other manufacturers were giving up. (Probably as a result of the "500" series recall, Firestone's share of the OE market had dropped from 26 to 21.5 percent.) With about 28 percent of the worldwide market, Goodyear was the world's largest tire company, though its radial-tire sales were only two-thirds those of Michelin, which was second-largest overall. Goodyear's relatively strong position enabled it to expand and modernize its operations.

Firestone's financial structure was relatively weak. Nevin sold off the company's plastics business, reduced employment by 18,000 people, and halted the proliferation of tire types. To satisfy original-equipment manufacturers, its own Dayton-Sieberling division, and numerous private brands, Firestone had once produced as many as 7,289 sizes, widths, and styles of tires. Now the company slashed this number to about 2,900. In 1980, Firestone closed its two plants in the United Kingdom and sold its retail stores there. Though it continued to serve the European market from plants in France and Italy, its main focus was the United States. (In 1979, Firestone had made $133 million, following 1978's loss of $148 million, largely attributable to the recall. The company lost $106 million in 1980.) With plants running at about 90% capacity, Firestone was repositioning itself for a permanently altered industry.

Firestone expected the number of miles driven each year—a major determinant of replacement-tire demand—to continue to grow, but at a decelerating rate. At the same time, the company believed average tire-life would increase substantially.

The Direct Forecast of Replacement Passenger Tires

Traditionally, tire companies forecast OE sales of passenger tires, which depended on car and light-truck production, separately from replacement demand, producing total industry forecasts and then projecting their shares of each market. The most direct way to forecast industry replacement-tire shipments was based on shipment history, which was reliably reported by the Rubber Manufacturers Association. In 1969, Firestone developed a regression model to express the number of replacement tires required in a given year as a function of the total new tires sold in five previous years, the OE tires sold in the forecast year, and a trend. The model took the form shown in Table A.

Table A

$$\text{Repl ship}(t) = B_0 + B_1{*}\text{OE ship}(t) + B_2{*}Tot(t-1) + B_3{*}Tot(t-2) + B_4{*}Tot(t-3) + B_5{*}Tot(t-4) + B_6{*}Tot(t-5) + B_7{*}t + \text{error}$$

where

$t = 0$ in 1954, 1 in 1955, etc., and

Repl ship(t) = replacement-tire shipments in year t

OE ship(t) = OE tire shipments in year t

$Tot(t-1)$ = OE plus replacement-tire shipments in year $t-1$, etc.

The model was fit to data on replacement shipments from 1955 through 1968, using data on OE shipments in those years and on replacement shipments from 1950 through 1968. See Exhibit 3 for the data, the equation output and fit, and the coefficient values.

Using this method to forecast future replacement-tire demand required forecasts of OE tire shipments as input. Ordinarily, Firestone developed a rolling five-year forecast of OE industry shipments, but for the 1969 study this forecast, as shown in Exhibit 4, Column 3, was extended an additional five years, through 1978.

The forecast model implied that the demand for replacement tires was generated by the wearing out of tires put into service earlier (possibly during the current year) plus an increase in replacement tires' share of the total tire market. As a result of the mix of drivers, driving conditions, and tire constructions, some tires would wear out in less than one year, others in their first year, others in their second, and so on. And as cars lasted longer, or as the number of miles driven per car per year increased, replacement tires would account for an increasing percent of the total (OE plus replacement) market. A simple lag model with a time trend was adequate to express this relationship, as long as the mix of tires and the mix of driving habits remained stable, so that the fraction of one-year-old tires or two-year-old tires needing replacement would be roughly the same from year to year. Constant regression coefficients would reflect this stability assumption. If, however, driving habits or the tire mix changed significantly, or the trend in increasing car lives or miles driven per car per year reversed, the model would predict poorly. Forecasts made with this model are shown in Exhibit 4.

Decomposition into Factors Generating Demand

Following the first OPEC oil crisis in late 1973, the task of forecasting became complicated by effects of escalating fuel costs, as well as the growing use of radials. In consequence, Firestone abandoned its direct forecasting methodology and developed a new methodology based indirectly on changes in tire technology and the economic environment. Since demand for replacement tires was a function of miles driven on passenger tires and of tire-life, Firestone forecasters determined how many tires had worn out in each recent year, and related that figure to number of miles driven and estimated tire-life for that year. They then forecast miles driven and average tire-life to determine how many tires would wear out in future years. Not every tire that wore out led to the purchase of a new replacement tire. Some tires were replaced with spare tires, some with retreads, and some (on light trucks) with truck tires; the tires on a car being scrapped were not replaced at all. Each of these categories had to be analyzed and forecast, then subtracted from the total number of tires wearing out.

Firestone excluded snow tires from forecasts, in order to distinguish the mileage (and hence wear) given to conventional and snow tires. Because snow tires were used only part of the year, they constituted a kind of customer-owned inventory. As the "all-weather" radial became the standard tire, fewer snow tires were in use, and the snow-tire proportion of miles driven changed. The trend in conventional replacement-tire demand emerged more accurately if snow-tire shipments (and miles driven on snow tires) were eliminated.

In 1980, after careful analysis of all these elements, Firestone released the forecast that diverged so materially from the rest of the industry's expectations.

The Controversial Forecast

The reaction to Firestone's replacement-passenger-tire forecast was one of surprise and some skepticism. In response to questions about their assumptions and their data, the forecasting staff (Dillard B. Feltner, manager—sales forecasting; Harry H. Hollingsworth, manager—management information; and Robert L. Kritzberger, senior sales forecaster) circulated an in-company explanation of their methodology and results. (It was entitled "What? Only 105 Million Replacement Passenger Tires in 1985?") Their detailed analysis of the forecast's major variables, "passenger-tire miles" and "miles per tire," forms the basis for the following discussion.

1. Passenger Tire Miles

The forecast of the first major variable affecting replacement-tire demand—the number of miles driven on passenger tires—incorporated several subsidiary models and forecasts. Basic data for vehicle-miles of travel came from the Department of Transportation (DOT), using information provided by the individual states. Twenty-three states and the District of Columbia estimated highway mileage by monitoring sections of different types of roads, and counting various types of vehicles on them over specified time periods. From these samples, total miles traveled by cars, trucks, buses, and so on, were estimated. Eleven states used motor-fuel tax receipts to determine total fuel consumption, and then multiplied that total by an assumed number of miles per gallon to yield a miles-driven figure. Sixteen states used both methods. The DOT employed these state totals, information about the federal fleet, and taxi and bus data from industry associations to develop estimates of miles driven every year by various types of vehicles.

a. The passenger-car-miles model — Data Resources, Incorporated (DRI), to whose services Firestone subscribed, related the DOT data to three independent variables—car-operating costs, personal income, and cars in use—to develop a forecasting model for miles driven by passenger cars. See Table B. DRI tracked the costs of car operation and developed forecasts based in part on oil and gasoline supply and prices, which were monitored and forecast by DRI's energy service. DRI also forecast personal income, using historical data published by the Department of Commerce, and the DRI 800-equation model of the United States economy. The final variable of the model, cars in use, was itself the output of another forecasting model, this one developed jointly by DRI and Firestone. Its independent variables included car sales (forecast by DRI), population over 15 (from census data), and scrappage rates by age of car (forecast by Firestone). Firestone had forecast that the number of cars in use would increase by slightly over two million (net) each year through 1986.

Table B

Passenger-car miles = *f*(Car operating costs, personal income, cars in use)

Cars in use = *f*(Car sales, population over 15, scrappage rate)

b. Adjustments to the passenger-car-miles model — Since both the historical data from the DOT and the forecast from DRI were for *car*-miles, certain adjustments were necessary in order to forecast shipments of conventional (nonsnow) passenger tires in the replacement market. Firestone forecasters revised actual and forecast figures to exclude snow-tire miles, and to include passenger-tire miles driven on light trucks. (DOT and DRI figures were associated strictly with passenger cars, whereas Firestone was also interested in any other vehicles using passenger tires. In fact, 68% of light trucks came with passenger tires as original equipment, and about 40% of replacement tires for light trucks were also passenger, rather than truck, tires.) The derivation of the passenger-tire-miles forecast, incorporating these refinements, is shown in Exhibit 5. Actual values are given for 1961-1980, and forecasts for 1981 through 1986.

As a final comment on tire-miles driven, the forecasters stressed

> ...the likelihood of at least one more energy shock during the next five years, such as the oil embargo in 1974 or the Iran Revolution in early 1979. The car-miles data make it clear that miles are depressed at such times and most observers expect the Iran-Iraq war, or some other development, to bring about another shock.

Since DRI's car-miles model depended in part on costs of automobile operation, it would reflect an energy shock, but of course it could not predict one.

2. Miles Per Tire

Many independent factors determined how many miles a tire could be driven before it wore out. Individual drivers influenced tire wear significantly by the speed at which they turned corners, or the care they exercised to maintain front-end alignment and tire balancing. Driving conditions differed regionally, varying with the weather, the straightness of the roads, and the quality of road maintenance. Cars themselves affected tire-life: heavier cars with power brakes and power steering put more stress on tires. And the construction of the tire contributed materially to variation in tire-life; radials, for example, lasted about twice as long as typical bias-ply tires under comparable driving conditions. (Exhibit 6 compares radial and bias tire constructions.) Firestone's forecasters wanted to forecast each factor that caused variation in tire-life independently, if possible. As a first step, they tried to assess the tire-life potentials of bias-ply, belted-bias, and radial modes of tire construction.

a. The Tire-Life Index (TLI) — Tires wear out for various reasons. Some are damaged by running over curbs or sharp objects; others suffer sidewall or other premature failure; others simply wear their tread down so much that further driving is dangerous. Firestone attempted, through several types of research, to learn how long tires lasted before being discarded. The company did its own track testing to estimate tread wear. It also carried out "parking-lot studies" in which tire tread on recent-model cars was measured and correlated with odometer readings. (A 1976 parking-lot study concluded that the median life of a radial tire was 31,700 miles; see Exhibit 7.) Tires turned in to Firestone dealers by people buying replacements were analyzed to see how much tread remained, and what caused the tire to need replacement; consumer surveys asked people how many miles their tires had lasted. In addition, Firestone personnel discussed the question with other members of the tire industry.

Using this information, the forecasters estimated how many miles would be provided by the average new bias, belted-bias, and radial tire put into service each year. First, they took into account the estimated lives of different lines (e.g., premium, first line, second line) within each construction type and the share accounted for by each line. Then they converted these estimated mileage figures to an index with a common base: the life of an average new bias tire during the period 1961-63. Exhibit 8 shows the index of tires going into service in any given year: the Tire-Life Index (TLI) for each construction is a weighted average of TLIs for all grades of that construction. Thus the TLI for bias tires (column 6) declines between 1969 and 1979 because the proportion of second- and third-line bias tires shipped increases in the later years.

The TLI reflected variations in tire construction, rather than the influences, for example, of heavier or lighter cars. In some cases, however, particular constructions were associated with particular applications: front-wheel-drive cars, for instance, used radial tires exclusively. Table C shows that front-wheel-drive cars will account for a rapidly increasing percentage of radials going into service.

Table C

Percentage of Radial Passenger Tires going into Service that will be on Domestic Front-Wheel-Drive Cars

Year	On New Cars	Replacements	Total
1979	7.8%		3.4%
1980	10.1%		4.7%
1981	18.4%		8.4%
1982	26.0%	5.2%	14.5%
1983	34.3%	7.4%	19.3%
1984	37.4%	13.2%	23.7%
1985	44.2%	22.6%	31.8%
1986	47.5%	29.7%	37.2%

When Firestone's forecasters learned that the X-body style car appeared to give greatly increased tire-life, they took account of this association in developing the TLI for radial tires. As the sidebar explains, the longer tire-life seemed to be due to the use of oversized tires, rather than front-wheel drive per se. In summary, Feltner and his colleagues wrote, "Our estimate of the effect of front-wheel-drive cars significantly increases our forecast of future tire lives."

To forecast the composite Tire-Life Index, Firestone had to anticipate trends in car design that would alter the number of miles a given tire would last; hence forecasters estimated the number of cars of various types that were likely to enter the new-car market in coming years. From 1979 to 1986, they expected to see a tenfold increase in the proportion of radial tires entering service that would be on domestically produced front-wheel-drive cars (see Table C).

The figures in column 12 of Exhibit 8 represent the Tire-Life Index of an average tire going into service in the indicated year. As radials became a greater proportion of total tires shipped, this "composite Tire-Life Index" rose. To develop the index of the average tire coming out of service (column 13), Firestone used a polynomially distributed lag (PDL) on the composite index of tires going into service. The lag here functioned analogously to the coefficients in Firestone's earlier direct forecast of replacement-tire shipments: it reflected the relationship between tire shipments in any one year and the rate at which those tires would be replaced in future years.

Using the composite Tire-Life Index, Firestone's forecasters could consider how many tires went into service, or out of service, in terms of "standardized" tires. A standardized tire was the equivalent of a new conventional bias-ply tire with Tire-Life Index of 100. If a radial had twice the life of a bias-ply tire, one new radial wearing out was the equivalent of two new bias tires wearing out, or of two standardized tires. By multiplying the number of tires wearing out by their average TLI, the forecasters could express the tires wearing out in a given year in terms of a common denominator, the standardized tire.

b. Forecasting the tire-construction mix— The composite Tire-Life Index (that is, the index of an average tire for a particular year) was changing rapidly, as the mix of radial, belted-bias, and bias tires changed. As radials increased their market penetration, the composite TLI rose, because radials lasted almost twice as long as bias-ply tires. Firestone's forecasters expected that radials would follow the same pattern of penetration in the United States market as they had in other countries. Harry Hollingsworth remarked that "the pattern of industry radial growth is common to all countries, the only difference being when this growth will occur." Firestone developed composite OE and replacement radial-penetration curves, based on information from non-United States markets. Exhibit 9 shows curves representing the pattern of radial penetration in various replacement markets, along with a summary curve. The summary curve implies, for example, that a country with 50% radial penetration could anticipate over 80% penetration four years later. Using such curves, the company's forecasters could predict fairly correctly the future penetration of radials into both the OE and replacement markets by locating a nation's present position on the appropriate curve, and looking ahead the required number of years. Then they could forecast the changing proportion of radial tires in both the OE and the new-replacement-tire mix (that is, replacement tires exclusive of retreads). This procedure yielded forecasts of the composite TLI for tires entering service in a given future year. Once they had settled on composite TLI numbers for the forecast period, the PDL was applied to the relevant actuals and forecasts, generating composite TLIs for tires going out of service during the forecast period.

Trade journals reported Goodyear engineers as showing average tread lives of 60,000 miles on GM X-body cars as compared to 40,000 miles on other rear-wheel-drive cars equipped with radial tires. Firestone data also showed almost unbelievable increases in tread life. Firestone believed that the reason was oversized tires rather than front-wheel drive. Tires on X-body GM cars were one-half size larger than needed for the following reasons:

1. Cosmetic—the size required looked too small
2. To meet standard bumper height requirements
3. To provide sufficient road clearance

c. Miles per standardized tire — tire-life depended not only on construction, but also on driving habits, car size, economic conditions, and so on. Throughout the 1960s, cars grew heavier and engines more powerful, while power brakes and power steering became standard equipment. The new cars went faster, stopped more suddenly, and could be turned with little or no rolling motion.

Tires experienced increased friction and stress, wearing out more rapidly. Average standardized-tire-life decreased steadily. (See column 3, Exhibit 10, for tire-life by standardized tire, the basic tire unit.) After 1968, standardized-tire-life remained fairly constant at about 19,000 miles, with some variation due to external events such as the imposition of the 55-miles-per-hour speed limit. Most clearly evident were the effects of recessions, when people tended to drive their cars longer before buying either new tires or new cars. The higher figures for tire-life in 1970, 1974, and 1980 reflected the depressed economic climate of those years.[30]

Firestone expected tire-life to lengthen in the mid-eighties because cars were growing lighter and less powerful, reversing the trend of the 1960s. Contrary to general expectations, they did not wear out faster than did larger tires, even though the tires for these lighter cars were smaller and made more revolutions per mile. At the same time, the composite TLI of tires going out of service (column 5, Exhibit 10) was also increasing. The increase in average tire-life reflected both the increasing share of radials in the mix of tire constructions and changes in car technology.

d. Tires wearing out once — Firestone's forecasters had assembled historical data and developed forecasts for the major variables, tire-miles driven and tire-life; they could forecast the number of standardized tires that would go out of service each year. The last column of Exhibit 10 converts standardized tires used (Column 4) to actual tires used by dividing by the Composite TLI (Column 5) and multiplying by 100. Firestone was forecasting that 160.1 million tires would wear out in 1985. But would they be replaced?

3. Replacement Tires

To estimate demand for new passenger replacement tires, the next step was to estimate all the alternative ways of replacing (or not replacing) the worn-out tires (columns 3-6 in Exhibit 11). Their totals could then be subtracted from the number of tires wearing out (column 2, identical to column 6 in Exhibit 10).

a. Tires on scrapped vehicles (column 3) — As tire-life increased, there were fewer tire replacements during the life of an average car. Thus scrapped cars accounted for a growing percentage of worn-out tires in general.

b. Tires replaced by spare tires (column 4) — Consumer research showed that most spare tires replaced a worn-out tire some time between the fourth and sixth year of a car's life. Because the worn-out tire was then kept as the spare, this replacement did not lead to the sale of a new replacement tire. Car manufacturers, however, were now providing temporary spare tires as standard equipment. A worn-out tire would thus generate a replacement-tire sale, and Firestone was forecasting a drop in the replacement of worn-out tires by spare tires.

---

[30] In determining miles per standardized tire, Firestone's forecasters adjusted the tire-wearing-out figure to reflect "estimated change in distributor inventories and net effect of the Firestone '500' recall." They found that the adjustments had no significant effect on their results, and consequently the figures used in the memo and in this case do not include those adjustments.

c. Tires replaced by retreads (column 5) — The Firestone forecasters noted that sales of retreads had been artificially depressed:

> Starting about 1970, there were not enough belted-bias casings available in the new tire sizes to keep up with normal retread demand. The same thing was repeated for radial tires starting in about 1973.

They expected the retread share of the market to rise as more casings became available.

d. Tires replaced with light-truck tires (column 6) — The forecasters estimated the proportion of passenger tires in use on light trucks that would be replaced with light-truck tires. As trucks became more popular, truck tires were expected to account for an increasing fraction of replacements.

e. Total replacements (column 9) — The totals in columns 3-6 were subtracted from the number of tires going out of service, producing the figures in column 7; the snow-tire data and forecasts (Column 8) provided the final adjustment, resulting in the historical replacement shipment data and forecasts through 1986 (Column 9). For 1985, Firestone forecast shipments of 105 million replacement passenger tires, and for 1986 the forecast was even lower, 102 million tires.

Technical Concerns

Firestone's forecasters also worried about some of the incidental techniques they used. Hollingsworth explained:

> Some of the most important relationships keep changing. We don't have quite the right way of dealing with that. For instance, we use a distributed-lag model to relate the Tire-Life Index of tires entering service to the Tire-Life Index of tires going out of service. The polynomially distributed lag [PDL] is derived through regression. But as radials come to dominate the mix, the relationship between the index of tires entering service and those going out must alter. Tires will last longer, and the PDL ought to show that — it might make our forecast lower yet! Even though we refigure the PDL using current data, our equation won't be entirely correct. The PDL captures an average situation for the available history, and therefore is not exactly right for the forecast. We have that problem, of changing relationships, a lot — it came up in the sixties, as cars grew heavier and more powerful, affecting tire-life directly.

Hollingsworth expressed concern over several other technical issues. He wondered if the use of averages to estimate the impact of tire-life on replacement-tire demand might be concealing effects that would emerge in a less summarized treatment of the data. Often, too, it was hard to distinguish among several causal factors that acted together to produce similar results. Thus, it was not possible to know how much of the lessening tire-life in the 1960s was due to the positive economic climate, how much to changes in car design, and how much to the building of new, faster roads. Similarly, multiple factors worked together to depress replacement demand in the late 1970s: lighter, less powerful cars; greater operating costs; diminished economic optimism; and much more durable tires. Disentangling their individual effects was highly desirable, but might be impossible at the present time.

A retired tire-company executive, musing on the problems of forecasting, commented:

> You know, for years we underforecasted. Cars got more and more powerful; people drove more and more, using cheap gas; families acquired second and even third cars. Roads were straighter, and you covered the miles faster. More cars, more weight, more miles, more people — we couldn't keep up with the reality, and in the '60s our demand forecasts were always low. Then everything changed at once, and we had a decade of overforecasting, partly in reaction to all the underforecasting. Now where are we in that cycle? Nobody really knows yet.

Dillard Feltner and Harry Hollingsworth discuss the forecasts:

DF: We're going to get some searching questions about these figures, you know.

HH: Probably. But that's the way the numbers come out. If people ask, we can explain everything we've done, and I don't think we're very vulnerable in any of our contributory analysis.

DF: We do depend on several outside forecasts — for instance, DRI's forecast for car miles. If they are wrong, we may be wrong too. But the last report from Chase Automotive Service showed a lower car-miles forecast than any that service has previously made — in fact, it's even lower than the latest DRI forecast, which has come up a bit. And the DRI car-miles forecast has been one of the principal reasons our forecast of industry replacement passenger tires for 1985 has been lower than anyone else's. The fact that Chase is now lower than DRI supports our low number.

HH: People do ask me why our forecast should be so different from Goodyear's. I don't know how Goodyear got its figure, of course, but I do know there are plenty of uncertainties in our data. We can't be sure that our Tire-Life Index is precisely right, because we don't have complete information about other companies' tires — that is, about how long their first-, second-, and third-quality tires actually last. And the whole question of front-wheel-drive cars is so new that we can't be too confident about our conclusions. We're making a large adjustment for the increased radial-tire-life on front-wheel-drive cars.

DF: The fact is, no matter how finely you decompose the process, you've still got to rely on other forecasts, and on judgmental adjustments. And those can differ enormously.

Exhibit 1

Replacement Passenger-Tire Shipments (millions)*

Shipments forecast by Rubber Manufacturers Association

Forecast Made in:	1973	1974	1975	1976	1977	1978	1979	1980
Nov-69	159							
Nov-70	156	164						
Nov-71	154	161	169					
Nov-72	153	158	163	168				
Nov-73		155	161	164	167			
Nov-74			145	147	150	152		
Nov-75				140	145	149	151	
Nov-76					145	146	149	153
Nov-77						140	142	144
Nov-78							145	143
Nov-79								133

Actual Shipments

1973	1974	1975	1976	1977	1978	1979	1980
148	131	130	133	140	148	135	120

* Both actuals and forecasts include non RMA shipments, as estimated by the RMA Statistical Committee.

Exhibit 2

U.S. Passenger-Tire Shipments (millions)

Original Equipment

	RMA	Imports & non–RMA	Total
1970	37.5	0.3	37.8
1971	48.6	0.5	49.1
1972	51.3	0.9	52.2
1973	56.0	1.0	57.0
1974	43.3	0.6	43.9
1975	39.3	0.5	39.8
1976	49.9	2.0	51.9
1977	55.7	2.1	57.8
1978	55.0	2.0	57.0
1979	48.2	2.0	50.2
1980	34.9	1.8	36.7

Replacement

	RMA	Imports & non–RMA	Total
1970	129.6	3.8	133.4
1971	135.0	4.9	139.9
1972	141.3	6.2	147.5
1973	142.0	6.6	148.6
1974	123.9	8.0	131.9
1975	122.4	7.4	129.8
1976	123.0	8.5	131.5
1977	129.2	9.5	138.7
1978	135.2	11.1	146.3
1979	121.9	12.4	134.3
1980	106.9	11.4	118.3

Source: Firestone Tire & Rubber Company

Exhibit 3

Direct Forecast of Replacement Passenger Tires: Data and Regression Fit (millions)

Actual Shipments				*Regression Inputs*										
Year	**Repl**	**OE**	**Total**	**Repl**	**OE**	**Tot-1**	**Tot-2**	**Tot-3**	**Tot-4**	**Tot-5**	**t**	**Y(est)**	**Residuals**
1950	47.1	36.7	83.8	47.1	36.7								
1951	34.2	26.7	60.9	34.2	26.7	83.8							
1952	45.5	24.1	69.6	45.5	24.1	60.9	83.8						
1953	45.9	33.1	79.0	45.9	33.1	69.6	60.9	83.8					
1954	47.0	29.7	76.7	47.0	29.7	79.0	69.6	60.9	83.8		0		
1955	50.1	42.6	92.7	**50.1**	42.6	76.7	79.0	69.6	60.9	83.8	1	50.13	-0.03
1956	53.2	30.9	84.1	**53.2**	30.9	92.7	76.7	79.0	69.6	60.9	2	51.47	1.73
1957	56.6	32.7	89.3	**56.6**	32.7	84.1	92.7	76.7	79.0	69.6	3	58.17	-1.57
1958	61.6	23.4	85.0	**61.6**	23.4	89.3	84.1	92.7	76.7	79.0	4	62.69	-1.09
1959	66.8	29.8	96.6	**66.8**	29.8	85.0	89.3	84.1	92.7	76.7	5	66.25	0.55
1960	68.5	36.3	104.8	**68.5**	36.3	96.6	85.0	89.3	84.1	92.7	6	69.26	-0.76
1961	73.3	30.4	103.7	**73.3**	30.4	104.8	96.6	85.0	89.3	84.1	7	71.23	2.07
1962	78.4	37.5	115.9	**78.4**	37.5	103.7	104.8	96.6	85.0	89.3	8	76.49	1.91
1963	79.0	41.9	120.9	**79.0**	41.9	115.9	103.7	104.8	96.6	85.0	9	81.76	-2.76
1964	88.2	42.5	130.7	**88.2**	42.5	120.9	115.9	103.7	104.8	96.6	10	88.58	-0.38
1965	94.9	51.4	146.3	**94.9**	51.4	130.7	120.9	115.9	103.7	104.8	11	94.95	-0.05
1966	101.8	47.4	149.2	**101.8**	47.4	146.3	130.7	120.9	115.9	103.7	12	101.65	0.15
1967	108.5	40.8	149.3	**108.5**	40.8	149.2	146.3	130.7	120.9	115.9	13	110.55	-2.05
1968	121.3	49.4	170.7	**121.3**	49.4	149.3	149.2	146.3	130.7	120.9	14	119.04	2.26

Regression Number1

Dependent Variable: Repl

	Constant	OE	Tot-1	Tot-2	Tot-3	Tot-4	Tot-5	t
Regr. Coef.	(8.164)	0.02626	0.01888	0.1504	0.2199	0.2214	0.1627	1.419
Std. Error	10.741	0.12980	0.12075	0.1239	0.1205	0.1467	0.1103	0.803
t value	(0.8)	0.2	0.2	1.2	1.8	1.5	1.5	1.8

# of obs =	***14***	*Deg of F =*	***6***
R-squared =	***0.9947***	*Resid SD =*	***2.329***

Exhibit 4

Direct Forecast of Replacement Passenger Tires:
Regression Forecast Computation (millions)

				Data for Computing Regression Forecast						
Year	Repl*	OE*	Total	OE	Tot-1	Tot-2	Tot-3	Tot-4	Tot-5	t
Actual										
1968	121.3	49.4	170.7	49.4	149.3	149.2	146.3	130.7	120.9	14
Forecast										
1969	0.0	45.7	45.7	45.7	170.7	149.3	149.2	146.3	130.7	15
1970	0.0	48.9	48.9	48.9	45.7	170.7	149.3	149.2	146.3	16
1971	0.0	51.0	51.0	51.0	48.9	45.7	170.7	149.3	149.2	17
1972	0.0	52.7	52.7	52.7	51.0	48.9	45.7	170.7	149.3	18
1973	0.0	53.4	53.4	53.4	52.7	51.0	48.9	45.7	170.7	19
1974	0.0	56.8	56.8	56.8	53.4	52.7	51.0	48.9	45.7	20
1975	0.0	58.9	58.9	58.9	56.8	53.4	52.7	51.0	48.9	21
1976	0.0	61.3	61.3	61.3	58.9	56.8	53.4	52.7	51.0	22
1977	0.0	61.9	61.9	61.9	61.3	58.9	56.8	53.4	52.7	23
1978	0.0	63.2	63.2	63.2	61.9	61.3	58.9	56.8	53.4	24

Regression Number1
Dependent Variable: Repl

	Constant	OE	Tot-1	Tot-2	Tot-3	Tot-4	Tot-5	t
Regr. Coef.	(8.164)	0.02626	0.01888	0.1504	0.2199	0.2214	0.1627	1.419
Std. Error	10.741	0.12980	0.12075	0.1239	0.1205	0.1467	0.1103	0.803
t value	-0.8	0.2	0.2	1.2	1.8	1.5	1.5	1.8

# of obs = ***14*** *Deg of F =* ***6***
R-squared = ***0.9947*** *Resid SD =* ***2.329***

* Actuals for 1968; forecasts for 1969-1978. OE shipments were forecast directly by Firestone; replacement shipments were calculated from the regression shown in Exhibit 3.

Exhibit 5

Calculation of Passenger–Tire Miles on Conventional Tires, 1961–86 (billions of miles)

	(1)	(2)	(3)	(4)	(5)
	Year	*Car Miles*	*Tire Miles*	*Net Adjustments*	*Passenger–Tire Miles on Conventional Tires*
ACTUAL					
	1961	603	2,412	29	2,441
	1962	627	2,508	17	2,525
	1963	642	2,568	(1)	2,567
	1964	674	2,696	(19)	2,677
	1965	706	2,824	(45)	2,779
	1966	745	2,980	(54)	2,926
	1967	766	3,064	(61)	3,003
	1968	806	3,224	(75)	3,149
	1969	850	3,400	(85)	3,315
	1970	891	3,564	(85)	3,479
	1971	939	3,756	(81)	3,675
	1972	986	3,944	(68)	3,876
	1973	1,017	4,068	(49)	4,019
	1974	991	3,964	(23)	3,941
	1975	1,028	4,112	4	4,116
	1976	1,076	4,304	4	4,308
	1977	1,119	4,476	54	4,530
	1978	1,171	4,684	102	4,786
	1979	1,142	4,568	159	4,727
	1980	1,130	4,520	189	4,709
FORECAST					
	1981	1,135	4,540	206	4,746
	1982	1,149	4,596	224	4,820
	1983	1,167	4,668	254	4,922
	1984	1,188	4,752	291	5,043
	1985	1,202	4,808	316	5,124
	1986	1,217	4,868	332	5,200

*Col(3)=4 * Col(2)*
Col(5)=Col(3) + Col(4)

Exhibit 6

Bias (Diagonal) vs. Radial Tire Construction

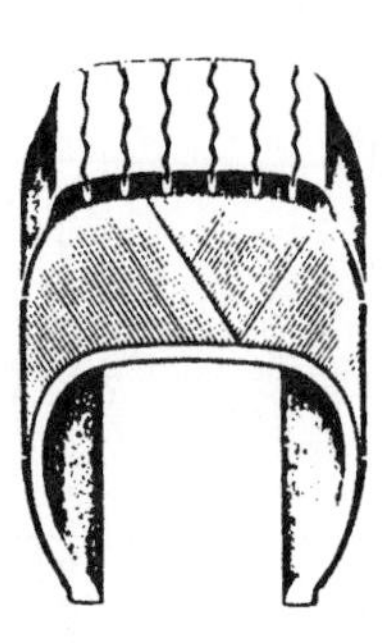

Diagonal Ply

The diagonal-ply tire has two or four layers of criss-crossed body plies. This conventional construction provides flexibility and two-directional strength in both tread and sidewall areas. It is a very serviceable construction and one that has been a standard in the tire-making industry for many, many years.

Radial Ply

The body cords of this type of tire run radially from bead to bead, at 90 degrees to the direction of travel. This radial-ply body provides flexible sidewalls that effectively cushion road shock with a minimum of cross-ply friction (and heat) since the ply cords don't criss-cross. Two or more stabilizing belts encircle the tire to add support to the body and to help hold the tread grooves open for effective traction. The belts also enable the elements of the tread pattern to resist the normal tendency to squeeze closed during road contact, effectively prolonging tread life.

Exhibit 7

Distribution of Projected Treadwear*

Decile	Miles
1	18,271
2	21,477
3	24,352
4	28,377
5	31,663
6	35,688
7	41,224
8	49,078
9	61,384

* Based on parking lot measurements of 1825 tires and a curvilinear projection. The curvilinear approach recognizes that radials (but not bias or belted bias tires) wear faster early in their lives, so that a tire that is one-fourth worn will last more than four times as many miles as it has already traveled.

Exhibit 8

Passenger Tires Going into Service and Tire-Life Index* (Tires in Millions)

(1)	(2)	(3)	(4)	(5)	(6)	(7)	(8)	(9)	(10)	(11)	(12)	(13)
New Tires									Retreads		Composite TLI	
Year	*Radial TLI*	*Radial% of New Tires*	*Belted TLI*	*Belted% of New Tires*	*Bias TLI*	*Bias% of New Tires*	*TLI, New Tires Only*	*New Tires as % of Tot Tires Going into Service*	*TLI*	*% of Tot Tire Going into Service*	*Going into Service*	*Coming Out of Service***
Actual												
1961					99	100.0%	99.0	78.6%	74	21.4%	93.7	92.9
1962	182	0.5%			100	99.5%	100.4	79.9%	74	20.1%	95.1	93.3
1963	182	0.5%			100	99.5%	100.4	80.9%	74	19.1%	95.4	94.0
1964	182	0.9%			101	99.1%	101.7	82.8%	74	17.2%	97.0	94.6
1965	182	1.2%			102	98.8%	103.0	85.2%	74	14.8%	98.7	95.4
1966	182	2.2%			102	97.8%	103.8	85.8%	74	14.2%	99.5	96.6
1967	182	2.7%			102	97.3%	104.2	86.3%	74	13.7%	100.0	97.9
1968	182	3.6%	123	7.2%	102	89.2%	106.4	86.9%	75	13.1%	102.3	99.0
1969	182	4.2%	123	26.3%	101	69.5%	110.2	88.1%	75	11.9%	106.0	100.2
1970	182	5.2%	123	44.5%	97	50.3%	113.0	87.5%	76	12.5%	108.4	102.1
1971	175	6.8%	134	47.4%	95	45.8%	118.9	88.9%	77	11.1%	114.3	104.7
1972	174	10.3%	131	46.9%	97	42.8%	120.9	90.2%	78	9.8%	116.7	108.3
1973	178	17.7%	134	44.8%	99	37.6%	128.8	90.5%	78	9.5%	123.9	112.2
1974	178	30.8%	133	33.8%	93	35.4%	132.7	89.3%	80	10.7%	127.1	116.8
1975	170	41.0%	131	28.3%	91	30.7%	134.7	89.1%	83	10.9%	129.1	121.4
1976	175	43.6%	136	26.8%	87	29.6%	138.5	90.2%	85	9.8%	133.3	125.4
1977	177	47.1%	133	25.4%	88	27.4%	141.3	91.7%	89	8.3%	136.9	128.7
1978	181	51.2%	129	23.5%	88	25.4%	145.3	91.7%	91	8.3%	140.8	132.1
1979	183	59.9%	129	17.1%	88	23.0%	151.9	91.9%	97	8.1%	147.5	135.8
1980	183	65.3%	125	12.8%	88	22.0%	154.9	81.3%	100	18.7%	144.6	140.2
Forecast												
1981	183	70.4%	121	11.7%	88	17.9%	158.7	90.1%	101	9.9%	153.0	144.7
1982	185	74.3%	118	9.5%	88	16.2%	162.9	90.1%	104	9.9%	157.1	148.9
1983	188	78.4%	114	7.8%	88	13.8%	168.4	90.1%	106	9.9%	162.2	152.6
1984	191	83.5%	110	6.6%	88	9.9%	175.5	90.1%	109	9.9%	168.9	156.6
1985	198	87.6%	108	5.5%	88	6.9%	185.5	90.1%	114	9.9%	178.4	161.6
1986	203	90.9%	107	4.5%	88	4.6%	193.4	90.1%	120	9.9%	186.1	168.0

*For Tire-Life Index, new bias in 1961-63=100. Tires going into service, including those on new vehicles, and Tire-Life Indexes estimated and forecast by Firestone Sales Forecasting section.

**Estimated by Firestone Sales Forecasting section using polynomial distributed lag which weights TLIs of tires going into service during previous four years.

Exhibit 9

Radial Tires Penetration Curves: Replacement Tires

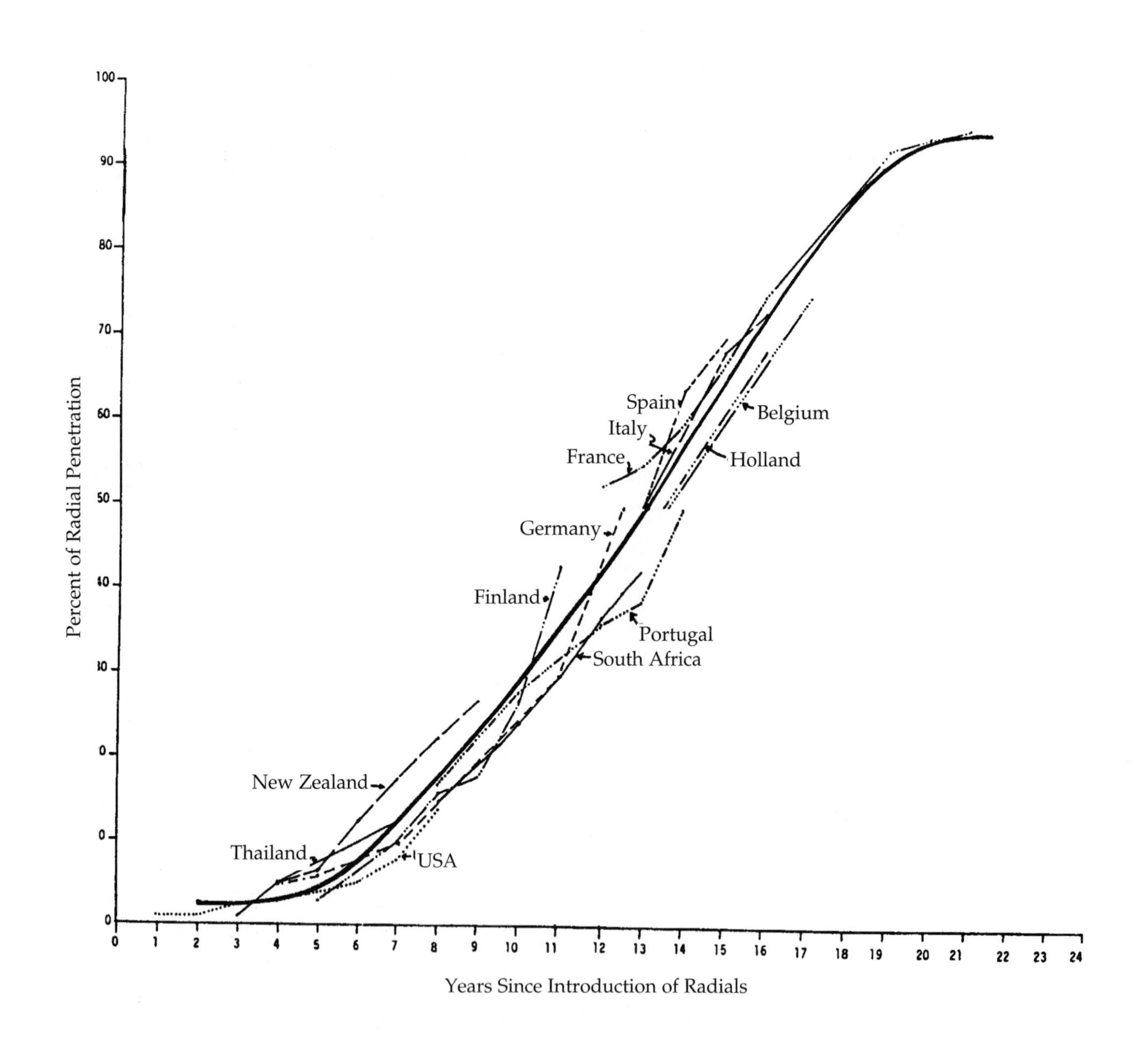

Exhibit 10

Tires Wearing Out

(1) Year	(2) Tire-Miles Driven on Conventional Passenger Tires* (billions)	(3) Miles Per Standardized Tire (thousands)	(4) Standardized Tires (millions)	(5) Composite TLI (Tires going Out of Service)**	(6) Actual Tires used (millions)
ACTUAL					
1961	2,441	22.5	108.5	92.9	116.8
1962	2,525	22.4	112.7	93.3	120.8
1963	2,567	22.3	115.1	94.0	122.5
1964	2,677	21.7	123.4	94.6	130.4
1965	2,779	20.8	133.6	95.4	140.0
1966	2,926	20.4	143.4	96.6	148.5
1967	3,003	20.5	146.5	97.9	149.6
1968	3,149	19.0	165.7	99.0	167.4
1969	3,315	18.7	177.3	100.2	176.9
1970	3,479	19.5	178.4	102.1	174.7
1971	3,675	18.9	194.4	104.7	185.7
1972	3,876	18.3	211.8	108.3	195.6
1973	4,019	18.8	213.8	112.2	190.5
1974	3,941	19.6	201.1	116.8	172.2
1975	4,116	18.8	218.9	121.4	180.3
1976	4,308	18.4	234.1	125.4	186.7
1977	4,530	18.4	246.2	128.7	191.3
1978	4,786	18.3	261.5	132.1	198.0
1979	4,727	18.7	252.8	135.8	186.1
1980	4,709	20.2	233.1	140.2	166.3
FORECAST					
1981	4,746	19.5	243.4	144.7	168.2
1982	4,820	19.0	253.7	148.9	170.4
1983	4,922	19.0	259.1	152.6	169.8
1984	5,043	19.4	259.9	156.6	166.0
1985	5,124	19.8	258.8	161.6	160.1
1986	5,200	19.9	261.3	168.0	155.5

* From Exhibit 5, Column (5)
** From Exhibit 8, Column (13)

Exhibit 11

Derivation of New-Replacement Tire Sales from Tires Wearing Out (Millions)

(1)	(2)	(3)	(4)	(5)	(6)	(7)	(8)	(9)
		Not Replaced by New Passenger Replacement Tire						
Year	*Total Tires Going Out of Service**	*On Scrapped Vehicles*	*Replaced with Spare*	*Passenger Retreads*	*Light Truck Tires*	*Passenger Tires Replaced with New Conventional Replacement Tires*	*Snow Tires*	*Total*
ACTUAL								
1961	116.8	18.3	7.6	26.0	0.6	64.3	8.8	73.1
1962	120.8	19.5	6.4	26.4	0.8	67.7	10.8	78.5
1963	122.5	21.2	6.3	25.9	0.9	68.2	10.8	79.0
1964	130.4	23.7	5.1	24.9	0.8	75.9	12.6	88.5
1965	140.0	28.0	6.4	23.1	0.9	81.6	14.0	95.6
1966	148.5	29.1	7.0	22.8	1.1	88.5	14.5	103.0
1967	149.6	26.3	6.4	22.1	1.2	93.6	16.3	109.9
1968	167.4	27.4	7.5	24.1	1.4	107.0	16.5	123.5
1969	176.9	30.6	8.2	22.6	1.3	114.2	18.2	132.4
1970	174.7	26.8	8.7	23.1	1.2	114.9	18.5	133.4
1971	185.7	30.5	10.0	22.4	1.5	121.3	18.2	139.5
1972	195.6	34.9	9.8	20.6	1.5	128.8	19.1	147.9
1973	190.5	29.8	9.0	20.7	1.5	129.5	18.6	148.1
1974	172.2	23.8	10.6	19.9	1.8	116.1	16.2	132.3
1975	180.3	30.2	10.5	19.7	2.2	117.7	12.5	130.2
1976	186.7	36.6	9.3	18.9	2.7	119.2	12.5	131.7
1977	191.3	35.0	11.2	18.3	2.3	124.5	14.1	138.6
1978	198.0	36.7	12.2	17.3	2.2	129.6	16.6	146.2
1979	186.1	35.0	13.0	16.3	2.9	118.9	15.7	134.6
1980	166.3	29.1	10.1	15.5	3.3	108.3	9.0	117.3
FORECAST								
1981	168.2	32.0	9.6	15.2	3.7	107.7	11.5	119.2
1982	170.4	36.6	11.0	15.6	3.1	104.1	11.0	115.1
1983	169.8	38.6	11.5	15.6	2.3	101.8	10.5	112.3
1984	166.0	40.2	9.3	15.8	2.6	98.1	10.0	108.1
1985	160.1	40.3	5.9	15.8	2.9	95.2	9.5	104.7
1986	155.5	40.0	3.6	16.0	3.1	92.8	9.0	101.8

* From Exhibit 10, Column (6)

DATA RESOURCES, INCORPORATED: NOTE ON ECONOMETRIC MODELS

Econometric forecasts such as those produced by Data Resources, Inc. (DRI) provide information on major aspects of the United States economy such as Gross National Product (more recently Gross Domestic Product) and its components, inventory levels, price levels, industrial production, and financial measures. Companies seeking to anticipate demand or costs for their own products, or interest rates for specific debt instruments, have much narrower concerns. Because macroeconomic trends clearly have an effect at the company level, however, individual companies often choose to link their demand or cost forecasting with macroeconomic forecasts, either through models developed for them by DRI or other macroeconomic forecasting firms, or through models they develop on their own. In either case, the macromodel drives the company's own forecasting. Exhibit 1 is a simple diagram showing DRI's conception of this linkage.

The following discussion summarizes the concept of econometric modeling, the way econometric models are used to forecast, and the necessary judgmental inputs (in the form of assumptions about values for exogenous variables and add factors)[31] that keep the forecast from being merely a mechanical output of the model and the data.

Econometric Models

An econometric model is a set of equations relating economic variables to each other, either behaviorally or by definition, so as to approximate the actual economic structure of the real world. Jan Tinbergen is generally credited with being the first to employ econometric modeling, and Lawrence Klein was its major proponent in the United States during the late forties and early fifties. (Klein is largely responsible for the model run by the Wharton Econometric Forecasting Associates.) By the early sixties the data base consisting of the National Income and Products Accounts of the United States Department of Commerce was sufficiently detailed, accurate, and stable to provide an adequate set of variables for large models, consisting of 100 to 200 equations. At the same time the availability of high-speed computers and the development of sophisticated algorithms made it possible to obtain solutions to such models. Model size has continued to increase, in order, as explained by Otto Eckstein, "to model the economic processes more fully as inputs to institutional decision making." More equations permit more variables to be used and economic relationships to be incorporated, and result in more disaggregated analyses and forecasts.

Harvard Business School case 9-183-097. This case was prepared by Alice B. Morgan, Research Associate, under the supervision of Professor Arthur Schleifer, Jr.

[31] See below for further discussion of exogenous variables and add factors.

Specification

A model-builder makes explicit his assumptions about economic relationships through the mathematical form of the model's equations. For example, an extremely simple model of the U.S. economy might involve just four variables:

- Consumption (*C*),
- Income (*Y*) ,
- Investment (*I*) ,
- Government Expenditures (*G*) .

In any one period, these variables might be related via two equations:

1) $C = A + BY + E$,

2) $Y = C + I + G$.

The first equation states that *C* (consumption) is a linear function of *Y* (income) plus *E*, a "disturbance" term. A and B are parameters of the first equation—numbers that represent the true (but unknown)[32] numerical relationship between *Y* and *C*. Equation 1 is a **behavioral** equation, in the sense that it describes the way consumption has behaved as a function of income in the past (and presumably will in the future); it is also a **stochastic** equation in the sense that even if *Y* were known with certainty in any one period, the value of *C* deduced from Equation 1 would be uncertain because of the disturbance *E* and the fact that the values of *A* and *B* are uncertain. Equation 2, on the other hand, is a definition: *Y* (income) is by definition the sum of *C* (consumption), *I* (investment) and *G* (government spending). Equation 2, having no error term, is nonstochastic. In this two-equation model, *I* and *G* must be determined outside the model. They are "exogenous" variables, while the variables *C* and *Y*, whose values are determined by the model, are "endogenous."[33]

The model expressed by Equations 1 and 2 is a "static" model: all the relationships bear on a single time period.

The following is an example of a dynamic model:

3) $C_t = A + BY_{t-1} + E_t$,

4) $I_t = I_{t-1} + H(C_t - C_{t-1}) + E'_t$,

5) $Y_t = C_t + I_t + G_t$.

The *t* subscripts indicate a particular time period, and *t*–1 indicates the preceding period. *C, Y, I* and *G* are defined as above, but the equations involve lagged as well as current values of these variables. Both *E* and E'_t are disturbance terms.

Equations 3 and 4 are behavioral or stochastic, while 5 is a definition. This model implies that consumption in a given time period varies linearly with income in the preceding period (Equation 3), and that the change in investment from the previous period is proportional to the change in consumption from the previous period (Equation 4).

The endogenous variables (those whose values are determined by the

---

[32] Although A and B are unknown, the model builder usually assumes that A is close to 0 and B is positive but less than 1: consumption should tend to increase roughly in proportion to increases in income, but consumption will generally not exceed income.

[33] With a little algebraic manipulation we can express the two endogenous variables in terms of the exogenous variables plus a disturbance term. Assuming $B \neq 1$,
1') $C = A/(1-B) + [B/(1-B)](I+G) + E/(1-B)$,
2') $Y = A/(1-B) + (I+G)/(1-B) + E/(1-B)$.
This is called the "reduced form" of the original equations.

model) are C_t, I_t, and Y_t. Variables whose values must be supplied to the model are Y_{t-1}, C_{t-1}, I_{t-1} and G_t. These variables are called "predetermined" in the econometrics literature, and they in turn can be partitioned into two subgroups: the "lagged endogenous variables" Y_{t-1}, C_{t-1}, and I_{t-1}, whose values are available from data already in hand; and the "exogenous" variable G_t whose value must be supplied to the model judgmentally.[34] (In "static" models like the one represented by Equations 1 and 2, the predetermined variables I and G are necessarily exogenous.)

Estimation

Next, the model builder must use data on income and consumption to estimate the parameters A, B, and (in the second model) H through regression analysis. The parameter estimates are denoted with lower-case letters, and whereas Equations 1–5 expressed structural relationships, Equations 1*–5* express regression estimates of those structural relationships:

1*) $C = a + bY + e$,

2*) $Y = C + I + G$,

3*) $C_t = a + bY_{t-1} + e_t$,

4*) $I_t = I_{t-1} + h(C_t - C_{t-1}) + e'_t$,

5*) $Y_t = C_t + I_t + G_t$.

The residual e in Equation 1* represents the difference between the actual value of C and the value estimated by the equation in which the residual appears, and similarly for the other residuals.

Verification and Testing of an Econometric Model

Before a model is used to forecast, it has to prove itself as a set of equations that make economic sense, that behave reasonably when solved using actual data, and that are likely to continue to make sense in the future. DRI accordingly tested its model using several types of simulation. One way was to introduce extreme values for some of the exogenous variables. Intuitively reasonable forecasts generated under these "exogenous shocks" tended to validate the model. Another way was to use the model and past data to "predict" values already in hand. Still another series of tests consisted of running the model out to a distant time horizon. Such simulations permitted assessment of the model's stability. If the model passed these and further validation tests, it could be treated as a reliable first cut at the future.

---

34 The reduced form, in which each endogenous variable is expressed in terms of predetermined variables only, can be shown to be

3′) $C_t = A + BY_{t-1} + E_t$,

4′) $I_t = AH + I_{t-1} + BHY_{t-1} - HC_{t-1} + (1 + H)E_t + E'_t$,

5′) $Y_t = A(1 + H) + I_{t-1} + B(1 + H)Y_{t-1} - sHC_{t-1} + G_t + (1 + H)E_t + E'_t$.

It might at first appear that because the lags are for one period only, known values of the lagged endogenous variables permit forecasts only one period ahead, but this is not true. If t is the period which is to be forecast, and if we are now in period t–2, C_t can be computed in Equation 3′ by using Equation 5′ to compute Y_{t-1} and substituting this value into Equation 3′: C_t will then depend on I_{t-2}, Y_{t-2}, C_{t-2}, and G_{t-1}, all of whose values, except for G_{t-1}, are known in period t–2. The values of I_t, and Y_t can similarly be computed from data available in period t–2, and this process can be rolled backward to provide forecasts any number of periods ahead based on known values of the lagged endogenous variables. Notice, however, that in all cases the value of the exogenous variable cannot be known at the time of the forecast.

Exogenous Variables

The exogenous variable G_t in the dynamic model represented by Equations 3–5 must either be known (the government might have announced its spending plans) or forecasted (on the basis of political, social and economic conditions). Typical exogenous variables in macroeconomic models include policy variables whose values depend on government monetary and fiscal plans; prices determined outside the United States such as OPEC's oil price; and others which are treated as exogenous for particular modeling reasons, e.g., the number of car-miles driven in a given year. Many variables can be either exogenous or endogenous, according to the preference of the modelers as reflected in the form and number of equations in the model. The policy variables in particular, because they are eventually determined by political and psychological considerations as well as economic and social ones, pose serious forecasting difficulties. In November 1981, for example, the DRI model forecasted a $130 billion budget deficit for 1983. Faced with this prediction, most of DRI's senior forecasting staff felt that some governmental action would be taken to prevent what at that time appeared to be so great a deficit. The staff examined the size, possible variations, and associated effects of many exogenous variables representing tax actions, government spending, government financing, and monetary policy; and individuals offered their views about appropriate and likely changes in some of these. Allen Sinai, DRI's senior vice president—Financial Sector, recalled:

> When we looked at the deficit forecasts we knew we couldn't go with those numbers. We had to think ourselves into the roles of government policy makers and forecast what they would do. I asked myself, "If I were Paul Volcker, the Chairman of the Federal Reserve, what would I want written on my tombstone?" My answer was "Here lies the man who ended inflation," and it gave me some idea what kind of decisions Volcker would make.

Once the model is in place, the exogenous variables forecasted, and all relevant data series updated, the forecast can be generated.

Forecasting with the DRI Model of the United States Economy

As new data came in, DRI ran its current econometric models, checking forecasts against actuals. Periodically, existing models were refit to include the most recent data; such refitting changed parameter estimates and hence forecasts. In addition to refitting an already specified model with new data, DRI and other econometric modeling firms periodically respecified their models, as economic theories were refined and as new data series became available. DRI had incorporated basic Keynesian assumptions—the circular flow of income and expenditure—into the heart of its model, but the model included as well a full representation of the economy's financial system. More recently, the company had added equations reflecting supply-side theory, and the model's structure was continually being analyzed and evaluated.

Chris Probyn, one of the model's architects, explained that DRI usually made significant changes in the model twice a year. Wholesale refitting was necessary in July when the Department of Commerce released revised data for the past several years, and some respecification might also occur at that time. More far-reaching respecification was usually accomplished early in the calendar year.

DRI had constructed its very large model so as to facilitate rapid and economical forecast solutions. Exhibit 2 shows, in simplified form, how major parts of DRI's model related to each other. The three-dimensional blocks at the top of the page represent exogenous variables; the arrows from these blocks show where in the model such variables appear. The flat shapes on the lower half of the page represent the model's sectors; the links connecting the sectors are shown as arrows. The model was structured so that, at these links, the forecasters could break into the interconnected flow of output from one equation to another with minimal disruption. Chris Probyn noted:

> We can interrupt the model solution at the links, and by "excluding" or "exogenizing" a few variables (somewhere between four and ten), we can work with just part of the model, temporarily ignoring the rest of it. To do this, we stipulate certain values for the exogenized variables—values that, in the full solution, would come from other parts of the model. Then we "tune" the unlinked sector, based on our exogenized-variable values, so that the equations are giving mutually consistent results. It's the model's capacity to be treated this way that makes rapid forecasting possible.

Exhibits 3 and 4 give some data about DRI's model as of spring 1981. Exhibit 3 provides information on the number and types of equations in the model's sectors (the more important sectors appear as the flat shapes in Exhibit 2). This exhibit also indicates the location of the model's 128 exogenous variables. In Exhibit 4, economic categories appear in the first column, and the theory underlying the model's treatment of each appears in column 2. The connections with other elements in the model are indicated in the final column.

Forecasting at DRI was an ongoing process, with those individuals who made exogenous forecasts constantly gathering information that would enable them to do so sensibly. At the same time, data from industry and government sources were accumulated and banked in the computer, and information from DRI staff who worked closely with government and business personnel was also steadily gathered and discussed. There were several subsidiary models run during the month, but the major macroeconomic forecast was produced in a frenzy of activity as soon as the Bureau of Economic Analysis of the U.S. Department of Commerce released the National Income and Product Account data it published every month. (This generally occurred between the 19th and 21st.) Banking the data to update DRI's time series required about 5 to 8 hours; it included checking data for consistency and updating series DRI tracked which depended on more than one government statistic (for example, many of the ratio series). The model was then solved for an initial forecast, and critical variables were examined in the light of current data and anticipated trends. Values for certain variables were chosen (i.e., those variables were exogenized) and individuals responsible for tuning specific sectors of the model—the consumption equations or the investment block—then started work. Their efforts and the necessary comparison and cross checking took place over the next few days. Outputs were scrutinized by appropriate personnel: for example, Allen Sinai checked all the financial forecasts. Any variable whose forecast appeared unreasonable was analyzed and revised as necessary. Otto Eckstein also reviewed the complete forecast, providing a final opportunity for questioning and revision. The forecast was usually ready for release less than one week after the government had made its data public.

Add Factors

Because each sector of the model used variables which appeared in other sectors, it was necessary to cross check constantly to be certain that results were internally consistent, and to make adjustments when they were not. Such adjustments were termed "add factors," and they might be required for any of a large number of reasons.

If an equation was systematically under- or over-forecasting, it could be corrected with an add factor. To determine the need for such an add factor, the "null solution" of the equation was run. The null solution can best be explained through an example. First, using data from, say, 1970-77, the equation's parameters were estimated. Then the equation was solved using later known values for both endogenous and predetermined variables (e.g., with data from 1978-81), and also using the coefficients estimated in the aforementioned period of fit (1970-77). This "null solution" thus provided a time series of values for the residual term ("*e*" in the examples above).

DRI ran the null solution for at least 16 quarters of known data. Equations showing biased errors, or correlated errors, could be adjusted through the use of add factors—alterations in the constant term of the equation.[35] As Otto Eckstein explained:

> There may be genuine serial correlation, so the add factors simply phase out the error terms according to formula. Policy, such as the legislated paths for the future prices of natural gas, may be a constraining variable, and add factors must therefore override the model solution. Some errors may be explained by nonrecurring events, such as harsh weather, strikes, or supply disruptions. But in the most difficult case, the surprising terms indicate economic change.
>
> Add factors are also the vehicle for introducing the information content of data not built into the model. Survey evidence on business and household plans, leading indicator signals, and even fragmentary evidence from direct field contact with the actors in the economy will add to forecasting performance. The model is an information-processing device, and all usable evidence, including the enormous body of data with positive information content not included in the model structure, should be considered. For example, it has long been known that filtered results from investment surveys beat any equation for the first few quarters. The construction of add factors is a means of giving weight to such evidence.

Consistency within or between blocks of equations, and with other DRI models, might also demand add factors. These would be incorporated as the month's forecast was developed.

---

[35] Since the errors were simply the differences between actual and forecast, statistical tests would determine whether they conformed to the standard "ordinary-least squares" regression assumption that they were independently drawn from identical distributions with zero mean. If the evidence seemed to contradict this assumption, appropriate adjustments in the constant term would be made; these adjustments were called "add factors."

Exhibit 1

Linkage Between DRI Macromodel and Company Forecasts

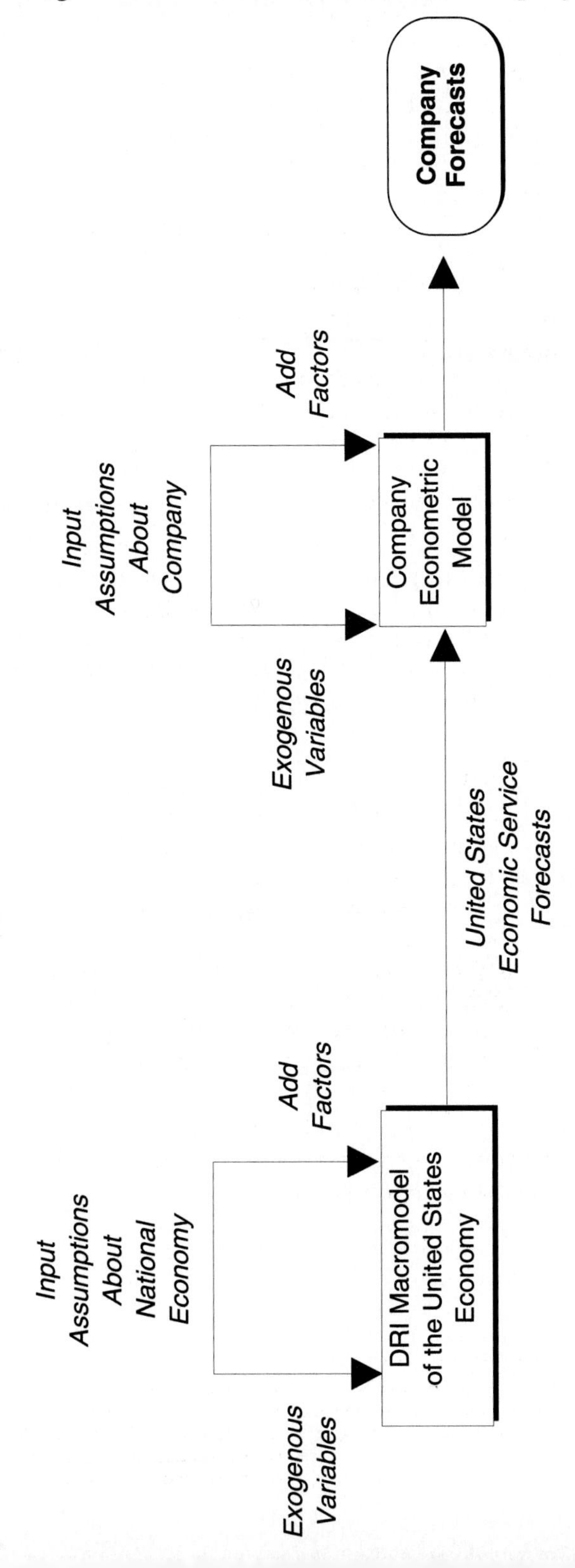

Exhibit 2

The DRI Model of the United States Economy

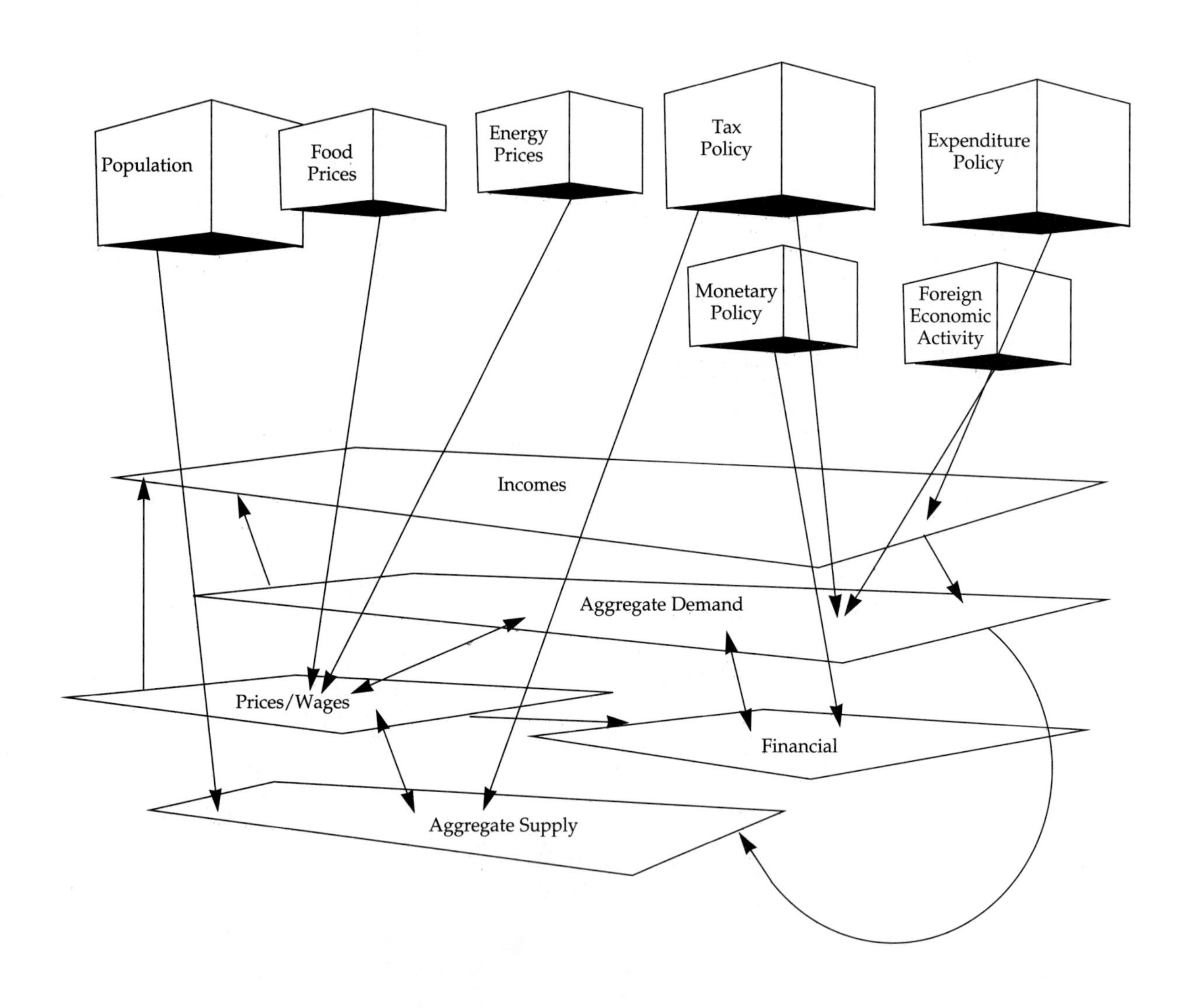

Exhibit 3

The DRI Model

	Stochastic Equations	Nonstochastic Equations	Total Equations	Exogenous Variables
Final GNP Demand	64	148	212	83
Consumption	19	30	49	5
Housing	8	12	20	2
Business Fixed Investment	7	22	29	11
Inventories	6	6	12	4
Government	10	38	48	34
Foreign	14	40	54	27
Income	15	37	52	6
Wages, Salaries and Supplements	0	6	6	3
Corporate Profits	3	6	9	–
Interest	3	1	4	1
Other	9	24	33	2
Financial	112	81	193	46
Monetary and Reserve Aggregates	8	14	22	8
Interest Rates and Stock Prices	26	1	27	18
Commercial Bank Loans and Investments	6	1	7	2
Flow of Funds—Households	20	12	32	–
Flow of Funds—Nonfinancial Corporations	25	33	58	10
Flow of Funds—Mortgage Activity	10	12	22	8
Flow of Funds—Government	3	1	4	–
Flow of Funds—Commercial Banks, Savings and Loan Associations, Mutual Savings Banks, Life Insurance Companies and Others	6	1	7	–
Equity Market, Inflation Expectations, and Others	6	3	9	–
Consumer Installment Credit	2	3	5	–
Supply, Capacity, Operating Rates	6	6	12	5
Prices, Wages and Productivity	57	37	94	14
Population				
Employment, Unemployment, and the Labor Force	9	1	10	14
Industry	112	96	208	6
Production	59	17	76	4
Investment	24	43	67	–
Capital Stock	0	32	32	–
Employment	29	4	33	2
TOTAL	**375**	**406**	**781**	**128**

Source: DRI

Exhibit 4

Economic Theory and Specifications in the DRI Model

	Theoretical Foundations	Extensions
Households Consumption	Utility maximization	
	Temporary and permanent income, real and financial assets, relative prices	Variance of income, debt burden, demographic structure, consumer confidence (modeled from macro risks of inflation and unemployment)
Labor Supply	Unemployment rate, wages	Demographic composition of the labor force
Wages	Price expectations, unemployment	Temporary and permanent price expectations
Firms	Profit maximization	
Fixed Investment	Rental price of capital, stock adjustment	Long- and short-term output expectations, surprises in actual output, cost of capital by financial sources, debt burden, balance sheet optimum, pollution abatement requirements
Inventory Investment	Stock adjustment to sales expectations	Errors in sales expectations, capacity utilization, delivery conditions, debt burden
Production	Variable coefficient input-output relations, supply from production functions including energy	Effects of several capacity constraints on output and price
Employment	Output, wage rates productivity trends	Cyclical productivity swings
Pricing	Material cost, unit labor cost, demand-supply disequilibrium, exchange rate	Vendor performance, stage of processing
Financial Institutions	Profit-maximizing portfolio behavior	
Portfolio Decisions	Balance sheet, expected own and alternative rates of return and opportunity costs	Modeling of flow-of-funds of households, corporations, financial institutions
Interest Rates	Price expectations, supply and demand of liquidity, sectoral borrowing demands	Segmented short- and long-term markets, competitive equity returns, interrelated portfolio adjustment dynamics
Central Bank	Exogenous in policy parameters	
State and Local Governments	Utility maximization for spending and taxes subject to budget constraint	Optimal revenue combination, demographic structure
Federal Government Spending	Real full employment values as policy variables	Policy levers for major fiscal instruments
Taxes	Income distribution, activity levels	Rates as policy variables
Rest-of-World Exports	Activity levels and relative prices abroad, exchange rate	World grain reserves, exchange rate response to balance of trade constraint
Imports	Relative prices, exchange rate, input-output relations	Capacity utilization, excess demand, real income

5 CAUSAL INFERENCE

INTRODUCTION

In Chapter 4, *Forecasting with Regression Analysis*, we saw how you could use regression analysis to forecast a future value of a dependent variable. In this chapter we take up a second use of regression: causal inference. We use regression to infer by how much a deliberate intervention that changes the value of some independent variable will *cause* the value of the dependent variable to change. For example, we may be interested in how much our company's stock price will change if we increase our dividend by $1, or how much gross domestic product changes if the Fed lowers interest rates by 1%, or how sales will change after an increase of $1 million in advertising expenditures. Sometimes you can infer these causal effects by examining past data. When such data are "observational" (not obtained from carefully controlled experiments), you can use the regression model to estimate those effects.

Many textbooks in statistics assert that you cannot infer causation from observational data; rather, you can only infer statistical association. They state that you might observe in a representative sample of data that whenever x increases by one unit, y increases by three units, on average, but from that you cannot conclude that the expected value of y will necessarily increase by three units (or at all) if you deliberately cause x to increase by one unit.

Contrast this with what managers, scientists, and economists do. A manager of a fast-food chain might observe that restaurants located near a McDonald's restaurant do better than those that are not near a McDonald's, and as a result will seek new restaurant sites near McDonald's. A doctor may observe that patients who smoke incur heart and lung disease at a higher rate than nonsmokers, and therefore advise her patients not to smoke. An economist may observe that high inflation is associated with low unemployment and vice versa, and recommend that to control inflation the government should take steps to increase unemployment. These are all violations of the textbook rule.

In this chapter, we show how you can sometimes infer from observational data how changing the value of one variable causes the value of some other variable to change. You will see that although you cannot "prove" causation, you may be able to make reasonable causal inferences. But you can easily make incorrect inferences unless you are very careful.

Harvard Business School note 894-032. This note was prepared by Professor Arthur Schleifer, Jr.

WHAT IS CAUSATION?

An assertion that reducing the price of a product from $18 per unit to $17 *causes* demand to increase from 22,000 units per month to 23,000 makes sense only in terms of an ideal experiment—one that can never be performed in practice. The experiment involves two scenarios. In the first, price is set at $18. The second scenario is identical to the first in all respects except that the price is set at $17. Under both scenarios, the monthly demand is observed. If demand was 22,000 units per month when price was $18, and 23,000 when price was $17, we can conclude that the price change *caused* the change in demand.

"Identical in all respects . . ." means just that. The two scenarios cannot take place in different periods of time, or in different geographical regions. Although it is common practice to assert that an increase in demand that followed a reduction in price was "caused" by the price reduction, such an assertion involves measurement of demand in two different periods, between which other factors that affect demand may have changed.

Given that the "ideal" experiment can never be performed in practice, we can never measure exactly how much a change in the value of one variable causes the value of another variable to change. Our challenge is to find methods that come as close as possible to mimicking the ideal experiment. But before discussing such methods, let's explore what we would learn if we could actually carry out this two-scenario experiment.

The price reduction—the variable whose value we deliberately change—is called a **treatment** or an **intervention**. What is the effect of that treatment? Although we have focused on just one effect that was of particular interest to us—the change in demand—it should be clear that the price change causes not just this one effect, but many effects. An increase in demand of 1,000 units next month means that some customers who would not have purchased at the old price decided to purchase as a result of the intervention. This, in turn, may mean that some of them will not purchase some other product, and that some will reduce their savings or increase their debt. Perhaps a competitor will respond to our price reduction by reducing his price as well (something he would not have done had we not lowered our price), and this too might have an impact on the demand for our product.

Some of those effects may have little to do with our "bottom line," but others may. If lowering the price of one item in our product line diverts demand away from other items, then the total effect of the intervention is the increase in demand for the item whose price was reduced, less the decrease in demand for substitute products. This is more appropriately measured in dollars than in units. A single intervention usually can change the values of many variables, but one of them—the net change in dollar sales across all items in our product line—is the one we select to be the **dependent variable**, the one that most appropriately measures the total effect of the intervention.

WHAT IS AN EXPERIMENT?

The nearest we can come in the real world to measuring the true effect of a treatment is to conduct an experiment by finding "matched pairs"—pairs of individuals (or experimental units) that are as alike as possible in all respects—and to apply the treatment to one member of the pair and not to the other. If the pairs were truly matched in all respects, we could achieve with matched pairs what the "ideal experiment" does with a single individual or experimental unit:

measure the effect by measuring the difference in their responses. Unfortunately, from a practical point of view, it is not possible to find individuals who are exactly alike in all respects. (Even identical twins have almost certainly been exposed to different environmental influences.) Thus "matched pairs" may be alike in many respects, but they may differ with respect to other, unmeasured variables that have an effect on the dependent variable. If these unmeasured variables happen to be correlated with the treatment, the observed treatment effect will include a **proxy effect** for these other variables.

Even the choice of which experimental unit gets the treatment and which does not may make it difficult to sort out effects. For example, the average effect on longevity of giving up smoking (the "treatment") may be different for people who voluntarily give it up than for people who are forced to give it up. In the extreme, we could imagine a situation in which those who voluntarily give up smoking are the only ones whose longevity is increased. If this were true, we would observe that those who gave up smoking lived longer than those who did not, but we would also discover that applying coercion or providing incentives to nonvolunteers to give up smoking would provide no benefits to them.

Unwanted proxy effects of unmeasured variables can be eliminated by using a random device to choose which experimental unit in a matched pair receives the treatment. Random assignment of treatments assures that, on average, whatever unmeasured variables affect the dependent variable will not be correlated with the treatment. Thus, the treatment will not capture unwanted proxy effects.

There are situations where this random assignment can be carried out relatively easily. Returning to our price-reduction problem, suppose the context is that of a direct-mail company. The company could prepare two sets of catalogs, both identical except for the item whose price reduction was under consideration. One set of catalogs would show the standard price; the other set, the reduced price. Matched pairs of customers could be selected based on the recency, frequency, and monetary value of their previous purchases, and for each pair the catalog with the reduced price could be assigned at random. In drug testing, it is routine to assign the drug to one member of a matched pair, and a placebo to the other, with the determination of who gets what decided by a randomizing device (e.g., the flip of a coin).

There are other situations where random assignment is virtually impossible, either because it is too hard to implement or because it is socially unacceptable. In the smoking experiment, it would be impossible to justify and enforce a policy in which some people, chosen at random, were instructed to keep smoking, while others were told to stop smoking. In dealing with the economy, we cannot segment the population into two groups, and take measures to increase unemployment among only one group, then observe the difference in the rate of inflation in the two segments. Even in the pricing example, the mail-order company's management might find it unacceptable to have two catalogs with different prices in circulation. We are often reduced to relying on observational data.

Observational Data

When we seek to estimate from observational data the effect of a "treatment"—an independent variable whose value we will be able to manipulate in the future—on a dependent variable, the estimation problem is made more difficult by the presence of other independent variables that may also affect the dependent variable, and may be correlated with the treatment variable.

An independent variable may be correlated with the treatment for one of four reasons:

1. There may be no causal relationship between the two variables, but they might be correlated by chance alone.
2. The independent variable may **affect** the treatment.
3. It may be **affected by** the treatment.
4. It and the treatment may both be affected by some other variable: they may be correlated due to a "common cause."

An Example

We may, for example, be interested in learning by how much changes in the posted speed limit on highways (the treatment) affect motor-vehicle death rates—deaths per thousand drivers per year (the dependent variable). The reason for our interest is that if we discover that reducing the speed limit reduces death rates, we might want to propose legislation to lower the speed limit.

Suppose we have a cross section of the fifty states in the United States, with data on each state's maximum speed limit and motor-vehicle death rate. Even if lowering the speed limit really reduced the death rate, a scatter diagram of speed limit vs. death rate might show that, in the data, death rate declines as speed limit increases. How could this be? It might be that states with very high death rates are states where bad weather makes driving conditions hazardous, where drivers drive long distances, where driving under the influence of alcohol is prevalent, etc. These states may have lowered the speed limit to reduce the carnage, but still have higher death rates than states in which driving is safer, but the speed limit remains high. If this story is correct, then low speed limits may reduce the death rate but also proxy for variables that increase the death rate.

If those other variables—weather conditions, miles driven per capita per year, alcohol consumption, etc.—are included, along with speed limit, as independent variables in a regression model, then the regression coefficient on speed limit will show how death rate varies with speed limit when the other variables in the model are held constant. Speed limit will no longer proxy for these other variables. If lower speed limits reduce the death rate, then the regression coefficient on speed limit will be negative. Weather conditions, miles driven, and alcohol consumption are examples of other independent variables that *affect* death rate and that are *correlated* with speed limit. The correlation occurs because these variables have *caused* the speed limit to be lowered in states where they are major contributors to highway deaths. These variables *should* be included in the model to eliminate their unwanted proxy effects on the treatment variable.

Suppose we discover that states whose citizens have pronounced concerns for public safety tend to have both low speed limits and rigorous automobile-inspection standards. This is a case where the treatment and another variable (automobile inspection) that may affect death rates are correlated because they are both affected by another variable that is a **common cause**—concern for public safety. Clearly, a measure of the rigor of inspection standards should be included as an independent variable; otherwise, speed limit will capture the unwanted proxy effect of inspection on death rate.[1]

[1] If inspection and speed are perfectly correlated, it will be impossible to sort out to what degree each contributes to lowering the death rate. This is an example of **collinearity**, a problem that occurs frequently in regression. Collinearity is not a defect in the methodology; it is a defect in the data. The only way you can sort out the effects in this extreme case is to change the relationship between speed limit and inspection in the data.

Now suppose that reduction in the posted speed limit causes drivers to drive slower, on average, and it is the actual reduction in speed driven, not the posted speed limit, that causes the death rate to decrease. Should we include average speed driven as an independent variable in our model? Clearly not: if we did, the regression coefficient on posted speed limit would show its relationship to death rate when average speed driven remained constant, and since actual speed driven, not posted speed limit, causes fatal accidents, the coefficient would indicate that the effect of posted speed limit on death rate was zero, and thus make it appear that changing posted speed has no effect on death rate. To assure that the regression coefficient correctly captures the causal relationship, we want posted speed limit to *include* the "good" proxy effect of driving speed, and thus we want to *exclude* driving speed from the regression model. Driving speed is an example of a variable that affects the dependent variable but is *affected by* the treatment variable. Such a variable should be excluded from the model, so that the treatment variable will capture its proxy effects.

Finally, consider the case where some other variable, say the average age of cars in the various states, affects death rates, but has no causal relationship to posted speed limits. Nonetheless, average age may be correlated with speed limits in the sample data: even variables that have nothing to do with one another are seldom perfectly uncorrelated in observational data. In this case, failure to include average age of cars as an independent variable in the model will cause speed limit to carry an unwanted proxy effect for age of cars. We should, therefore, *include* age of cars as an independent variable.

Which Independent Variables Should Be Included?

Based on this example, we can state the following rules. When you want to estimate the effect that a treatment or intervention will have on a dependent variable, you should:

- include as an independent variable any variable that you believe might affect the dependent variable and that is correlated with the treatment variable because (a) it affects the treatment variable, or (b) both the dependent variable and the treatment variable are affected by a common cause, or (c) the correlation occurred purely by chance.
- exclude from the model any variable that you believe might affect the dependent variable and that is correlated with the treatment variable because it is affected by the treatment variable.

If a variable affects the dependent variable but is uncorrelated with the treatment variable, whether you include it or not makes no difference in the regression coefficient for the treatment variable: there are no proxy effects from an uncorrelated variable. As a matter of practice, you should include any variable that affects the treatment variable; at the very least, it will improve the fit of the model. In a sample of observational data only rarely are two variables completely uncorrelated.

The consequences of these rules may seem counterintuitive. A variable that affects the dependent variable and is correlated with the treatment variable must be included in the model; the higher the correlation, the more important it is to include it. Including such a variable will not greatly improve the fit (increase R^2, decrease RSD). Nevertheless, omitting a variable like this distorts the apparent effect of the treatment variable by causing it to pick up the unwanted proxy effects of the omitted variable.

On the other hand, omitting an independent variable that affects the dependent variable and is affected by the treatment variable assures that the treatment variable captures its "good" proxy effects. Nevertheless, omitting such a variable invariably results in a poorer fit (lower R^2, higher RSD).

In both cases, what is good for proper causal inference is bad for forecasting. This seemingly counterintuitive result is resolved by recognizing that causal inference involves correctly estimating a particular regression coefficient, while forecasting involves providing a good fit to past data. What is good practice for dealing with one of these problems is not necessarily good practice for dealing with the other.

How to Identify the Relevant Independent Variables

Given that you should include any independent variable that might affect the dependent variable and that is correlated with the treatment variable, but is not affected by it, how do you decide what variables to include? The answer depends on your understanding of what causes what, and on your ingenuity in finding ways of measuring crucial variables. Sometimes you have to settle for a variable that is correlated with such a crucial variable. In our speed-limit example, you may have trouble obtaining data on drunk-driving convictions in a state, but probably can easily get statistics on alcohol consumption. For many reasons this may be an imperfect measure of driving under the influence, but it may be good enough for our purposes. Think about how you would measure weather conditions that are dangerous for drivers, or amount of driving per person.

A useful tool for depicting causal relationships is the **influence diagram**. Figure 5.1 shows such a diagram schematically. A treatment variable and a dependent variable are shown. Other variables are classified by type. Type A variables affect the dependent variable directly as well as indirectly through their effect on the treatment variable. Type B variables affect the dependent variable and are correlated with the treatment variable by virtue of a common cause. Type C variables affect the dependent variable and are correlated with the treatment variable by chance. Type D variables affect the dependent variable but are uncorrelated with the treatment variable. All of these variables should be included as independent variables in the regression model.

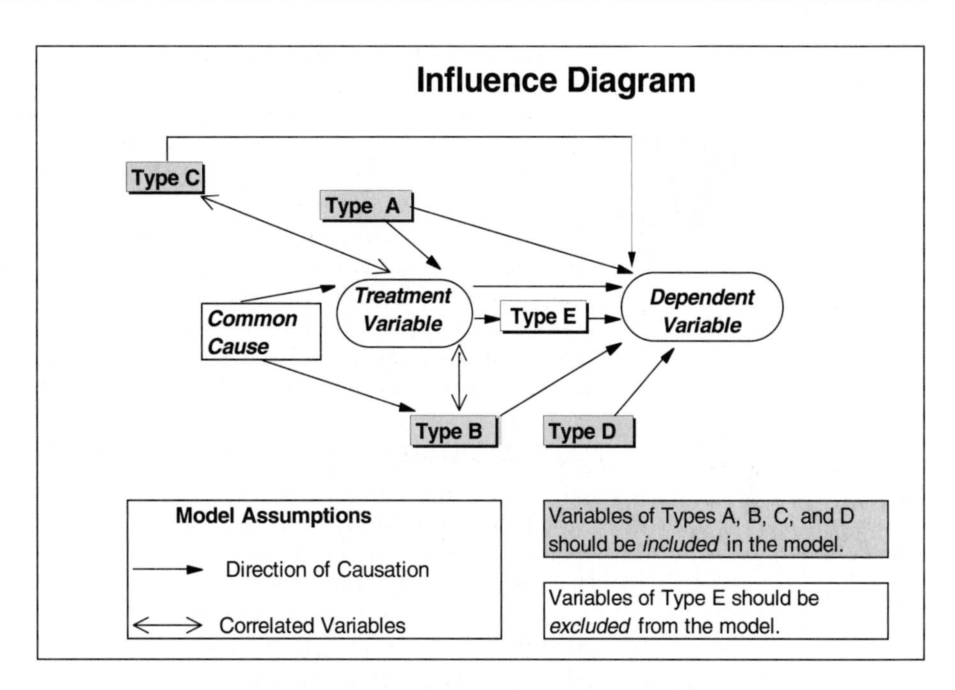

Figure 5.1

Type E variables, on the other hand, affect the dependent variable directly, but are affected in turn by the treatment variable. They should not be included in the model.

It is not always clear which way the causation goes. Advertising expenditures by your competitor may be correlated with your advertising. The correlation may be due to a common cause (seasonality, business conditions), in which case your competitor's advertising should be included as an independent variable. On the other hand, your competitor may simply be reacting to your advertising, raising expenditures when you raise yours, and vice versa, in which case it should be excluded. About all you can do under such circumstances is perform the regression with and without the competitive-advertising variable, and weight the resulting regression coefficients on the treatment variable by the probability you assign to the two competing causal models.

Exercise on Causal Inference

In 1976 the Federal Trade Commission (FTC) launched an investigation to provide insight into the question of whether test preparation centers such as Stanley H. Kaplan have a reasonable basis for claiming they help students increase their Scholastic Aptitude Test (SAT) scores.

The data base KAPLAN.XLS contains observations on 246 high school students who took the test twice. Some of these students were coached by Kaplan or some other test preparation center between the first and second administration of the test. The variables consist of SAT scores, demographics, high-school-performance indicators, and whether or not coaching occurred.

What do you conclude about the effectiveness of coaching?

THE GOTHAM GIANTS

On an afternoon in January 1980 Harold Glasshofer, the owner of the Gotham Giants major-league baseball team, sat down to review the report which had just been completed by his consultant. The report contained an analysis of the effects of television broadcasts and promotions on attendance at the Giants' games. Glasshofer hoped that the consultant's analysis would be useful in his upcoming negotiations with WQJY, a local television station.

The Economics of Major-League Baseball

The most important source of income for a major-league baseball team is ticket sales, which typically account for 50%-70% of total revenue. For "home games" (games played at the Giants' stadium in Gotham), the Giants receive 80% of the ticket revenue and the opposing team receives 20%.

Television and radio contracts also provide substantial income. Currently, two major television networks broadcast games nationally or regionally, each showing one game per week for part of the season. The networks pay Major League Baseball for the rights to broadcast the games, and this revenue is divided among all the teams. At the local level, typically one TV station will purchase from the local team the rights to broadcast some (or most) of that team's games.

Concessions are a third source of income for major-league baseball teams. For the upcoming 1980 season, J.H. Pierce and Company had purchased from the Giants the right to sell refreshments and souvenirs at the Giants' home games. As part of the agreement, J.H. Pierce and Company would pay the Giants $0.75 for each person in attendance. The principal expenses associated with operating a team are player salaries, which have been growing rapidly in recent years.

Effects of Television on Attendance

Late in 1979 Glasshofer decided to re-evaluate his strategy concerning the televising of games. It was generally believed that televising a home game would lower attendance, because people would find it easier and cheaper to watch the game on TV than go to the ballpark. Glasshofer was therefore surprised and delighted to find that in 1979 the average attendance had been substantially *higher* for televised games than for non-televised games (Table A).

Table A

	Average Attendance	Number of Games
TV	36,784	40
No TV	27,140	39

Harvard Business School case 9-182-153. This case was prepared by Professor William S. Krasker.

Since the Giants received $12,500 from station WQJY for each televised game, Glasshofer was tempted to televise as many games as WQJY would agree to. In December 1979, as the negotiations with WQJY were nearing, he hired a consultant to study the situation further and make a recommendation. Glasshofer also asked the consultant to analyze the effect of "special events."

Special Events

A special event is a promotion designed to boost attendance at a particular game. A "give-away" is the most common special event. Every person who attends the game that day is given a complimentary gift, such as a baseball cap or jacket. Though special events always boost attendance, they also cost money (see Exhibit 2 for the costs of certain give-aways), and Glasshofer was not sure if the special events yielded a profit.

Seasonal Patterns in Attendance

The major-league baseball season starts in April and ends in the first week of October. Attendance is generally low early in the season, particularly in cities like Gotham where the spring-time temperatures can be uncomfortably low. Though attendance always increases as the season progresses, the size of the increase depends partly on how well the team is playing. Attendance is rarely high in September unless the team is still in contention for the championship.

Nearly half of the Giants' ticket sales are to season-ticket[2] holders. These sales occur before the start of the baseball season. The remaining ticket purchases for a particular game occur mainly during the month preceding that game. The exceptions to this general rule are games against the Giants' arch-rivals. Tickets for those games are often purchased several months in advance (Exhibits 3 and 4).

Exhibit 1

Attendance and Ticket Prices in Recent Years

	1975	1976	1977	1978	1979	1980
Attendance (millions)	1.29	2.01	2.10	2.34	2.54	–
Average ticket price*	$3.57	3.63	5.00	5.25	5.60	5.85

* Calculated by dividing revenue from a full stadium by the number of seats.

Exhibit 2

Cost of Give-aways

Item	Variable Cost
Jacket	$1.89
Cap	.96
Bat	2.04
Ball	1.14
Batting Glove	1.39
T-Shirt	1.49

The fixed cost of a give-away is approximately $20,000 which includes promotion, administrative costs, and additional ushers.

[2] A season ticket was originally a pass which entitled the bearer to a special seat at all of a team's home games. In recent years teams have begun to offer season tickets for certain subsets of the games, for example, all weekday night games.

Exhibit 3

1980 Giants Schedule

MONTH	DATES	DAY OF WEEK*	OPPONENT**
April	18 19 20	F S Su	8
April	21 22 23	M T W	12
April	25 26 27	F S Su	7
May	9 10 11	F S Su	6
May	12 13 14	M T W	5
May	16 17 18	F S Su	4
May	26 27 28	M T W	9
May	30 31 June 1	F S Su	11
June	16 17	M T	1
June	18 19	W Th	3
June	20 21 22	F S Su	2
June	23 24 25	M T W	13
June	27 28 29	F S Su	10
July	15 16 17	T W Th	6
July	18 19 20	F S Su	5
July	21 22 23	M T W	8
August	4 5 6	M T W	4
August	8 9 10	F S Su	12
August	11 12 13	M T W	7
August	28 29 30 31	Th F S Su	1
September	1 2 3	M T W	2
September	4 5 6 7	Th F S Su	3
September	16 17 18	T W Th	11
September	19 20 21	F S Su	13
September	22 23 24 25	M T W Th	10
October	2 3 4 5	Th F S Su	9

* M – Monday, T – Tuesday, W – Wednesday, Th – Thursday, F – Friday, S – Saturday, Su – Sunday

** To maintain confidentiality, the 13 other teams are referred to by number only.

Exhibit 4

Data on the Giants' Home Games in 1979

VARIABLE	DESCRIPTION
OBS	Observation Number
MONTH	4 = April, 5 = May, ..., 9 = September
DAY	Calendar date
DAY OF WEEK	1 = Monday, 2 = Tuesday, ..., 7 = Sunday
ATTENDANCE	Attendance at the game
TV	1 if game was televised
TEMPERATURE	Air temperature in degrees Fahrenheit
TYPE OF DAY	0 = clear skies, 1 = cloudy, 2 = rainy
OPPONENT	Opposing team, numbered 1 through 13
STAR PITCHER	1 if one of the Giants' star pitchers was playing*
SPECIAL EVENT	1 if special event
NIGHT GAME	1 if the game was played at night
GAMES BEHIND	Games behind for the Giants**

In addition, several dummy variables derived from the variables listed here are used in the regressions which follow. Examples are

WEDNESDAY (1 if DAY OF WEEK = 3), AUGUST (1 if MONTH = 8), OPPONENT 7 (1 if OPPONENT = 7), APRIL-MAY (1 if MONTH = 4 or MONTH = 5), and RAINY DAY (1 if TYPE OF DAY = 2).

* The pitcher is the most important of the nine players on a baseball team. Due to the strain on a pitcher's arm, an individual cannot pitch every day, and therefore a team must have several pitchers. Two of the Giants' pitchers were regarded as outstanding.

** Games Behind is a measure of how far the Giants are behind the first-place team. It is a measure of the Giants' success to date and is defined as

$$\text{GAMES BEHIND} = \frac{(W-WS) + (LS-L)}{2}$$

where

W = Number of wins for the first-place team

WS = Number of wins for the Giants

L = Number of losses for the first-place team

LS = Number of losses for the Giants

As the season progresses, this becomes a closely-watched statistic, especially if the Giants have a chance of finishing first in their league.

EXHIBIT 4 CONTINUED

DAY OF GAME		DAY OF WEEK (1 = Mon)	NIGHT GAME (1 = night)	ATTENDANCE	TV? (1 = yes)	SPECIAL EVENT (1 = yes)	TEMP. (DEG F)	TYPE OF DAY	STAR PICHER (1 = yes)	OPPONENT	GAMES BEHIND
Month	Day	Wkday	Nite	Attend	TV	Spec	Temp	Weath	Star	Opp	Behind
4	5	4	0	52,719	1	1	55	1	1	7	0.0
4	7	6	0	17,387	1	0	49	1	0	7	1.0
4	8	7	0	26,954	0	0	64	1	1	7	2.0
4	17	2	0	20,135	0	0	54	1	1	1	1.5
4	18	3	1	25,562	0	0	62	0	1	1	0.5
4	19	4	0	21,201	0	0	64	0	0	1	0.5
4	20	5	1	26,651	1	0	64	0	0	12	0.5
4	21	6	0	25,530	1	0	70	0	0	12	0.0
4	22	7	0	35,250	1	0	70	1	1	12	1.0
5	4	5	1	17,705	1	0	72	1	0	9	4.0
5	5	6	0	30,167	1	0	68	0	1	9	4.5
5	6	7	0	46,750	1	0	68	1	0	9	4.5
5	7	1	1	14,065	0	0	75	0	0	11	4.5
5	8	2	1	15,981	1	0	81	0	0	11	4.5
5	9	3	1	14,738	0	0	94	0	0	11	4.5
5	10	4	0	14,394	0	0	94	0	1	11	3.5
5	11	5	1	37,998	0	0	79	1	0	10	3.5
5	12	6	0	28,783	0	0	61	2	0	10	4.5
5	13	7	0	30,083	1	0	72	1	0	10	4.5
5	14	1	1	15,650	1	0	68	2	0	5	4.0
5	15	2	1	18,876	0	0	73	1	1	5	4.5
5	16	3	1	43,843	1	0	78	1	1	5	4.0
6	1	5	1	33,230	0	0	83	0	1	3	4.5
6	2	6	1	53,539	1	1	73	1	0	3	3.5
6	3	7	0	55,073	1	1	66	1	0	3	4.5
6	4	1	1	30,164	1	0	66	2	1	6	3.5
6	5	2	1	24,988	0	0	83	0	0	6	3.5
6	6	3	1	34,075	1	0	74	1	1	8	3.5
6	7	4	0	20,722	0	0	79	1	0	8	3.5
6	19	2	1	36,211	0	1	82	0	1	13	8.5
6	20	3	0	32,129	1	0	83	0	0	13	9.5
6	21	4	0	20,078	0	0	77	0	0	13	10.0
6	22	5	1	33,776	0	0	73	1	0	4	9.5
6	23	6	1	25,818	1	0	81	1	1	4	9.5
6	24	7	0	55,049	1	1	65	0	0	4	10.0
6	29	5	1	53,306	1	0	78	0	1	2	9.5
6	30	6	0	50,253	1	0	77	1	0	2	11.0

EXHIBIT 4 CONTINUED

DAY OF GAME		DAY OF WEEK *(1 = Mon)*	NIGHT GAME *(1 = night)*	ATTENDANCE	TV? *(1 = yes)*	SPECIAL EVENT *(1 = yes)*	TEMP. (DEG F)	TYPE OF DAY	STAR PICHER *(1 = yes)*	OPPONENT	GAMES BEHIND
Month	**Day**	**Wkday**	**Nite**	**Attend**	**TV**	**Spec**	**Temp**	**Weath**	**Star**	**Opp**	**Behind**
7	1	7	0	51,246	1	0	87	1	0	2	12.0
7	2	1	1	51,211	1	0	79	1	1	2	12.0
7	3	2	1	35,158	0	0	87	1	0	7	11.0
7	4	3	1	20,084	0	1	71	2	0	7	11.0
7	5	4	0	31,878	1	0	69	1	1	7	10.0
7	19	4	1	22,648	0	0	88	0	0	9	11.0
7	20	5	1	30,481	0	0	85	0	1	9	10.5
7	21	6	0	50,084	1	1	82	1	1	9	11.5
7	22	7	0	40,156	1	0	86	0	0	11	11.5
7	23	1	1	20,674	1	0	88	0	0	11	11.5
7	24	2	1	33,497	0	0	92	0	0	10	11.5
7	25	3	1	47,449	1	0	91	1	1	10	12.0
7	26	4	0	43,141	0	0	87	1	1	10	12.0
8	3	5	1	51,151	0	1	87	0	0	1	14.0
8	4	6	1	46,407	1	0	84	1	0	1	15.0
8	5	7	0	54,478	0	0	94	0	1	1	16.0
8	6	1	0	36,314	1	0	90	0	1	1	15.0
8	7	2	1	33,513	0	0	83	0	0	3	14.0
8	8	3	1	20,048	1	0	94	1	0	3	14.0
8	9	4	1	21,535	0	0	89	0	0	3	13.0
8	13	1	1	24,977	0	0	78	0	1	12	15.0
8	14	2	1	24,125	0	0	82	1	0	12	14.0
8	15	3	1	25,905	1	0	73	1	1	12	14.0
8	16	4	0	22,036	0	0	78	0	0	8	14.0
8	17	5	0	30,372	0	0	78	0	0	8	14.0
8	18	6	0	38,695	0	0	68	2	0	8	14.0
8	19	7	1	47,723	1	0	83	0	1	8	14.0
8	30	4	1	30,717	1	0	91	0	0	6	14.5
8	31	5	1	35,229	1	0	88	0	1	6	15.5
9	1	6	0	30,130	0	0	83	0	0	6	14.5
9	2	7	0	34,008	1	0	82	1	0	6	14.5
9	3	1	0	46,298	1	1	86	1	1	2	14.5
9	4	2	1	37,259	0	0	90	0	0	2	15.0
9	5	3	1	38,644	1	0	86	0	1	2	14.5
9	15	6	0	30,050	1	0	77	0	1	5	16.0
9	16	7	0	40,192	1	1	79	0	0	5	15.5
9	25	2	1	15,699	0	0	68	0	0	4	17.0
9	26	3	1	16,354	0	0	81	0	1	4	16.0
9	27	4	1	12,111	0	0	76	0	0	4	16.0
9	28	5	1	17,647	0	0	75	1	0	13	15.5
9	29	6	0	30,016	0	0	79	1	1	13	15.5
9	30	7	0	21,641	0	0	71	2	1	13	14.5

Regression Number 1
Dependent Variable: Attend

	Constant	TV
Regr. Coef.	27,140	9,644
Std. Error	1,779	2,501
t value	15.3	3.9

# of obs =	***79***	*Deg of F =*	***77***
R-squared =	***0.1619***	*Resid SD =*	***11,112***

Regression Number 2
Dependent Variable: Attend

	Constant
Regr. Coef.	29,992
Std. Error	1,308
t value	22.9

# of obs =	***79***	*Deg of F =*	***77***
R-squared =	***0.1983***	*Resid SD =*	***10,868***

Regression Number 3

Dependent Variable: Attend

	Nite	Constant	TV	Spec	Temp	Star	Behind	Cloudy	Rainy
Regr. Coef.	(3,833)	4,373	6,973	14,451	248.6	3,263	229.8	3,999	(1,194)
Std. Error	2,290	10,752	2,280	3,315	148.8	2,208	253.5	2,420	4,537
t value	(1.7)	0.4	3.1	4.4	1.7	1.5	0.9	1.7	(0.3)

# of obs = ***79*** *Deg of F =* ***70***

R-squared = ***0.4352*** *Resid SD =* ***9,568***

Regression Number 4

Dependent Variable: Attend

	Nite	Constant	TV	Spec	Temp	Star	Behind	Fri	Sat	Sun	Cloudy	Rainy	April-May	June-August	Opp 1,10	Opp 2	Opp 3,5,6,7,8
Regr. Coef.	414.8	8,600	3,438	12,518	44.74	3,725	55.30	5,708	8,158	12,758	953.3	(5,032)	1,935	8,870	10,311	17,345	3,671
Std. Error	1,906.8	8,799	1,737	2,348	107.90	1,518	289.21	2,263	2,393	2,529	1,682.8	3,211	4,363	2,586	2,288	2,979	1,793
t value	0.2	1.0	2.0	5.3	0.4	2.5	0.2	2.5	3.4	5.0	0.6	(1.6)	0.4	3.4	4.5	5.8	2.0

# of obs = ***79*** *Deg of F =* ***62***

R-squared = ***0.7712*** *Resid SD =* ***6,470***

Regression Number 5

Dependent Variable: Attend

	Nite	Constant	TV	Spec	Star	Fri	Sat	Sun	Rainy	April-May	June-August	Opp 1,10	Opp 2	Opp 3,5,6,7,8
Regr. Coef.	600.7	13,140	3,544	12,273	3,754	5,568	8,231	12,939	(5,926)	970.9	8,775	10,616	17,635	3,695
Std. Error	1,830.2	2,658	1,685	2,232	1,479	2,201	2,317	2,463	2,819	2,466.4	2,158	2,198	2,872	1,721
t value	0.3	4.9	2.1	5.5	2.5	2.5	3.6	5.3	(2.1)	0.4	4.1	4.8	6.1	2.1

# of obs = ***79*** *Deg of F =* ***65***

R-squared = ***0.7694*** *Resid SD =* ***6,344***

NOPANE ADVERTISING STRATEGY

Nopane was a mature proprietary drug product which had been marketed for a decade. Nopane's marketing program had undergone little change for some time when a new product manager, Alison Silk, assumed responsibility for the brand.

Silk undertook a careful review of the brand's history and available marketing research information. There were 12 competing brands, four of which, accounting for 60% of the market, were nationally distributed and supported by media advertising. Nopane was an important but not dominant brand among the four, with a 15% share of market. No consumer promotion (dealing, couponing, etc.) was used to any appreciable extent and price cutting was negligible, but all four major brands were priced above the level of the remaining brands in the category, and all four advertised heavily.

After working on Nopane for several months, Silk became convinced that sales could be increased by repositioning the brand. The brand's advertising agency prepared and tested some new approaches in focus group interviews. The results were quite favorable.

Encouraged, Silk authorized the agency to produce television commercials, to be aired on local TV stations, representing two different advertising strategies, one emphasizing what was labeled an "*emotional*" appeal, the other a "*rational*" approach. Silk planned a market test to reveal which one to use in a national roll-out at a later stage. Silk was also uncertain about what level of advertising was needed to support the strategy change. After consulting with the Marketing Research Manager, Silk proposed that an advertising experiment be conducted to address the following issues:

1. Do the "emotional" and "rational" copy alternatives differ in their effectiveness?
2. What level of advertising should be used for Nopane for the coming fiscal year?

Experimental Design

Two elements of the advertising campaign were to be systematically varied: copy and media spending. Two copy treatments ("emotional" and "rational") and three levels of advertising intensity were to be tested. Expressed in six-month expenditures per 100 "prospects" (potential customers) in a geographical area, the levels to be tested were $2.50, $4.75, and $8.00. The company had divided the U.S. into two segments; Segment A consisted of states lying along the east and west coasts of the United States while the rest of the country comprised Segment B. The two segments contained about equal numbers of total prospects. Twelve sales territories (out of a total of 75) were selected at random from the region designated as Segment A and another twelve were selected from Segment B. The complete experimental design, therefore, provided 24 observations (2 segments * 2 copy executions * 3 levels of media expenditure * 2 test territories).

Harvard Business School case 9-893-005. This case was prepared by Professor David E. Bell.

Sales measurements were obtained based on point-of-sale information in each of the territories. The experiment was run for six months, a period known from prior investigations to be sufficient for the long-term response to advertising to become clear. Arrangements were made to monitor competitive advertising activity in each of the 24 test territories.

Experiment Results

Exhibit 1 shows the results of the experiment. The 24 rows in the table each reflect one test sales territory. Each row records the segment, copy type, Nopane advertising expenditure, unit sales of Nopane, and competitor advertising expenditure in a territory. The last two columns represent segment and copy as dummy variables (1 = A, 0 = B for segment, 1 = Emotional, 0=Rational for copy type) for the purpose of running regressions.

Silk first regressed Nopane sales against the other variables in the table. (See Regression 1, below.) She was particularly interested, of course, in seeing how sales responded to Nopane advertising expenditure ("Ad Dollars") and to the two ad types ("Dum Copy").

Regression Number 1

Dependent Variable: SALES

	Ad Dollars	Constant	Competition	Dum Segm	Dum Copy
Regr. Coef.	1.477	32.59	(0.5652)	0.3514	2.134
Std. Error	0.338	2.53	0.1622	1.4005	2.027
t value	4.4	12.9	(3.5)	0.3	1.1

# of obs =	***24***	*Deg of F =*	***19***
R-squared =	***0.5889***	*Resid SD =*	***3.411***

Meeting with the Division Vice-President

On February 7, 1992, Alison presented her conclusions to the Division Vice-President, Stanley Skamarycz, whose approval would be needed for a change in advertising strategy. Alison showed him the experimental data and her regression. She was surprised by his response. "These results are worthless!" he said. "It's clear that our competitors have run interference on your experiment. As you can see, they systematically varied their own advertising strategy across our different test territories."

"It could be," remarked Alison, "that their response mirrors the response they will follow if we go national."

"I doubt it," responded Stanley. "I'd be prepared to bet that no matter which copy we choose, or what media expenditure we pick, they'll spend at a rate of about $19 per 100 prospects per sales territory per six-month period. I say that because that is what they have always done in the past."

Further Analysis

Back in her office, Silk ran a new regression (Regression 2) that seemed to confirm Skamarycz's theory that the amount of competitor advertising in a sales territory had depended on whether the "emotional" or "rational" advertising strategy had been used by Nopane.

She wrote two hypotheses on a piece of paper:

Regression Number 2
Dependent Variable: COMPETITION

	Ad Dollars	Constant	Dum Segm	Dum Copy
Regr. Coef.	0.8515	9.713	0.9167	9.083
Std. Error	0.4251	2.726	1.9196	1.920
t value	2.0	3.6	0.5	4.7

# of obs =	***24***	*Deg of F =*	***20***
R-squared =	***0.5711***	*Resid SD =*	***4.702***

Silk's Hypothesis — Nopane's competitors will react to our national strategy (whatever it might be) in the same way as they did in the test.

Skamarycz's Hypothesis — Whatever our national strategy, we can expect our competitors to spend an average of $19 per 100 prospects per sales territory per six-month period.

Concerned that these two hypotheses might have profoundly different implications for a national advertising strategy for Nopane, she decided to re-run Regression 1 without the "Competitor" variable (see Regression 3).

Regression Number 3
Dependent Variable: SALES

	Ad Dollars	Constant	Dum Segm	Dum Copy
Regr. Coef.	0.9959	27.10	(0.1667)	(3.000)
Std. Error	0.3848	2.47	1.7376	1.738
t value	2.6	11.0	(0.1)	(1.7)

# of obs =	***24***	*Deg of F =*	***20***
R-squared =	***0.3263***	*Resid SD =*	***4.256***

Exhibit 1

Results of Experiment

SEGMENT	COPY	ADVERTISING DOLLARS (per 100 Prospects)	UNIT SALES NOPANE (per 100 Prospects)	COMPETITOR ADVERTISING $ (per 100 Prospects)	DUMMY VARIABLES SEGMENT 1 = A 0 = B	FOR COPY 1 = Emotional 0 = Rational
		AD DOLLARS	SALES	COMPETITION	DUM SEGM	DUM COPY
A	Emotional	2.50	26	16	1	1
A	Emotional	2.50	26	20	1	1
A	Emotional	4.75	31	23	1	1
A	Emotional	4.75	32	24	1	1
A	Emotional	8.00	24	25	1	1
A	Emotional	8.00	31	33	1	1
A	Rational	2.50	25	20	1	0
A	Rational	2.50	26	14	1	0
A	Rational	4.75	32	19	1	0
A	Rational	4.75	35	12	1	0
A	Rational	8.00	40	13	1	0
A	Rational	8.00	38	15	1	0
B	Emotional	2.50	33	16	0	1
B	Emotional	2.50	30	15	0	1
B	Emotional	4.75	35	24	0	1
B	Emotional	4.75	26	25	0	1
B	Emotional	8.00	30	28	0	1
B	Emotional	8.00	25	34	0	1
B	Rational	2.50	26	14	0	0
B	Rational	2.50	25	18	0	0
B	Rational	4.75	30	16	0	0
B	Rational	4.75	33	10	0	0
B	Rational	8.00	38	11	0	0
B	Rational	8.00	37	12	0	0

Explanation of Numerical Variables

Ad Dollars	=	Number of dollars per 100 prospects spent on Nopane advertising
Sales	=	Unit sales of Nopane per 100 prospects
Competition	=	Number of dollars per 100 prospects spent by competitors advertising their own products
Dum Segm	=	1 if sales territory is in Segment A
	=	0 if sales territory is in Segment B
Dum Copy	=	1 if "emotional" copy used
	=	0 if "rational" copy used

Note: All sales figures and advertising expenditures are per 100 prospects per six-month period.

LINCOLN COMMUNITY HOSPITAL

The five-person Executive Committee of the Board of Trustees of the Lincoln Community Hospital, a 180-bed not-for-profit hospital in Sparta, New York (population: 135,000), met Saturday, January 19, 1985, to develop an understanding of the factors driving hospital costs. Lincoln had been incurring an operating deficit (the difference between the hospital's revenues for services rendered and its operating costs) for each of the last six years, its financial position was precarious, and the Executive Committee was under increasing pressure to bring in a for-profit management team. An understanding of hospital costs would be a relevant input to a decision on whether to change management.

Lincoln's case was typical of the predicament facing many not-for-profit community hospitals in the mid-1980s. Faced with rising costs (in part because of the increasing complexity of modern medical care) and lower revenues (primarily because of cost containment by third-party reimbursements of hospital bills, e.g., the government, Blue Cross/Blue Shield, and other medical insurers), many of these hospitals were experiencing persistent and chronic operating deficits. Most not-for-profit community hospitals did not have many beds and could not take advantage of the supposed scale economies in running a hospital. Being independent, these hospitals were also denied the benefits of any scale economies that might result from the management of a large chain of hospitals. Furthermore, they were often the only hospital in the community and could not control the complexity of the mix of cases they took in. Finally, these hospitals had to contend with criticisms alleging that they were not run as efficiently as for-profit hospitals.

The debate at the Executive Committee meeting was heated and inconclusive. Finally, at 9:30 p.m., Dr. Otto Planck, the Committee's chairperson and a former director of the hospital, said: "Let's call it a day. It's clear we have conflicting notions as to what drives hospital costs. Why don't we look at some data to examine the merits of these notions? Since we are all in agreement that Lincoln should not become affiliated with a hospital chain, I will examine the data available on operating costs for other independent hospitals and report back to you next Saturday."

For his analysis, Dr. Planck looked at data for the year 1983 on costs and related variables for 494 independent for-profit and not-for-profit hospitals in the United States. Dr. Planck's analysis is documented in the five regressions in Exhibit 1; the variables used in these regressions are described in Exhibit 2.

Harvard Business School case 9-191-149. This case was prepared by Professor Anirudh Dhebar. It was adapted from the case *Bellevue Hospital* by Professor Richard F. Meyer, which in turn was based on Regina E. Herzlinger and Williams S. Krasker, *Who Profits from Nonprofits?* Harvard Business Review 65, #1 (January - February, 1987), pp. 93-106. The data have been disguised for instructional purposes.

Exhibit 1

Regression Outputs

Regression Number 1
Dependent Variable: Cost

	Constant	Beds	Patient-Days
Regr. Coef.	(4,722,847)	119,655	114.7
Std. Error	351,921	4,972	21.9
t value	(13.4)	24.1	5.2

# of obs =	**494**	Deg of F =	**491**
R-squared =	**0.9153**	Resid SD =	**5,036**

Regression Number 2
Dependent Variable: Occupancy

	For-Pft	Constant
Regr. Coef.	30.39	189.7
Std. Error	6.51	5.7
t value	4.7	33.1

# of obs =	**494**	Deg of F =	**492**
R-squared =	**0.0424**	Resid SD =	**60.40**

Regression Number 3
Dependent Variable: Cost/bed

	Beds	For-Pft	Constant	Occupancy
Regr. Coef.	182.6	(15,188)	30,889	225.4
Std. Error	9.1	2,552	4,132	17.3
t value	20.1	(6.0)	7.5	13.0

# of obs =	**494**	Deg of F =	**490**
R-squared =	**0.5383**	Resid SD =	**23,150**

Regression Number 4
Dependent Variable: Cost/bed

	Beds	Casemix	For-Pft	Constant
Regr. Coef.	(25.39)	120,947	11,600	(43,379)
Std. Error	20.73	10,845	3,145	10,829
t value	(1.2)	11.2	3.7	(4.0)

# of obs =	**494**	Deg of F =	**490**
R-squared =	**0.5040**	Resid SD =	**23,990**

Regression Number 5
Dependent Variable: Cost/bed

	Beds	Casemix	For-Pft	Constant	Occupancy
Regr. Coef.	(17.38)	117,442	4,308	(82,031)	220.3
Std. Error	17.26	9,031	2,664	9,387	14.9
t value	(1.0)	13.0	1.6	(8.7)	14.8

# of obs =	**494**	Deg of F =	**489**
R-squared =	**0.6570**	Resid SD =	**19,970**

Exhibit 2

Variables in Dr. Planck's Data Base

Year of study: 1983

Number of hospitals: 494

VARIABLE NAME	DESCRIPTION	MINIMUM	AVERAGE	MAXIMUM
NATURAL VARIABLES				
Cost	Total Operating Cost	$755,773	$14,885,419	$88,524,179
Beds	Number of beds	16	136	562
Patient-days	Number of Patient-days	1,647	28,958	140,846
Casemix	Average complexity of cases in hospital	0.640	1.074	2.097
For-Profit	1 if for-profit hospital; 0 if not-for-profit	0.000	0.775	1.000
TRANSFORMATIONS				
Cost/Bed	Operating Cost/ Number of beds	$18,276	$92,031	$195,735
Occupancy	Occupancy rate (Patient-days/Beds)	33.1	213.3	364.1

Multiplicative Regression Models

Introduction

The standard regression model relates a dependent variable y to an independent variable x using the relationship:

$$y = b_0 + b_1 * x + \text{residual} \quad ,$$

so that the estimated value of y is given by:

$$y_{est} = b_0 + b_1 * x \quad .$$

For given values of b_0 and b_1, this implies that the relationship between x and y_{est} is *linear.* Furthermore, we have assumed that if the regression is correctly specified, the residuals will be indistinguishable. Thus, the frequency distribution of the residuals tells us all we can learn about what the value of a residual on a new observation will be; knowing the value of x (or any other variable not included in the model) on such an observation will not enable us to improve on our prediction of the residual.[1] In particular, the mean and standard deviation of the residuals should not depend on the value of x or of any other variables.

This is often a satisfactory form of model to use when y and x are difference-scale variables, or when x is a dummy variable. It may be good enough even when x and/or y are ratio-scale variables that do not vary over a very large range. For example, we were able to derive a reasonable relationship between weight and height using a linear model.

An Example

Business and economic data often involve ratio-scale variables that cover a substantial range of values. If this is true the simple linear model may be substantially incorrect for a number of reasons. To see why, let's look at Figure 6.1, which is a scatter diagram relating the yield per acre for an agricultural product (avocados) to its price. The data pertain to California, where most avocados in the United States are grown, and cover the twenty-five years from 1950 through 1974.

Harvard Business School note 9-893-013. This note was prepared by Professor Arthur Schleifer, Jr.

[1] Remember that we use the distribution of residuals to make probabilistic forecasts—see Figure 4.3 in Chapter 4, *Forecasting with Regression Analysis.*

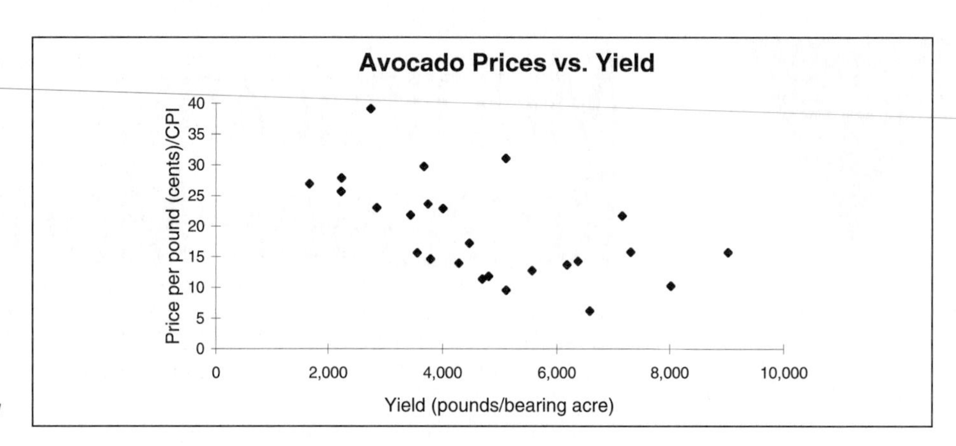

Figure 6.1

Crop yield varies considerably from one year to the next largely because of weather and crop diseases. In addition, yields are affected by soil treatment (fertilizers), and the age of the trees. In years of low yield, prices tend to be bid up by people who are willing to pay a premium to obtain avocados that are in short supply. By contrast, in years of high yield, a market surplus tends to drive the price down. Of course, many other factors affect the price of avocados as well. A growing population, the number of acres planted with avocado trees, greater affluence (avocados are a luxury food item), inflation,[2] changes in consumer taste and eating habits, attempts to influence consumers by advertising and promotion, changes in distribution, and competition from foreign growers and from alternate luxury fruits all affect price. Nonetheless, one's general impression from Figure 6.1 is that as yield increases, price tends to decrease, albeit with many exceptions.

A regression with price as the dependent variable and yield as the independent variable results in an estimated constant term $b_0 = 30.59$, and an estimated regression coefficient $b_1 = -0.00241$; the residual standard deviation is 6.65. Figure 6.2 is the same scatter diagram as Figure 6.1 with the estimated regression line superimposed.

The regression line is downward sloping, conforming to our belief that high yields drive prices lower, and the fit is fairly good ($R^2 = 0.33$). We shall return later in this chapter to a more sophisticated analysis of this problem, but for now let's just look at the implications of what we have.

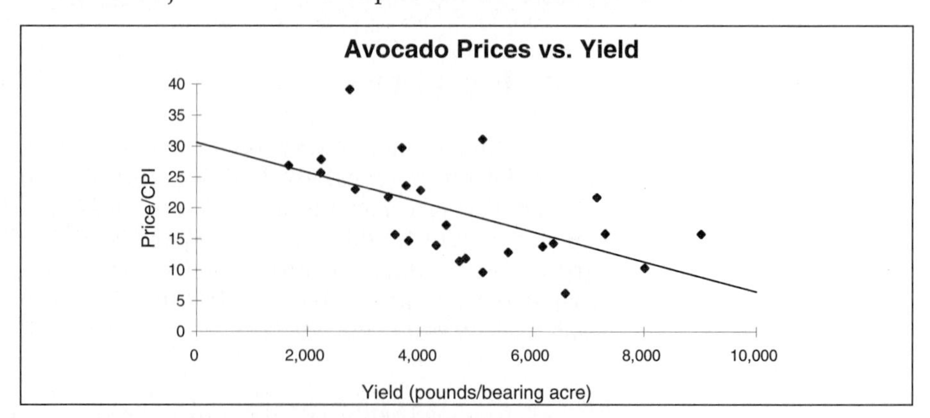

Figure 6.2

[2] Inflation effects can be taken into account by expressing price in constant cents per pound. This has been done in Figure 6.1 by dividing each year's current price by the value of the consumer price index in that year.

PROBLEMS WITH THE LINEAR MODEL

We would feel quite comfortable forecasting the price in a year when the yield per acre was 7,500 pounds. By formula, estimated price = 30.59 – 0.00241 *7,500 = 12.52 cents per pound. But what if we want to extrapolate beyond the range of the data? Suppose we anticipate a banner year for crop yield: 13,000 pounds. Our formula then predicts that price will be 30.59–0.00241 * 13,000 = –0.74 cents! We don't even need to be this extreme, however, to run into negative price forecasts. For instance, a 95% confidence interval given a yield of 7,500 pounds would extend from 12.52 – 2*6.65 to 12.52 + 2*6.65, or from –0.78 to 25.82 cents per pound. Although high yields are likely to drive prices down, negative prices are patently absurd.

Now suppose yields were very low. Our model says that as yield approaches 0, estimated price will not rise above 30.59 cents per pound. Again, this seems absurd. Some restaurants and some premium food stores will be willing to pay much more to be among the few outlets that have avocados available. Thus we might expect prices to soar far above 30 cents per pound in a year when yields are very low.

Finally, we might anticipate a wider confidence interval when yields are low and prices are high than in the opposite situation. If our model gave an estimated price of $1.00 per pound when yields were very low, we might not be surprised to have actual prices differ from the estimate by 20%, from 80 cents to $1.20, for example. By contrast, if the estimated price in a high-yield year were 10 cents, we might still expect actual to differ from estimated price by 20%, but that would result in an interval from 8 to 12 cents, a much narrower interval than in the previous case. However, the regression model that we have used assumes that the residuals are indistinguishable. Therefore confidence limits will be based on the residual standard deviation of 6.65 regardless of the level of yield or the estimated price. Therefore, it cannot take into account our belief that confidence intervals should vary in width, depending on the value of the estimated price.

The simple regression model saying that estimated price = 30.59 – 0.00241 times yield implies that a 1,000 pound increase in yield will decrease price by 2.41 cents, whether we are going from a yield of 1,000 to 2,000 pounds, or from 7,000 to 8,000 pounds. A more satisfactory model might hypothesize that when the yield doubles, price will decrease by some fixed percentage. For example, when yield goes from 1,000 to 2,000 pounds it would have the same percentage effect on price as when it goes from 4,000 to 8,000 pounds. Such a model would also permit confidence intervals to be expressed as percentages of (instead of differences from) estimated price. Models of this sort are called **multiplicative**, as contrasted with the **linear** model we have been considering so far.

Notice that our dissatisfaction with the linear model stems not from lack of fit, or from data that dramatically disagree with assumptions of the model, but from quite theoretical considerations, based primarily on what the model implies for extreme cases. This kind of analysis can be very fruitful in discovering better ways of specifying regression models.

Figures 6.3A and 6.3B once again show the scatter diagram of price vs. yield. Figure 6.3A shows the linear regression estimate and 95% confidence limits. Figure 6.3B shows a multiplicative regression estimate and 95% confidence limits. Both figures have vertical scales that extend beyond the scales in Figures 6.1 and 6.2, but are comparable to one another. In the multiplicative model depicted in Figure 6.3B, notice that low yields result in estimated prices far in excess of 30 cents, that high yields lead to low but positive estimated prices, that the confidence intervals vary in width, and that even for high yields the lower confidence limit is never negative. In the remainder of this chapter we shall discuss how to formulate and interpret multiplicative regression models.

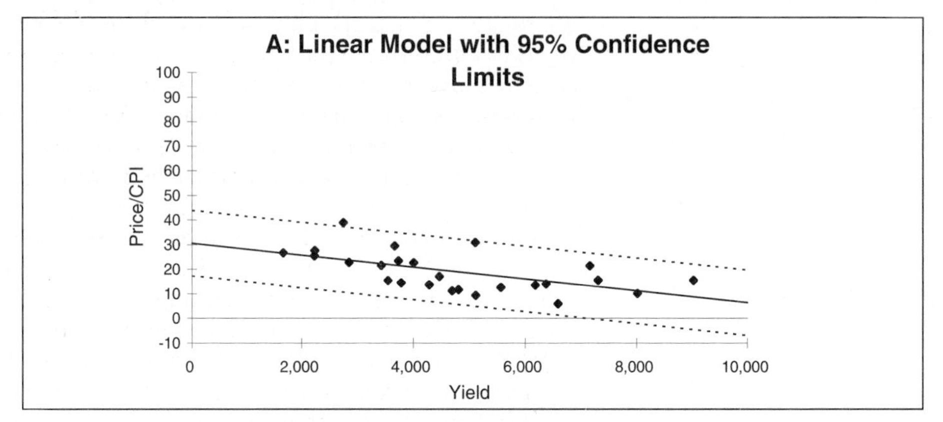

Figure 6.3A

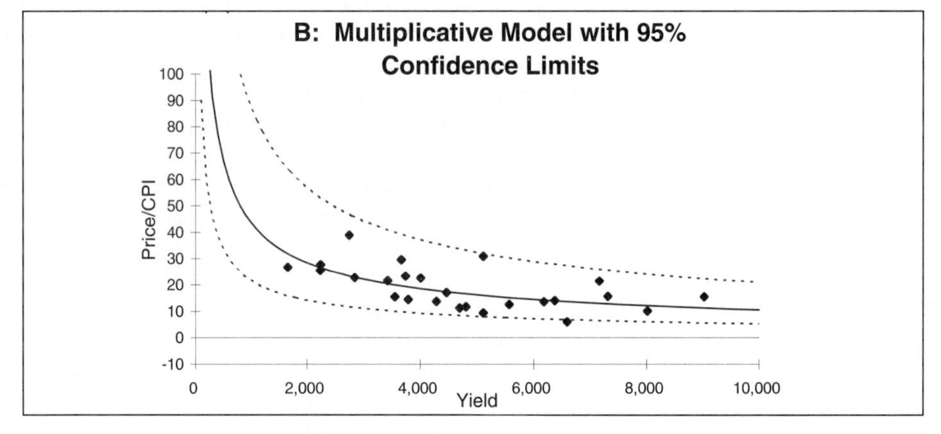

Figure 6.3B

Three Standard Multiplicative Models

There are three cases to consider:

1. If x is a ratio-scale variable, then as x increases by a fixed *percentage*,[3] y_{est} may increase or decrease by a fixed *amount*. For example, a 1% increase in x might cause y_{est} to increase by 3 units, so that as x went from 100 to 101, or from 500 to 505, or from 2000 to 2020, y_{est} would go up by the same 3 units.
2. If y is a ratio-scale variable, then as x increases by one unit, y_{est} may increase or decrease by a fixed *percentage*. For instance, a one-unit increase in x might cause y_{est} to increase by 2%. Thus, if $y_{est} = 100$ when $x = 3$ and $y_{est} = 102$ when $x = 4$, this model would imply that y_{est} would go from 682.679 to 696.333 (a 2% increase) when x went from 100 to 101.

[3] In all three examples, we assume that if there are other x's included in the model, their values are held constant while we vary the particular x in whose relationship with y_{est} we are interested, and that whatever that relationship is, it does not depend on the values at which the other x's were fixed.

3. If both x and y are ratio-scale variables, then as x increases by a fixed *percentage*, y_{est} may increase or decrease by a fixed *percentage* as well. By way of example, a one percent increase in x might be accompanied by a 0.5% increase in y_{est}. If y_{est} went from 400 to 402 when x went from 100 to 101, this model would imply that y_{est} would go from 1268.54 to 1274.88 (a 0.5% increase) when x went from 1000 to 1010 (a 1% increase).

In what follows we shall look at how to model each of these relationships. Remember that multiplicative relationships can be converted to additive ones by using logarithms (see Chapter 1, *Data Analysis and Statistical Description*), so it should be no surprise that logarithmic transformations will be the key to converting multiplicative models into ones that can be analyzed by regression. The interpretation of the output of such models is tricky, however, and the main point of this chapter is to provide you with such interpretations.

Here are all the mathematical facts you need to know. We shall be dealing with so-called *natural* logarithms (logarithms to the base $e = 2.71828...$). In Excel, you can find the natural logarithm of a number using the = LN function; on most pocket calculators there is an LN key that converts a number to its natural logarithm. If you know the natural logarithm of a number, and want to find the number itself, you use the = EXP function in Excel, or the EXP key on most pocket calculators. Thus, as you should verify, LN(2) = 0.6931, and EXP(0.6931) = 2.

We shall introduce data sets in which one or another of the three models described above is appropriate. While the main purpose of this chapter is to show cases where it makes sense to think of relationships between variables as multiplicative, the particular data sets will give us some opportunity to illustrate some other methodologically useful points as well.

Model 1: Life Expectancy

We start with an example in which one of the independent variables is clearly measured on a ratio scale, where a plausible model might specify that as it changes by a given multiplicative factor, the estimated value of the dependent variable changes by a given amount.

Data file LIFEEXP.XLS shows the life expectancy (in years) and income per capita of people in 101 different countries. The countries are further classified into three categories: industrialized, petroleum exporting, and lesser developed. The data are from 1974;[4] income per capita is given in thousands of 1974 U.S. dollars.

We might start by hypothesizing that income per capita positively affects life expectancy, and test this assumption by performing a regression with life expectancy as the dependent and income per capita as the independent variable. The output of such a regression is shown in Figure 6.4; it suggests that each additional $1,000 of income per capita increases longevity by 6.8 years on the average. The R^2 of 0.54 and the high t values may make it seem that this is an adequate model.

[4] See Ann Crittenden, "Vital Dialogue Is Beginning Between the Rich and the Poor," *The New York Times*, September 28, 1975, page E-5.

Regression Number 1

Dependent Variable: LIFE EXP.

	Income/Cap	Constant
Regr. Coef.	6.755	46.322
Std. Error	0.627	1.101
t value	10.8	42.1

# of obs =	***101***	*Deg of F =*	***99***
R-squared =	***0.5399***	*Resid SD =*	***9.018***

Figure 6.4

Do the implications of this model make sense? It's not surprising to find life expectancy going up with income per capita, but wouldn't we expect to find the effect on longevity of going from $1,000 to $2,000 per year to be greater than the effect of going from $4,000 to $5,000? Indeed, many of the industrialized countries have life expectancies in the 70s. Added income might increase these numbers somewhat, but there are probably diminishing returns. We run up against natural limits that cannot be easily overcome by the benefits that income provides—better medicine, nutrition, shelter, public safety, etc. In those lesser-developed countries with life expectancies in the 20s and 30s, on the other hand, a little additional income per capita might go a long way.

To test the idea of diminishing returns, let's compute values of y_{est} and the residuals, and produce a scatter diagram of these two variables, as shown in Figure 6.5. It indicates that there are no estimated values below 46, even though actual life expectancy was as low as 27. More importantly, the residuals tend to be large positive numbers for estimated life expectancies between about 50 and 70, and to be large negative numbers for very low and very high values of estimated life expectancy. This suggests that actual values of life expectancy exceed estimated values in the middle range, and are less than estimated values at the extremes—an indication of a **curvilinear** [5] relationship between income per capita and life expectancy.[6] According to our diminishing-returns hypothesis, this seems to be a reasonable way for the two variables to be related.

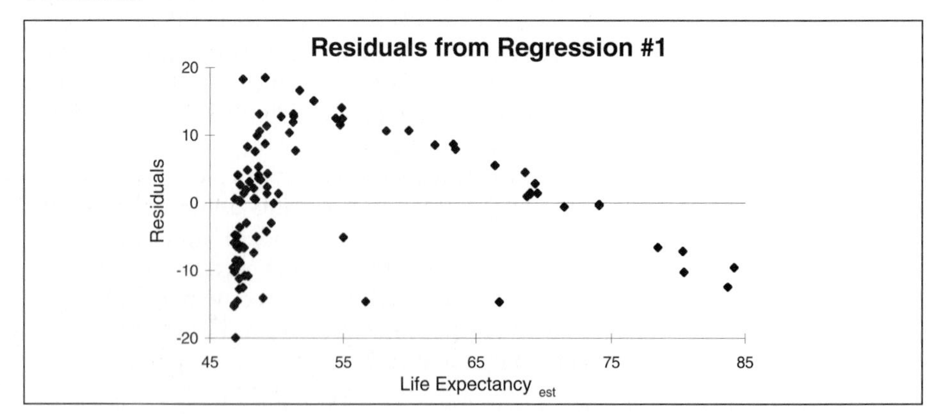

Figure 6.5

[5] We could try to take account of the curvilinearity by adding as a transformed variable the square of income per capita (see Chapter 4, "Forecasting with Regression Analysis"). But there are other characteristics of the relationship between income per capita and life expectancy that lead us to consider a different transformation.

[6] Had we produced a chart of life expectancy vs. income per capita at the beginning, we would have observed such a relationship, but not in as clear cut form as in Figure 6.5.

Further examination of the file shows that life expectancy ranged from 27 years (in Guinea) to 74.7 years (in Sweden), while income per capita ranged from $50 per year in Mali to $5,596 in Sweden. The ratio of high to low value was less than 3 for life expectancy, but more than 100 for income per capita.

Suppose we now state our diminishing-returns hypothesis more explicitly: a given *percentage* change in income per capita increases life expectancy by a fixed number of years. This implies that doubling income—going from $100 to $200 per year, for example—would result in the same increase in estimated life expectancy as going from $1,000 to $2,000, or from $2,000 to $4,000.

To see whether this hypothesis provides a better explanation of the relationship between the two variables, we must first perform a *logarithmic* transformation (using the = LN function in Excel) of income per capita. Why a logarithmic transformation? Because the *difference* between LN(100) and LN(200) is the same as the difference between LN(1,000) and LN(2,000), or the logarithms of any other number and twice that number. Similarly, differences between the logarithms of pairs of numbers differing by any other multiplicative factor would be the same, i.e., the difference between LN(100) and LN(110) is the same as the difference between LN(1,000) and LN(1,100). If life expectancy increases by the same amount every time income increases by a fixed percentage, it will increase by the same amount every time LN(income) increases by a fixed *amount*.

In Figure 6.6 the output of a regression with life expectancy as the dependent and LN(income per capita) as the independent variable is shown. The fit is clearly better (higher R^2), and a plot of the residuals of this regression against the values of y_{est} (Figure 6.7) shows that there is no discernible pattern (although there are a number of large residuals, leading us to wonder whether there are some additional explanatory variables that would account for them).

Regression Number 2
Dependent Variable: LIFE EXP.

	LN(inc/cap)	Constant
Regr. Coef.	8.411	60.76
Std. Error	0.491	0.80
t value	17.1	76.1

# of obs =	***101***	*Deg of F =*	***99***
R-squared =	***0.7474***	*Resid SD =*	***6.681***

Figure 6.6

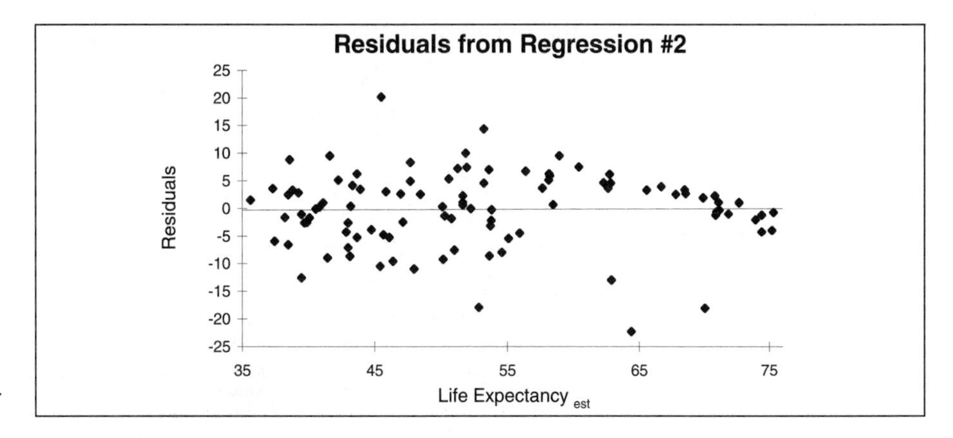

Figure 6.7

Interpretation of Output. The estimated regression coefficient of 8.411 in Figure 6.6 indicates that as LN(income) increases by one unit, estimated life expectancy goes up by 8.411 years, but that is of little value to a person who wants to understand the relationship of income to life expectancy. What you would like to know is how estimated life expectancy increases as income—not LN(income)—increases by a given percentage. The formula for estimated life expectancy, derived from the output in Figure 6.6, is:

$$y_{est} = 60.76 + 8.411*\text{LN(income)} .$$

We can use this formula to compute y_{est} for pairs of values of income that differ by 1%: 100 vs. 101, and 1,000 vs. 1,010. This is done in Table 6.1, which shows estimated life expectancy as a function of per capita income.

Table 6.1

INCOME ($000)	LN(INCOME)	y_{est}
0.100	–2.303	41.39
0.101	–2.293	41.48
1.000	0	60.76
1.010	0.00995	60.84

As income increases from 100 to 101, y_{est} increases from 41.39 to 41.48, an increase of 0.09 years. And as income increases from 1,000 to 1,010, y_{est} increases from 60.76 to 60.84, a difference of 0.08, essentially the same as before except for roundoff error. We conclude that a 1% increase in income is accompanied by a 0.08-year increase in estimated life expectancy. Furthermore, this increase is roughly equal to 1% of the regression coefficient on LN(income). This is not a coincidence. Indeed, in any regression model of the form:

$$y = b_0 + b_1*\text{LN}(x_1) + ... + \text{resid} ,$$

(where the "+ ... +" indicates that there may be other independent variables besides x_1), it can be shown that the effect on y_{est} of increasing x_1 by a factor k is $b_1*\text{LN}(k)$; doubling x_1, for example, would increase y_{est} by $b_1*\text{LN}(2) = 0.6931b_1$. If $k = 1.01$, representing a 1% increase in x_1, the increase in y_{est} is $b_1*\text{LN}(1.01) = 0.00995*b_1$, or approximately $0.01*b_1$. Thus we can approximate the effect on y_{est} of a 1% increase in x_1 directly from the regression output, without performing any additional calculations.

Additional Analysis. We noticed a number of large negative residuals in Figure 6.6, and upon scanning the data we see that many of these are associated with countries that are classified as petroleum exporting. Whereas most of the countries in the data base experienced relatively little change in income per capita over the years, the oil exporters experienced a significant jump in 1974 as a result of the OPEC price increase in late 1973. What affects life expectancy is not current income per capita, but rather average income per capita over a sufficiently long period of time to affect the public-health and associated infrastructures. In 1974 income-per-capita of petroleum exporters was probably far above that long-term average. Hence their estimated life expectancy will be too high, giving rise to negative residuals.

In Figure 6.8 we show the results of a regression having as independent variables, in addition to LN(income), two dummy variables representing lesser-developed countries and industrialized countries (petroleum exporters are the base case). The regression coefficients for the two dummies, compared with the base case, are positive (the regression "corrects" for oil exporters' estimated life expectancy being too high), although the difference between the regression coefficients for LDCs and industrialized countries is not sufficiently great to warrant putting them in separate categories. In addition, the R^2 is higher, and the RSD is lower: the additional analysis was clearly worthwhile.

Regression Number 3
Dependent Variable: LIFE EXP.

	LN(inc/cap)	Constant	Devel	LDC
Regr. Coef.	8.810	54.25	6.647	7.770
Std. Error	0.753	2.16	2.887	2.384
t value	11.7	25.1	2.3	3.3

# of obs =	***101***	*Deg of F =*	***97***
R-squared =	***0.7734***	*Resid SD =*	***6.393***

Figure 6.8

Model 2: Smoking and Death Rates

Figure 6.9 shows death rates for men as a function of age and smoking behavior, results derived from SMOKING.XLS. A similar chart could be produced for women. As we move from one five-year age bracket to another, the death rate goes up at an increasing rate in every smoking category. Does that mean, in this case, that the death rate goes up by a constant multiplicative factor as we go from one age group to the next? How does the rate go up as degree of inhalation increases? And how is the rate affected by gender?

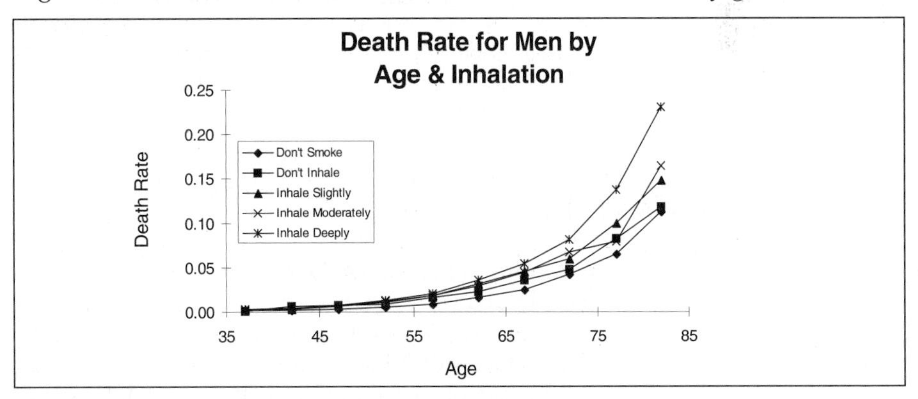

Figure 6.9

To answer these questions, we must first rearrange the data so that each age/inhalation/gender category represents an observation. If we code age using the midpoint of each bracket (37, 42, ... , 82), code inhalation as 0 for don't smoke up to 4 for deeply inhale, and code gender as 0 for men, 1 for women, we can create a four-column array of 100 observations. Column 1 contains death rates, and the other three columns contain coded values of age, inhalation, and gender. For convenience, that rearrangement has been performed in data file SMOKINGX.XLS.

If we believe that the effects of the independent variables on death rate are multiplicative, then the effects on the logarithm of death rate should be additive, in the sense that each five-year step in age should add a given amount to LN(death rate), and each increment in severity of inhalation should likewise add a certain amount. If the amounts added are the same for all age increments, or for all inhalation levels, the effects of these variables on LN(death rate) would be linear.

Figure 6.10 shows LN(death rate) as a function of age for men and women and the different levels of inhalation. Although no attempt is made to label the various "curves," it is quite clear that, except in a few instances (primarily cases where death rate is based on very few deaths and thus subject to considerable sampling error), the relationships are very close to linear.

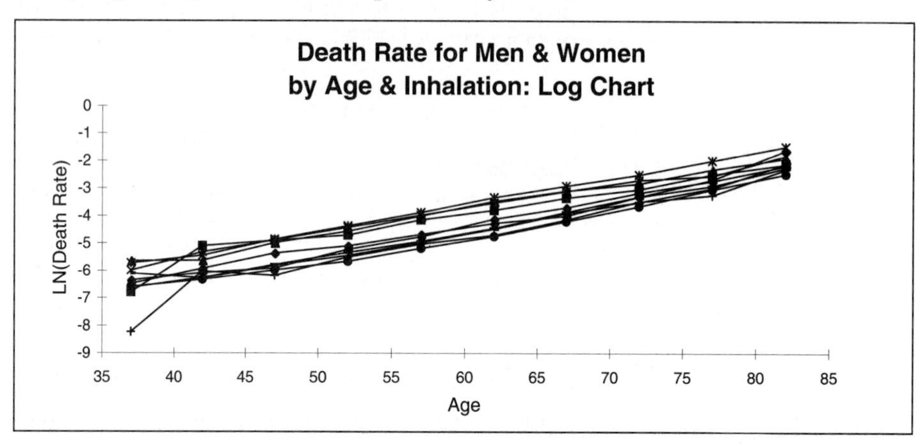

Figure 6.10

Figure 6.11 shows the results of a regression in which LN(death rate) is the dependent variable and age, inhalation, and gender are the independent variables. The R^2 is remarkably high (0.9710).

Regression Number 1
Dependent Variable: LN(DEATH RATE)

	Inhalation	Gender (M=0)	Age	Constant
Regr. Coef.	0.1484	(0.6724)	0.09462	(9.882)
Std. Error	0.0177	0.0500	0.00174	0.115
t value	8.4	(13.4)	54.3	(85.9)

# of obs =	***100***	*Deg of F =*	***96***
R-squared =	***0.9709***	*Resid SD =*	***0.2501***

Figure 6.11

Interpretation of Outputs. How do the various independent variables affect estimated death rate? From Figure 6.11, a one-year increase in age increases estimated LN(death rate) by 0.0946. Therefore, estimated death rate is increased by a factor of EXP(0.0946) = 1.099, i.e., by about 10% for each additional year of age. In a similar manner, each level of additional inhalation increases death rate by about 15%. The fact that a one-unit increase in x_1 increases y_{est} by a factor EXP(b_1) is true whether x_1 is a more-or-less continuous variable, like age, or a dummy variable, like gender. Thus the death rate for women is less than that for men by a factor of EXP(–0.6724) = 0.5105: the death rate for women is only around 50% of that for men of similar age and smoking behavior.

Additional Analysis. The linearity of LN(death rate) with respect to age is apparent from Figure 6.10, but are we justified in assuming that the ordinal variable that represents levels of inhalation causes estimated death rate to go up by roughly equal multiplicative increments? To answer this question, we created five dummy variables, each representing one of the levels of inhalation. Treating "don't smoke" as the base case, we replaced the "inhalation" variable with the other four dummies. The results of the regression are shown in

Figure 6.12. We see that the R^2 is very slightly improved and that the RSD is very slightly decreased. The regression coefficients for the dummy variables are plotted in Figure 6.13, and the straight-line relationship implied by the regression of Figure 6.11 is also shown. The "effects" of each additional level of inhalation do not increase in exactly equal steps, but they are not much different from the "effects" implied by the linear model of Figure 6.11. The slightly better fit hardly justifies the added complexity of description. Here is a case where an ordinal variable can be treated as if it were a difference-scale variable.

Regression Number 2

Dependent Variable: LN(DEATH RATE)

	Gender (M=0)	Age	Constant	Don't Inhale	Inhale Slightly	Inhale Mod.	Inhale Deeply
Regr. Coef.	(0.6724)	0.09462	(9.897)	0.1397	0.4033	0.4060	0.6086
Std. Error	0.0498	0.00173	0.120	0.0787	0.0787	0.0787	0.0787
t value	(13.5)	54.6	(82.6)	1.8	5.1	5.2	7.7

# of obs =	***100***	*Deg of F =*	***93***
R-squared =	***0.9721***	*Resid SD =*	***0.2490***

Figure 6.12

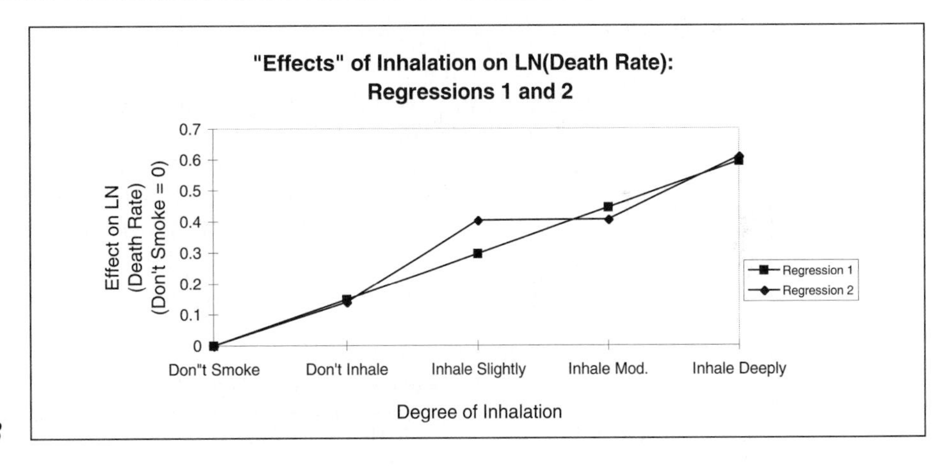

Figure 6.13

Model 3: Avocado Prices

We now return to a more complete discussion of avocado prices. Data file SUNRANCH.XLS contains data on California avocado prices,[7] production, and yield per acre, as well as disposable personal income, U.S. population, and the consumer price index from 1950 through 1974. Using these data, how can we forecast 1975 avocado prices?

California accounted for about 80% of the avocados produced in the U.S. in 1974. Newly planted avocado trees take about five years until they bear fruit. Yield per acre varies quite unpredictably from year to year, so that there are wide swings in production volume. Between 1950 and 1974 growers would increase their acreage when avocado prices were high; ultimately the increased supply of avocados would drive the price down, but better distribution procedures and heavy trade promotion would create sufficient demand to drive prices up again. Figure 6.14 shows prices and production as a function of time.

---

[7] Prices are in cents per pound, production in millions of pounds (for California only), disposable income in billions of dollars, yield in pounds per bearing acre, population in millions, and CPI is scaled so 1967=100%. Data from *Sun Ranch*, Harvard Business School case 9-185-076, by Professor Richard F. Meyer. Reproduced with permission.

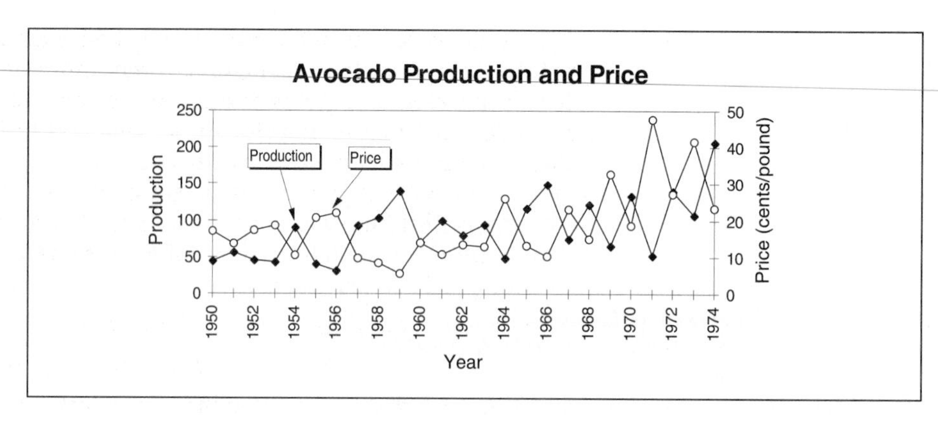

Figure 6.14

Figure 6.15 shows the output of a regression in which price was the dependent variable and production, income per capita (income divided by population) and population were the independent variables.[8] The high R^2 indicates a very good fit to the data. Yet the model implies that if production increased to 321.4 (millions of) pounds (quite far outside the range of the data, but not inconceivable if enough additional acres were planted and yields remained high), while income and population remained at 1974 levels, a point forecast of the price would be negative, and a 95% confidence interval on forecast price would include negative values even if production were only 301 million pounds. Furthermore, a plot of the residuals of this regression against y_{est} (Figure 6.16) shows that for values of y_{est} in the middle of its range the residuals tend to be negative, while for values of y_{est} at either extreme, the residuals tend to be positive: the model underestimates price for extreme values of y_{est}.

Regression Number 1
Dependent Variable: PRICE

	Constant	Pdtn	Pop	Inc/Cap
Regr. Coef.	25.25	(0.2129)	(0.1319)	15.67
Std. Error	8.34	0.0132	0.0589	1.23
t value	3.0	(16.1)	(2.2)	12.7

# of obs =	***25***	*Deg of F =*	***21***
R-squared =	***0.9594***	*Resid SD =*	***2.184***

Figure 6.15

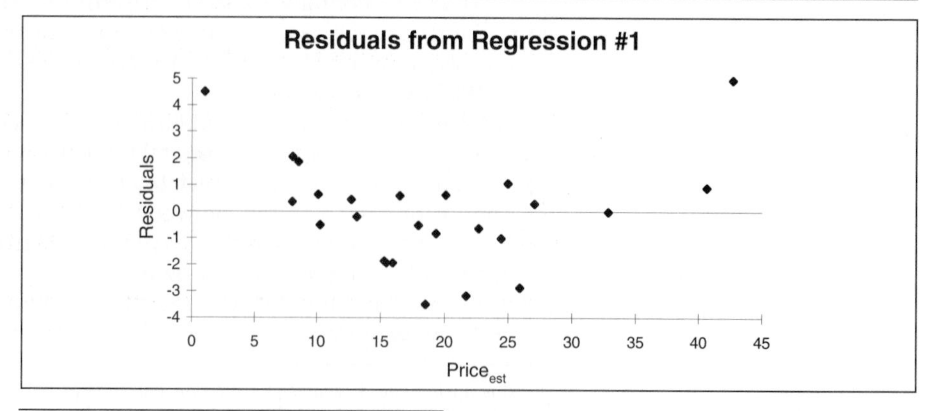

Figure 6.16

[8] Notice that we do not, in this or the next model, express price in constant cents (as we did in the introduction to this chapter). We expect any inflationary effects to be captured by the independent variable income per capita.

All of the variables in the model are ratio-scale variables, and a model that hypothesizes *percentage* changes in the independent variables causing *percentage* changes in y_{est} might make more sense; at least there would be no way for forecast values of price to be negative. In Figure 6.17 we show the results of a regression in which logarithms of all the variables are used. The fit as measured by R^2 is very slightly worse, but this time an examination of a scatter plot of residuals against values of y_{est} shows essentially random scatter. Even though the R^2 is not quite as high as in the nonlogarithmic model, this one is more satisfactory overall.

Interpretation of Outputs. In a model where:

$$\text{LN}(y) = b_0 + b_1 * \text{LN}(x_1) + \ldots + \text{resid}$$

it can be shown that an increase in x_1 by a factor k causes y_{est} [the estimated value of y, not of LN(y)] to change by a factor k^{b_1}; for example, doubling x_1 will change y_{est} by a factor 2^{b_1}. If $k = 1.01$, representing a 1% increase in the value of x_1, y_{est} will change by a factor 1.01^{b_1}. To a good approximation, if b_1 is between –5 and +5, a 1% increase in x_1 will cause y_{est} to change by b_1%. For example, using the regression of Figure 6.17, suppose we hold income and population constant at the 1974 levels of 984.6 and 215.47 respectively, so that income per capita is 984.6/215.47 = 4.5695, then ask what would happen to estimated price if production increased from its 1974 level of 207 to 209.07, a 1% increase. From the regression coefficients,

Regression Number 2
Dependent Variable: LN(PRICE)

	Constant	LN(Pdtn)	LN(Inc/Cap)	LN(Pop)
Regr. Coef.	20.36	(0.9280)	2.351	(2.942)
Std. Error	3.60	0.0578	0.207	0.730
t value	5.7	(16.1)	11.4	(4.0)

# of obs =	***25***	*Deg of F =*	***21***
R-squared =	***0.9589***	*Resid SD =*	***0.1094***

Figure 6.17

$$\begin{aligned} \text{LN(price)}_{est} &= 20.36 - 0.9280 * \text{LN(production)} \\ &\quad + 2.351 * \text{LN}(4.5695) \\ &\quad - 2.942 * \text{LN}(215.47) \\ &= 8.1253 - 0.9280 * \text{LN(production)}\ . \end{aligned}$$

Table 6.2 shows estimated price for production levels of 207 and 209.07:

Table 6.2

PRODUCTION	LN(PRODUCTION)	LN(PRICE)$_{est}$	PRICE$_{est}$
207.0	5.3327	3.17655	23.964
209.07	5.3427	3.16727	23.743

As we can see, a 1% increase in production caused a change in price of (23.743 – 23.964)/23.964 = –0.922%: – 0.922 is approximately equal in value to b_1.[9]

Forecasts. In a regression model involving a logarithmic transformation of the dependent variable, point forecasts can be computed by the method we have just used. To express forecast uncertainty in the form of a confidence interval, you can first compute a lower and upper confidence limit for LN(y), and then compute EXP of these limits to obtain similar limits for y itself. For example, we have already seen that the point forecast for LN(price) based on income per capita of 4.5695, population of 215.47, and production of 207 is 3.17655. A 95% confidence interval for LN(price) has a lower limit of 3.17655 – 2*0.1094 = 2.9578 and an upper limit of 3.17655 + 2*0.1094 = 3.3954, where 0.1094 is the value of the RSD in the regression of Figure 6.17. From these results, the 95% confidence limit for price extends from EXP(2.9578) = 19.26 to EXP(3.3954) = 29.83. (Remember that this interval, like all intervals derived in this way from regression output, is too narrow; in particular, it assumes that we know production for 1975, which depends on yield, and yield has been very hard to forecast in the past.) Because values computed using the EXP function are necessarily positive, a nice consequence of this analysis is that no confidence interval covers a negative price.

SUMMARY

The three models we have considered in this chapter are

$$y = b_0 + b_1 \text{ LN}(x_1) + \dots + \text{resid}\,, \qquad \text{(Model 1)},$$

$$\text{LN}(y) = b_0 + b_1 x_1 + \dots + \text{resid}\,, \qquad \text{(Model 2)},$$

and

$$\text{LN}(y) = b_0 + b_1 \text{ LN}(x_1) + \dots + \text{resid}\,. \qquad \text{(Model 3)}.$$

Model 1 was used to represent the relationship between life expectancy and income per capita in a sample of 101 countries; Model 2 to represent the relationship between death rates and age, smoking behavior, and gender in a sample of 1,000,000 adults; and Model 3 to represent the relationship between avocado prices and production, income per capita, and population in a sample of 25 years.

In Model 1, if x_1 increases by a *factor k*, y_{est} changes by b_1LN(k) *units*; if x_1 increases by 1%, y_{est} changes by approximately $0.01b_1$ units.

In Model 2, if x_1 increases by one unit, y_{est} changes by a *factor* EXP(b_1). If b_1 is between –0.2 and +0.2, y_{est} changes by a factor of approximately $1 + b_1$, or $100b_1$%.

In Model 3, if x_1 increases by a factor k, y_{est} changes by a factor k^{b_1}; if x_1 increases by 1%, y_{est} changes by a factor of approximately $1 + 0.01b_1$, or b_1%.

---

[9] Economists call the percentage change in one variable that accompanies a 1% change in another variable an **elasticity**. They usually look at how quantity demanded or supplied varies with price or other factors. Here we are looking at it the other way: how does price vary with quantity supplied (production) and other factors? Because the percentage change in price for a 1% change in production is –0.922%, whether production changes from 100 to 101 or from 300 to 303, a regression model in which both the dependent variable and the independent variables are expressed in logarithmic form (Model 3) is said to be a **constant-elasticity** model.

EXERCISE

Data file TIO2.XLS contains cost, capacity, and output data about DuPont's titanium dioxide business from 1955 through 1970. The standard learning-curve model asserts that every 1% increase in cumulative output will be accompanied by a constant percent decrease in unit manufacturing cost (measured in real, not nominal, dollars). For example, if manufacturing cost of an item were $100 when cumulative output was 1,000, the cost might drop to $99.50 (a 0.5% drop) when cumulative output increased to 1,010, and if this were so, we might expect a 0.5% drop in cost (in real terms) when cumulative output went from 2,000 to 2,020.

1. Use the data in file TIO2.XLS to see whether the manufacturing cost of titanium dioxide followed such a learning curve. If so, estimate by how much real costs decreased for each 1% increase in cumulative output. By how much did they decrease when cumulative output doubled?
2. In addition to learning effects, manufacturing costs are also influenced by economies of scale and capacity utilization. Scale is often measured in terms of plant capacity; from the data available, we can compute average plant capacity by dividing total capacity by the number of plants. Capacity utilization is simply the output in a given year divided by the capacity available in that year, usually expressed as a percent. Develop a model that relates cost to cumulative output, scale, and capacity utilization. Interpret the results of your model.

Barbara J. Key vs. The Gillette Company (A)

In 1975 Barbara J. Key, a former Gillette employee, brought suit against the company, charging that it had discriminated against her on the basis of her sex and race (Mrs. Key was black). She claimed relief and damages under Title VII of the Civil Rights Act of 1964. Subsequently, Mrs. Key reframed her complaint, bringing the action on the part of herself and other "exempt"[10] female employees of Gillette in southern Massachusetts, as there was already a class action suit pending against Gillette involving discrimination against blacks.

Barbara Key. Barbara Key worked for Gillette from May 13, 1968, to March 30, 1973. She had completed high school in Virginia and graduated from Hampton Institute, also in Virginia, with a major in business management. Gillette hired her as a compensation administrative assistant in the corporate personnel department at a starting yearly salary of $8,600; her position was categorized as Grade 8. In December 1968 she received a merit increase of $500, and in October 1969 she became a research assistant in the corporate personnel department, without any change in job classification or official title. Mrs. Key received a further merit increase in February 1970 of $800, and her grade level was changed to Grade 9, although her title remained the same. She was now earning $9,900.

Mrs. Key received merit increases of $600 in both 1971 and 1972, but no promotions or promotional salary increases. She did not receive a merit increase in February of 1973. When she left Gillette in March 1973, Mrs. Key was earning $11,100 per year.

In bringing suit on her own behalf, Mrs. Key made several claims of inequitable treatment in promotions, pay raises, and educational opportunities. She also claimed verbal harassment, and charged that she was treated differently from male employees in the matter of overtime, lunch breaks, vacation time, and punch-in and punch-out time. All these issues were investigated in the course of the legal process, which did not terminate until June 1982. In addition, however, other types of claims were essential to support Mrs. Key's assertion that Gillette's treatment was discriminatory on the basis of sex. She was required to demonstrate that Gillette's stated or unstated policies deprived women of opportunities in hiring and promotion, and affected the compensation, terms, conditions and privileges of their employment. For this claim, evidence was partly anecdotal, provided by personal testimony on behalf of the plaintiff and challenged by personal testimony on behalf of the defendant. But the main support for the class action issues, pro and con, necessarily derived from employee salary and promotion data.

The Gillette Company. Gillette was a large international consumer products company which developed, manufactured, and marketed a wide variety of consumer products, including razors and blades, personal care items, writing

Harvard Business School case 183-092. This case was prepared by Professor Stephen P. Bradley.

[10] Such employees are exempt from the Fair Labor Standards Act governing overtime pay.

instruments, and other products. In 1972-73 Gillette employed more than 33,000 people throughout the world, of whom 9,700 worked in the U.S. Approximately 5,000 employees were employed in Massachusetts, 1,300 to 1,700 in the "exempt" category during 1972-75. Gillette had three major facilities in Massachusetts: the Prudential Tower complex, where the company had its international headquarters; the Safety Razor Division in South Boston; and the Andover manufacturing facility, located in Andover. In 1973 Gillette relocated three major divisions—Personal Care, Papermate, and Appliances—from Chicago to Boston, and as a result hired many new employees. The company also entered into two new businesses in 1973-74, the calculator business and the digital watch business, and added some new products in its established divisions. These new businesses and products also necessitated additional hiring.

Exempt Job Grades at Gillette. Ninety-three percent of all Gillette's exempt employees were classified in Grades 8–20. A few sales jobs fell into job grades lower than 8, while certain high-level executive jobs were in grades above 20. Positions were classified and graded by the Compensation Department on the basis of how complex the necessary duties were, what education and experience were required, and the degree of responsibility and accountability the position carried. Newly created positions were graded before they were filled. From 1972 to 1975 Gillette hired 531 exempt employees, who held 243 different position titles in 9 separate functions.

Each job grade comprised a rather broad salary range, within which there were 5 distinct levels. The "minimum" level was the lowest salary a person in that grade level could be paid; there were also "25%" (one-fourth the spread between minimum and maximum), "control point" or "midpoint" (halfway between minimum and maximum), "75%" (three-fourths of the way between minimum and maximum), and finally "maximum"—the highest level of compensation paid in that grade. The full spread between minimum and maximum in any grade was 40%. Salary ranges for adjacent and nearly adjacent job grades often overlapped.

Within the various grades, differences in starting salaries depended on several variables. Gillette generally hired at or below the control point, although when market conditions required it, the company hired at higher levels within the grade. For example, if there was a shortage in a particular year for engineers, then higher salaries might have to be paid within grade for that occupational classification. Gillette claimed that it had, in fact, hired 35 engineers (out of 211 exempt employees) in 1974 under highly competitive conditions, more than half at or above control point. The company had to provide jobs that would appeal to people who had the opportunity to work in technologically more advanced firms in the Boston area, and needed to compensate them accordingly. In addition, Gillette routinely sought employees with particular backgrounds, filling most of its exempt positions through employment agencies (60%-80%) and newspaper advertising (10%-15%). Some positions, for example, required not only a certain level of education, but course work in specific areas: examples cited included product management and consumer advertising courses for those hired as product managers, or engineering or scientific undergraduate degrees (in addition to law degrees) for those hired as patent lawyers. People with appropriate educational or work experience might be hired at the high end of the in-grade salary range.

Promotion Procedures. Gillette did not have a formal promotion procedure by which individuals applied for publicly posted jobs. Instead, supervisors were asked to suggest appropriate personnel to fill positions as they became available: such positions were circulated among the supervisory staff, and all exempt employees were considered potential candidates. During the early 1970s, most supervisors were male, and Mrs. Key claimed that the company's subjective hiring practices made it difficult or impossible for women to know what promotions they might be eligible for, or to take steps to seek such promotions. Gillette had many more women in lower level job grades than in higher ones, and virtually none in important management positions.

Plaintiff's Analysis. As the basis for analysis, the plaintiff requested employee data from Gillette for the years 1968 to 1975 inclusive. Information was compiled on 1,850 full-time exempt employees who had worked at the Prudential, South Boston, or Andover facilities and who had never been transferred overseas. Data for each employee included:

- Name
- Employee Number
- Sex
- Birth Date
- Seniority Date
- Highest educational level achieved, in six categories: less than high school, high school graduate, college but no degree, college graduate, graduate or professional school but no degree, graduate or professional degree
- Marital Status

In addition, for each year the employee was employed by Gillette, the following information, representing status as of the end of the year in question, was included:

- Salary
- Salary code (reported salary is hourly, weekly, or annual)
- Group
- Division
- Termination Code (involuntary discharge, voluntary termination, still employed)
- Termination Date
- Class or Job Code
- Facility (Prudential, South Boston, Andover)

From the raw data supplied, the following information for each employee was computed:

1. Age in 1975
2. Seniority in 1975
3. Annualized Salary (hourly wages * 2,000; weekly wages * 52)

The Plaintiff's statistical expert also computed salary increase data. For each person employed in consecutive 2-year intervals (1968-69, ..., 1974-75), the percent increase of second year's salary over first year's salary were determined, and the average and median percentages for men and for women was obtained. These averages and medians, and the number of cases involved, are displayed in Table A for the years 1971-72 through 1974-75 only.

Table B shows, for each of the four years, average salaries for men and women, the difference between men's and women's average salaries, and the ratio of women's to men's average salary.

Table A

Average and Median Percent Salary Increases for Men and Women

	Salary Averages		Medians		Cases		
Interval	**Men**	**Women**	**Men**	**Women**	**Men**	**Women**	**Total**
1971-72	9.0%	8.4%	7.8%	6.9%	771	64	835
1972-73	10.0%	11.8%	7.3%	8.6%	780	70	850
1973-74	13.4%	14.0%	13.0%	13.7%	929	108	1037
1974-75	11.6%	13.6%	10.2%	12.5%	1022	142	1164

Table B

Salaries, Men vs. Women

	Salary Averages		Differences	Percents
Year	**Men**	**Women**	**Men–Women**	**Women/Men**
1972	$18,098	$12,414	$5,684	68.6%
1973	$19,083	$13,479	$5,604	70.6%
1974	$20,875	$14,282	$6,593	68.4%
1975	$22,753	$15,809	$6,944	69.5%

Barbara J. Key vs. The Gillette Company (B)

Barbara Key had brought suit against The Gillette Company for discriminating against her in salary, promotions, and educational opportunities. She contended that this discrimination was based on her sex, and accordingly had filed suit both on her own behalf and as a representative of the class of female employees. For discussion of the suit and the company, see Barbara J. Key vs. The Gillette Company (A).

To argue the matter of discrimination against women, both the plaintiff and the defendant presented testimony from expert witnesses analyzing data on men's and women's salaries. The testimony of both experts had certain elements in common: they both relied on regression analyses; they both ran separate analyses for the four years 1972 through 1975; they both used two forms of the dependent variable: annualized salary and the natural logarithm of annualized salary; and they both included among their independent variables a dummy variable for sex, coded 0 for men, 1 for women. Both experts agreed that the models with salary as the dependent variable implied that the effects on salary of the independent variables included in the model were "additive," while the models with log(salary) as the dependent variable implied "multiplicative" effects.

Output and Interpretation of Regression Models. Although some of the independent variables and the number of observations included in the analyses were different for the two experts, the basic outputs and interpretation of results were the same. Both witnesses were interested in estimating the salary differential attributable to being a woman given that other factors relevant to salary levels were "held constant." The differential in question might be additive (on the average, women are paid $5,000 per year less than men, for example) or multiplicative (on the average, women are paid 70% as much as men.) The model with annualized salary as the dependent variable is additive; the regression coefficient of the dummy variable for sex estimates the additive differential. The model with log (salary) as the dependent variable is multiplicative; the exponentiated value of the regression coefficient of the sex dummy estimates the multiplicative effect (the value of e_b is computed, where b is the value of the regression coefficient).

Both witnesses focused their analyses on the regression coefficients for the sex dummy variable, and the corresponding additive and multiplicative differentials. They were also interested in the goodness of fit of each of the regression models they ran, as measured by R^2.

Plaintiff's Analysis. The plaintiff's expert included the following independent variables (in addition to the dummy variable for sex):

- Age in 1975
- Seniority in 1975
- Dummy variables representing the six levels of education
- Dummy variables representing Gillette's three Massachusetts facilities

Harvard Business School case 183-093. This case was prepared by Professor Stephen P. Bradley.

Regarding the educational dummies, level 1 (less than high school) was taken as the base case, and five dummy variables, representing levels 2 through 6 were created; regarding the facilities dummies, Prudential was taken as the base case and two dummies, representing South Boston and Andover, were created.

Data on which the regressions were based included only those employees for whom information on the dependent variable and all of the independent variables was available. Because information on education was not routinely collected by the personnel department, a number of employees had incomplete records and were eliminated from the analysis. Table A shows the number of men and women who had complete records and who were included in plaintiff's analysis.

Table A

Number of Employees in Plaintiff's Data File

Year	Men	Women	Total
1972	694	45	739
1973	822	81	903
1974	940	130	1,070
1975	975	148	1,123

Table B shows values of the regression coefficient on the sex dummy variable and its standard error for each of the four years and for the two forms of the dependent variable. The value of R^2 for each regression is also shown. For the models in which log(salary) was the dependent variable, estimates of the multiplicative effect and its standard error are also shown.

Table B

Summary of Plaintiff's Regression Runs

Dependent Variable:
Salary or Log(Salary)

Independent Variables:
Sex
Age in 1975
Seniority in 1975
5 Education Dummies (Educ=2 thru 6)
2 Facility Dummies (S. Boston, Andover; Pru=Base Case)

Dep. Var.:	Salary			Log(Salary)			Multiplic. Effect	
Year	Coef.*	Std Err	R^2	Coef.*	Std Err	R^2	Estimate	Std Err
1972	(4,727)	1,107	0.310	(0.2635)	0.0427	0.409	0.768	0.033
1973	(4,308)	794	0.333	(0.2392)	0.0317	0.407	0.787	0.025
1974	(4,287)	647	0.367	(0.2327)	0.0252	0.439	0.792	0.020
1975	(4,572)	625	0.380	(0.2246)	0.0232	0.445	0.799	0.019

* Coefficient on Sex Dummy (Woman=1, Man=0)

Plaintiff's expert wrote:

> Under the assumptions that (a) these models correctly describe the relationship between salary and the included variables, and (b) any variables excluded from the model representing personal qualifications that affect salary are uncorrelated with sex, the implications of these models are as follows:
>
> Given any man and woman identical as regards age, seniority, educational level, and facility location, the best estimate of the woman's salary relative to the man's is that it will be:
>
> (1) lower than the man's by the amount of the additive effect (e.g., $4572 lower in 1975), given that the additive model is correct;

(2) a fraction (less than one) of the man's salary equal to the multiplicative effect (e.g., 79.9% of the 1975 man's salary) given that the multiplicative model is correct.

Defendant's Analysis. Defendant's analysis is shown in Table C. Data were taken from the end-of-year personnel files furnished by Gillette for 1972-75. The data included every employee whose personnel record was complete with regard to birth date, seniority date, facility, and job grade, who was exempt and who was not terminated during the year. Information about education level was not sought in the defendant's analysis, permitting defendant to obtain considerably more data (around 1,600 employees) than plaintiff.

Table C

Summary of Defendant's Regression Runs

DEPENDENT VARIABLE:	INDEPENDENT VARIABLES:
Salary or Log(Salary)	Sex
	Age in 1975
	Seniority in 1975
	2 Facility Dummies (S. Boston, Andover; Pru=Base Case)

1. REGRESSIONS WITH GRADE DUMMIES INCLUDED:

DEP. VAR.:	SALARY			LOG(SALARY)			MULTIPLIC. EFFECT	
Year	Coef.*	Std Err	R^2	Coef.*	Std Err	R^2	Estimate	Std Err
1972	(2,086)	1,666	0.38	(0.08)	0.03	0.71	0.92	0.03
1973	(1,650)	1,312	0.42	(0.06)	0.02	0.73	0.94	0.02
1974	(1,303)	971	0.52	(0.06)	0.02	0.81	0.94	0.02
1975	(1,256)	976	0.54	(0.06)	0.01	0.82	0.94	0.01

2. REGRESSIONS WITH GRADE DUMMIES EXCLUDED:

DEP. VAR.:	SALARY			LOG(SALARY)			MULTIPLIC. EFFECT	
Year	Coef.*	Std Err	R^2	Coef.*	Std Err	R^2	Estimate	Std Err
1972	(6,503)	1,835	0.18	(0.33)	0.05	0.26	0.72	0.04
1973	(6,804)	1,480	0.18	(0.32)	0.04	0.26	0.73	0.03
1974	(7,799)	1,173	0.20	(0.35)	0.03	0.33	0.70	0.02
1975	(7,625)	1,199	0.20	(0.32)	0.03	0.34	0.73	0.02

* Coefficient on Sex Dummy (Woman=1, Man=0)

The defense expert performed regression analyses to predict salary and the logarithm of salary as a function of sex, age, seniority, and facility, with and without dummy variables representing job grade.[11] When dummy variables for job grade were included (Table C1) the estimated additive salary differentials showed a deficit for women relative to men ranging from $1,256 to $2,086; these differentials were not statistically significant. Without the grade dummies, the ability of the regressions to explain salaries fell dramatically, as evidenced by the R^2 values in Table C2, while the estimated differential in salary showed deficits ranging from $6,503 to $7,799 and became statistically significant. The regressions on log(salary) with grade dummies included (Table C1) revealed differentials of about 92% to 94%, which, although fairly close to 100%, were nevertheless statistically significantly less than 100%. Regressions with grade dummies excluded (Table C2) revealed much larger differentials, ranging from 70% to 73%, which were also significantly less than 100%.

[11] Job Grade 8 was treated as the base case, and twelve dummy variables, representing Grades 9 through 20, were used.

INDEX

References to footnotes are indicated by the page number followed by a lowercase n.

D

E

F

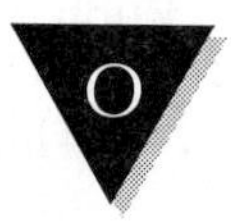

S

T

U

V

W